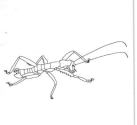

Heel-walkers & gladiators
MANTOPHASMATODEA p.132

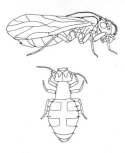

Psocids (booklice & barklice)
PSOCOPTERA p.134

Lice
PHTHIRAPTERA p.136

Bugs
HEMIPTERA p.138

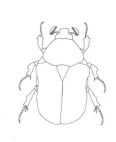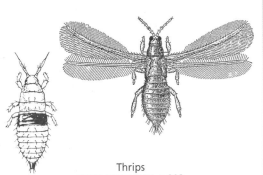

Thrips
THYSANOPTERA p.202

Dobsonflies & alderflies
MEGALOPTERA p.204

Lacewings, antlions & kin
NEUROPTERA p.206

Beetles
COLEOPTERA p.224

Stylopids
STREPSIPTERA p.308

Hangingflies & scorpion flies
MECOPTERA p.310

Flies
DIPTERA p.312

Fleas
SIPHONAPTERA p.360

Caddisflies
TRICHOPTERA p.362

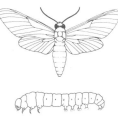

Moths & butterflies
LEPIDOPTERA p.368

Sawflies, wasps, bees & ants
HYMENOPTERA p.460

FIELD GUIDE TO
INSECTS
OF SOUTH AFRICA

Mike Picker • Charles Griffiths • Alan Weaving

Published by Struik Nature
(an imprint of Penguin Random House South Africa (Pty) Ltd)
Reg. No. 1953/000441/07
The Estuaries No. 4, Oxbow Crescent, Century City, 7441 South Africa
PO Box 1144, Cape Town, 8000 South Africa

Visit **www.penguinrandomhouse.co.za** and join the Struik Nature Club
for updates, news, events and special offers.

First published in 2002 by Struik Nature
Second edition published in 2004
Third edition published in 2019

3 5 7 9 10 8 6 4 2

Publisher: Pippa Parker
Managing editor: Helen de Villiers
Editor: Emily Donaldson
Designers: Dominic Robson & Neil Bester
Proofreader: Thea Grobbelaar

Reproduction by Hirt & Carter Cape (Pty) Ltd
Printed and bound in China by C&C Offset Printing Co., Ltd

MIX
Paper from
responsible sources
FSC
www.fsc.org FSC® C018179

Print: 9781775845843
ePub: 9781775845850

Cover (front): *Chrysis concinna*. **Back cover (top left):** *Sphaerocoris* sp.;
(centre right): *Amata cerbera*; **(bottom left):** *Heteracris* sp. **Page 3 (left to
right):** *Pseudocreobotra wahlbergii*; *Trichostetha fascicularis*; *Bagrada hilaris*.
Page 5 (top to bottom): *Eddara euchroma*; *Phymateus leprosus*;
Graptocoris aulicus; *Philoliche rostrata*; *Sigara* sp.; Rhipiceridae sp.

CONTENTS

INTRODUCTION

Insects are by far the most diverse group of organisms on earth, and must surely intrigue anyone with the slightest interest in the natural world. The global number of named insect species is approximately 1 million, which means they make up about 60 per cent of all known species on earth, or several times the number of all vascular plants (260,000) or of all vertebrates (60,000).

The total number of insect species (as opposed to just the number that are named) lies in the range of 5–30 million, but cannot be given with any precision: the number of known species is also constantly changing, as many thousands of new names are added to the list each year. It is a sobering thought that, given current rates of habitat destruction, millions of insect species will likely become extinct long before they are discovered and named.

INSECT DIVERSITY IN THE REGION

The insect fauna of South Africa currently comprises some 50,000 described species, a figure that underrepresents the actual number present in the region, which is estimated as 250,000 species. Many groups have never been thoroughly studied, and the fact that a new insect order was discovered in the region as recently as 2002 gives some indication of just how poorly surveyed the insect fauna of southern Africa is.

The known fauna represents a mixture of both indigenous or naturally occurring species, and of alien species – those that have been either deliberately or accidentally introduced to the region as a result of recent human activities. About 300 alien insect species are currently recorded as having been introduced to the region, and this number increases by a few species per year, not only because new species are constantly being introduced through various avenues, but also because species long-present in the region are only now being discovered. Although alien species represent less than 1 per cent of the total number of insect species in the region, they tend to be common and associated with human dwellings and with agricultural crops, so are often among the most frequently observed and destructive insects. Of the 300 introduced species, 219 were introduced accidentally. These include a wide range of domestic pests, such as several cockroach, louse and termite species, pests of stored products, and fleas. The majority are pests of agricultural and ornamental plants and were probably introduced along with the crop or garden plant on which they feed. The remaining 80 alien insect species were purposely introduced as biocontrol agents to inhibit the spread of alien plants or, more rarely, to control other alien insect species that have become crop pests. Many of these have proven to be extremely efficient and cost-effective agents in the battle to control alien invasive plants.

Ensign wasps are alien to South Africa and parasitize cockroach egg cases.

THE IMPORTANCE OF INSECTS

The critical importance of insects in natural ecosystems is best illustrated by imagining what would happen if all insects suddenly died out. Within days, thousands of insectivorous invertebrates, fish, amphibian, reptile, bird and mammal species would begin to starve to death, as would the larger predators that feed on them. Since most flowering plants rely on insects for pollination, these would also die out. Within a few decades, the only surviving large species would be wind- or water- pollinated plants and a reduced number of other invertebrates. In many ecosystems, insects such as termites are the most important recyclers of nutrients. In their absence the fertility of the soil would drop, causing the mass death of plant life. Interestingly, although both terrestrial and freshwater ecosystems would be dramatically transformed, the marine environment would hardly be affected, as insects are almost completely absent from the sea – the only habitat that they have failed to colonize successfully.

ECOLOGICAL AND ECONOMIC IMPACTS

Insects are key to life as we know it: predatory and parasitic insects control the density and population structure of many other life forms, including plants. Smaller organisms may fall prey to dragonflies, robber flies, predatory bugs, ladybeetles and the like. Larger vertebrates, such as birds and mammals, may be infected by diseases carried by tsetse flies, mosquitoes and other insects.

Parasitoids (parasitic insects) infect the larvae of other insect species, including important pest species, and may thus play a valuable role in regulating pest populations.

Insects are the most important plant pollinators, and are thus instrumental in the completion of the life cycle of many plant species. The majority of our fruit crops are dependent on insect pollination, so declines in dominant pollinators, such as bees, result in reduced fruit crops. The role of insects as pollinators is undoubtedly their greatest economic service to humans. The successful cultivation of a wide variety of fruit and vegetable crops depends on pollination by insects, mostly by honey bees, but other species as well. The value of these services is enormous. For example, the fresh fruit industry in South Africa, which is almost wholly dependent on bees as pollinators, generates an annual income of over R12 billion and employs approximately 460,000 people.

Insects also impact humans negatively through the transmission of major diseases, such as bubonic plague, sleeping sickness, dengue fever, leishmaniasis and malaria. Plague – which is transmitted through the bite of infected rat fleas – was responsible for the death of more than 75 million people during the 'black death' epidemic in Eurasia in the 14th century. Malaria, which is said to have killed more people than all the wars in history, is transmitted in the saliva of female *Anopheles* mosquitoes. Despite intense mosquito-control programmes, the disease still infected an estimated 214 million people in 2015, causing 438,000 deaths. Although malaria deaths have in fact declined by about 50 per cent over the past decade, there is concern that climate change will now result in the spread of the disease to new areas and that malarial parasites may develop growing resistance to the drugs currently used in treatment.

Anopheles mosquitoes transmit malaria.

The larvae of Indian Flour Moths contaminate stored food products.

Insects can cause widespread damage to agricultural crops. Massive costs are incurred in the manufacture and application of insecticides, and these chemicals have a negative ecological impact on non-target species, including humans. Many insect species attack manufactured products; these include termites and borer beetles, which cause extensive damage to wooden structures and can destroy entire buildings. Clothes moths and silverfish damage woollen and paper products, and cockroaches, beetles and moths infest stored food products, causing major losses to industry. In rural areas, where insecticide use is limited, entire stored grain crops can be destroyed by grain-feeding beetles and moths.

Not all insects are pests. Indeed, most are harmless, and many are hugely beneficial. Many insects that are predators or parasites of pest species are deliberately cultured and used as agents of biological control. Others generate by-products, which are the basis of large and profitable industries. Silk and honey are the most valuable of such products, but insects generate other useful substances, including beeswax, food dyes, shellac and the poisons used by the San (Bushmen) to tip their arrows.

A number of insects, including termites, stink bugs, locusts, beetle grubs and ants, are edible and are even regarded as delicacies. Mopane worms are a well-known local example and the basis of a substantial commercial enterprise. In rural parts of Africa and Asia insects such as caterpillars, termites and locusts are staple food items and crucial resources in times of drought or crop failure. Currently there is growing interest in insect food industries in the western world, as they are a well-balanced food source that has a far lower environmental impact when compared to the livestock industry.

Another, somewhat unusual, use of insects is in forensic entomology, where the insect species (mostly flies and beetles) that are attracted to, and then feed on, a corpse are used to provide accurate estimates of time of death. Different fly and beetle species are attracted to a corpse at various stages of its decomposition, so identifying the species of insect present on a cadaver gives an accurate estimate of the time of death. This is further refined by estimating the age of the maggots, and the temperature of the environment, allowing for very precise estimates of both date and time of death, information often crucial to solving murder investigations. Flies provide a very useful ecological function by feeding on dead animals, breaking down the corpse quickly, and thus both cleansing the environment and redistributing energy and nutrients to other levels in the ecosystem.

Mopane worms are just one example of the many edible insects.

Finally, insects are creatures of great beauty that have evolved novel and fascinating ways of surviving. By examining the insect world, we can observe a wide range of life histories, modes of locomotion, feeding habits and behaviour patterns. Moreover, this can be done simply and easily, without expensive trips to exotic locations, since even the smallest piece of natural bush or suburban garden harbours a wealth of insect life.

ComquatW/CICC by SA 3.0

THE INSECT BODY

The defining characteristics of insects are: a single pair of antennae, 3 pairs of legs, and a body divided into a distinct head, thorax and abdomen.

Insects are the only invertebrates to have evolved wings, although these are present only in adults and have been secondarily lost in many specialized groups (for information on the differences between insects and related groups, see pp.20–21).

The body of an insect is encased in a hardened exoskeleton made up of separate plates named tergites (those on top) and sternites (those below). These are joined together by sections of softer tissue to allow for movement. The exoskeleton provides a sturdy framework onto which the muscles can attach internally and also provides effective protection against predators and harsh environmental conditions. Being external, the skeleton can be moulded into a bewildering variety of shapes and textures and also helps prevent desiccation, allowing insects to inhabit the driest parts of the planet.

The external skeleton of insects accounts largely for their great success in surviving and evolving into the most diverse group, by far, of organisms on earth. The only real disadvantage of the exoskeleton is that it restricts growth, but juvenile insects overcome this by periodically shedding and producing another, bigger exoskeleton underneath, a process known as moulting (see p.12).

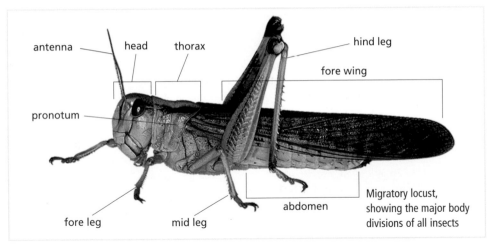

Migratory locust, showing the major body divisions of all insects

HEAD

An insect's head is made up of 6 fused segments on which are carried the mouthparts, eyes and a single pair of antennae. These structures are extremely variable in size and shape and are an important aid to identification, as detailed below.

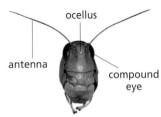

Eyes

Most insects have a single pair of prominent compound eyes, comprising many independent sensory structures called ommatidia, visible as hexagonal facets on the surface of the eye. The eyes may be so large that they meet in the middle of the head, or reduced to just a few facets: in rare cases they are completely absent. In addition to the compound eyes, most juvenile and many adult insects have up to 3 simple ocelli, each with a single lens, which cannot form an image but are very sensitive to changes in light intensity.

Mouthparts

Most insects have chewing mouthparts, consisting of (from front to back) an upper lip or labrum, a pair of prominent jaw-like mandibles, a pair of maxillae and a lower lip or labium. Long, stalk-like sensory organs called palps are attached to both the maxillae and labium, and these are often the most conspicuous visible component of the mouthparts.

However, this basic set of appendages is greatly modified in certain orders. For example, in bugs (Hemiptera), which are all fluid-feeders, the mouthparts have become elongated into a tubular piercing-and-sucking structure called a rostrum. Small bugs such as aphids can target individual plant cells, whereas larger bugs (e.g. twig wilters, p.146) inject a salivary secretion into plant tissue and then proceed to suck up the semi-digested plant fluids using a small pump at the base of the mouthparts, often resulting in the collapse of plant shoots. In butterflies and moths (Lepidoptera), the mandibles have disappeared, and the maxillae have developed into an elongate, coiled proboscis, used to suck up liquids. A third major modification occurs in flies (Diptera), where the mouthparts have become fused into either a blunt suction-pad-like structure modified to mop up liquids, or (in mosquitoes) into an elongate piercing tube.

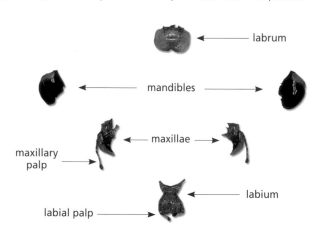

Antennae

These are elongate, segmented structures involved in both smell and touch. Their size and shape are extremely variable and are important aids to identification. Some of the recognized antennal shapes are shown in the box below. Many of the deviations from simple filiform antennae are related to increases in surface area to support the thousands of sensory cells (sensillae) that are used in smell. The plumose antennae of emperor moths (p.396) carry over 17,000 such sensillae, enabling male moths to respond to just a few molecules of the female sex pheromone.

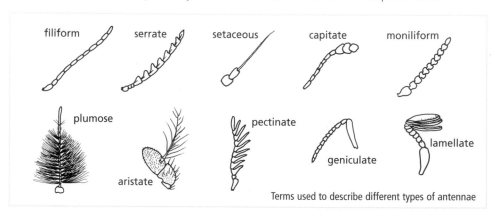

Terms used to describe different types of antennae

THORAX

This is made up of 3 segments, termed (from front to back) the prothorax, mesothorax and metathorax. To each of these segments is attached 1 of the 3 pairs of jointed legs. The last 2 thoracic segments also bear the wings, when these are present.

Legs

The legs of insects are divided into 5 main segments. The basal joint, or coxa, often looks as if it is part of the thorax and is followed by the tiny trochanter, which is easily overlooked. As a result, the third segment, the femur, is the first conspicuous segment and usually the bulkiest. Beyond this lies the thin, elongate tibia and the 'foot' or tarsus, which comprises up to 5 segments and usually ends in a pair of claws. In some cases (e.g. flies and heel-walkers, p.132) there is a membranous

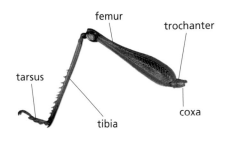

pad between the claws for additional purchase on slippery surfaces. Insect legs are highly variable in length and shape and may be specialized to perform a wide range of functions, including running, swimming, digging, jumping, clinging, producing sound, and capturing and grasping prey.

Wings

Most adult insects have 2 pairs of wings, but 1 or both pairs may be reduced or absent. Wings are always absent in primitive wingless or apterous orders, such as silverfish and bristletails, and have been secondarily lost in other, often parasitic, groups, such as fleas, lice, booklice and many bugs. One pair of wings may also be modified. For example, in flies the hind wings are transformed into small balancing organs (halteres), and in beetles the fore wings are modified into tough covers or elytra, which protect the flying wings. The position, size and pattern of supporting veins in the wings are very important features in insect classification, but because of the complexity of the system used to name these veins, and the cells between them, use of these characters is avoided in this guide wherever possible.

Butterflies (left) have two similar-sized pairs of wings. In flies (above) the second pair is greatly reduced.

ABDOMEN

The abdomen of an insect is usually made up of a maximum of 11 relatively soft segments and contains most of the digestive and reproductive organs. The only appendages are usually the terminal ovipositor of females (which can be very long and conspicuous in some wasps and crickets) and the complicated copulatory structures of male insects. These are derived from the last 1 or 2 abdominal segments. In many of the more primitive orders, such as silverfish (Thysanura), mayflies (Ephemeroptera) and stoneflies (Plecoptera), the tenth segment bears 2 or 3 long, thin, segmented sensory appendages called cerci. In earwigs (Dermaptera) and some diplurans (Diplura), these are modified into stout pincers. In the bugs and advanced insects with a pupal stage, the cerci have been lost.

11

LIFE HISTORY

Because the bodies of insects are enclosed within a rigid exoskeleton, in order to grow they need to moult, or shed their hard outer skin. A new layer of exoskeleton must be grown underneath before the old skin can be shed. When the insect crawls out of its old exoskeleton, the newly exposed skin is still soft and pale and can expand to accommodate the growing body. Once the skin has stretched sufficiently, it hardens and usually darkens through a chemical process. Insects thus grow in a series of discrete 'steps', called instars. The number of instars in the life cycle varies from group to group, but is normally between 4 and 8. It is only in the last, adult, instar that an insect is normally capable of reproduction and flight. Adult insects (apart from silverfish and bristletails) do not grow or moult. Any size differences between adults of the same species can be attributed to differences in food availability during the larval stages.

The life histories of insects can follow 2 very different patterns. In the more primitive groups (hemimetabolous insects), the eggs hatch into nymphs that resemble adults in both body form and habits, but are initially wingless. As they grow through successive moults, the nymphs gradually develop wing pads on the outside of their body, but these only become functional and expand to their full size after the last moult. These exopterygotes (exo = outside; pterygota = wings) include cockroaches, grasshoppers, mantids and bugs. This is demonstrated in the life cycle of the Green Stink Bug (*Nezara viridula*).

In the more advanced (holometabolous) insect orders, the eggs hatch into caterpillars or grub-like larvae, which are quite unlike the adults in both habits and appearance. After almost continuous feeding, these ultimately enter a separate, immobile, pupal stage, during which the entire body is broken down and reorganized into the adult form. This endopterygote (endo = inside; pterygota = wings) type of life history is typified by butterflies, beetles and flies. The advantage of this system is that the juvenile stages can specialize for rapid feeding and growth, and usually consume completely different resources from those used by the adults, so do not compete with them.

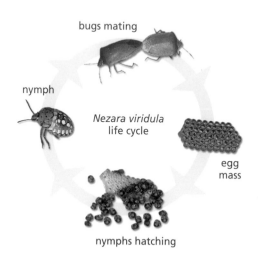

The nymphs resemble adults in the life cycle of hemimetabolous insects such as the Green Stink Bug (*Nezara viridula*).

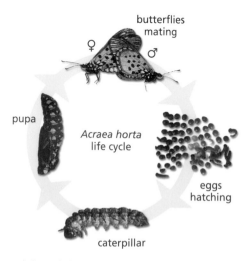

In holometabolous insects such as the **Garden Acraea** (*Acraea horta*), the larva is unlike the adult butterfly.

DISTRIBUTION PATTERNS

Some insect species have very wide natural distributions, occurring all over the region, and extending into the rest of Africa and even further afield. Some have been introduced to the region from distant lands, either accidentally or as biocontrol agents. Most insect species, however, have relatively restricted ranges, often associated with a particular habitat and vegetation type. Hence the distribution of vegetation types is a good proxy for insect distribution patterns.

The vegetation of South Africa can be divided simplistically into 5 major types. In the species descriptions in this guide, these vegetation types are used to describe an insect's preferred habitat.

Many insects are associated with **bushveld**. Such species have a very wide distribution, extending across the northern parts of South Africa and down the east coast. They are generally subtropical species whose presence in South Africa reflects a southern extension of a range that is primarily Afrotropical. The insect fauna of this region is diverse and contains many spectacular species.

The more arid **karoo**, divided into Nama karoo and succulent karoo (Namaqualand), has unique insect assemblages, with an above-average representation of beetles, grasshoppers, flies, wasps and lacewings, many emerging as adults only for brief periods in spring.

The **grasslands** of the Free State, much of Mpumalanga, Gauteng, Lesotho and the Eastern Cape have a distinctive insect fauna with many grass mimics. Termite species dominate these landscapes. The few surviving regions of **forest** in South Africa support rich butterfly and other insect communities.

The **fynbos** regions of the Western Cape, Cederberg and the southern Cape have a unique assemblage of insects of great evolutionary interest. Many are found only at the southern tip of Africa and southern parts of the continents that were joined 150 million years ago to form the supercontinent Gondwanaland. These groups have their closest relatives in New Zealand, Australia, South America and Madagascar. Their present distribution reflects a long history of passive migration on the slowly separating subcontinents of Gondwanaland. Some also occur in the Drakensberg mountains of KwaZulu-Natal and Mpumalanga.

Stream insects are a diverse and important group that respond to various water quality parameters, such as flow, oxygen content, temperature, nutrient content and pH. The nutrient-poor, acidic waters of the southwestern Cape are stained brown by humic compounds and support a distinctive insect fauna. Further inland, the neutral to alkaline turbid waters carry a higher sediment load. They too have their own distinctive aquatic insect fauna.

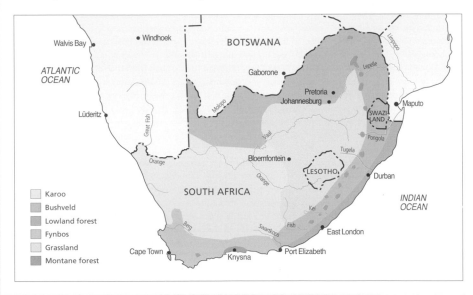

13

OBSERVING AND COLLECTING INSECTS

The main aim of this guide is to encourage readers to learn about and appreciate live insects in their natural habitats. However, some focused collecting and preservation of insects is necessary for scientific or educational purposes, or simply to identify species that are too small or elusive to be examined in the wild.

In light of the widespread and largely unregulated use of insecticides in South Africa, it might appear surprising that the collection of insects requires a collecting permit in all provinces. This is especially true of areas that fall under the control of South African National Parks, provincial or local nature conservation bodies and the Department of Water Affairs and Forestry. A few insect species (mostly certain rare or range-restricted butterflies and beetles) are also completely protected.

Limited collecting from private property or disturbed sites is unlikely to be questioned, but you should nevertheless check this with the appropriate nature conservation body in your region. If you do decide to make a collection, you should avoid killing more insects than you really need. One or 2 specimens of each species, perhaps a male and a female, are sufficient for most purposes.

Properly documented and conserved collections made by amateurs have often made a useful contribution to our knowledge of insects. The guidelines that follow (pp.15–19) may be supplemented by comprehensive instructions in books, such as that by Uys et al (p.507).

INSECT CLASSIFICATION

All animal species are classified at the broadest level into major groupings called phyla, each of which shares a unique body plan. For example, mammals, birds and reptiles all fall within the Phylum Chordata (animals with backbones).

Insects are classified within the large Phylum Arthropoda, members of which are characterized by a hard, jointed external skeleton and jointed legs. This group is split into several subphyla, which include the Chelicerata (spiders and their kin), Myriapoda (centipedes, millipedes etc.) and Crustacea (crabs, prawns, etc.). Insects are grouped as Class Insecta within another of the subphyla, Hexapoda ('six-legged'), which also includes the class Entognatha (internal mouthparts) made up of subclass Collembola and orders Diplura and Protura – primitive wingless groups that were once also considered to be insects, and remain their closest relatives. Two of these groups, the Diplura and Collembola, have been included in this book, as they are quite common, and easily confused with true insects. However, apart from sharing the 6-legged condition with insects, they have a very different body design from that of insects.

The phyla are in turn broken down into smaller groupings – classes, orders, families and genera (as well as a number of other, less frequently used groupings), each of which consists of more and more closely related species. Within the subphylum Hexapoda, there are 33 orders, 30 of which occur in South Africa.

Orders, in turn, are subdivided into families (the names of which always end with the suffix '-idae') and genera. An order may contain any number of families, from 1 to over 100, and a family, likewise, any number of genera.

Individual species within a genus are ultimately identified by a unique combination of 2 names, the binomial, which is always italicized. The first of these is the genus name and the second name that of the component species. For example, the House Fly has the binomial *Musca domestica*.

Few insect species have common names. Although common names may be familiar to many readers, they are not a reliable or standardized means of identification. Many species have multiple common names in different regions, as well as in each different language. Moreover, the same common name is often applied to a whole group of insects and, indeed, the same common name is sometimes used to describe entirely different species in different geographical areas. A pictorial guide to help readers assign specimens to the correct order is given inside the front and back covers of this book.

WHERE TO FIND INSECTS

Using a USB microscope in the field

One of the benefits of observing and collecting insects is that you can find a large variety of species wherever you are. However, the more different the habitats you explore, the more types of insect you are likely to discover. Moreover, many species are present seasonally, or are active only under particular weather conditions, or at specific times of day or night. You will thus come across different species at the same site at different times. Remember that insects span a wide range of sizes and you may miss much of the diversity if you restrict your attention to larger, more obvious, species. Get in close, or search with a magnifying glass. A 10x or 20x magnifying hand lens (obtainable from watchmaker suppliers) is an essential part of any serious insect collector's equipment. USB microscopes – small, inexpensive, portable digital microscopes that display live images onto a laptop computer – have also transformed our ability to observe tiny, live creatures in the field. Conventionally these microscopes are used indoors and mounted on a small, downward-looking stand. They are, however, easily modified to be held on a small tripod, or even a desk-lamp stand, and used to observe live subjects in the field. Collections of digital photographs made through such a microscope, or with conventional cameras (see p.18), are a very useful alternative to collections of dead specimens, offering a number of advantages (no need to kill subjects, no permit required, live colour and natural posture recorded, easy to store, can be copied and shared).

The most obvious places to look for insects are on flowers, fruits and leaves, or on and under the bark of trees or under stones and logs. Many insects are restricted to particular plants, so search as many plant species as possible. Dung and decomposing plant or animal material also have specific associated faunas, as do other specialized habitats, such as deserts, beaches, gravel plains or salt marshes. Many insects are aquatic, and a search in and around the wet margins of dams and streams will yield many species not found elsewhere. There are many other specialized habitats to look in, including wood, fungi, stored products and the bodies or nests of other creatures, many of which host a wide variety of parasitic insects. A number of nocturnal insects are attracted to light, and searching in the vicinity of outside lights, particularly in country areas, can be very productive.

COLLECTING METHODS

The various methods of collecting insects may be broadly divided into 2 categories: active hunting and baiting or trapping.

Active hunting

Visual detection is an effective method, and is the best way to observe an insect's habits and behaviour patterns. Once spotted, many insects can simply be collected in a small transparent container. Glass pill vials ('polytops') or similar small plastic canisters are ideal, but make a small air hole in the lid if you want to keep your insects alive for any length of time.

An insect net – useful for catching active and large species

Using a pooter to collect small insects

A pooter is another simple device that can be very useful for collecting small, fragile insects without damaging them. It consists of a small, resealable, transparent chamber with a tube at one end through which an insect can be suctioned up, and one at the other end, the tip of which is covered with mosquito netting to prevent accidental ingestion. The user simply 'vacuums' the insect up through one tube and into the collecting chamber by sucking on the other tube.

Larger or more active species need to be captured with an insect net. This should be at least 30cm across and made of fine netting, tough enough to withstand being dragged through vegetation (nylon curtain netting is adequate). The bag of the net should be at least twice as long as its diameter, allowing it to be folded over the frame after each stroke, trapping the insects within it. The rim may be made of metal rod, but where the netting folds over it, a strip of heavier reinforcing material (such as canvas) should be added to reduce wear. A stout wooden dowel or length of aluminium tubing, about 1m long, makes a good handle.

Hand nets can be used in many different ways. The most productive is sweeping, i.e. systematically swinging the net from side to side so it brushes through vegetation. This method yields many small, cryptic species, but a thorn bush or two can quickly reduce your net to tatters! Another good method is to hold the net below a branch to catch any insects that fall as the branch is shaken or tapped vigorously with a stick. Nets can also be used to sweep through water, or be placed across a stream to catch insects that you dislodge from vegetation or stones upstream. Nets used for beating, or in water, are normally more robust in construction than the usual 'butterfly net', and the bag should be constructed of shade cloth or very tough netting material.

Baiting and light trapping

Alternatively, insects can be collected with baits or traps. Lures to attract butterflies, moths, flies and beetles may be as simple as fermenting fruit, carrion or dung – or even a captive female, which can be used to attract potential mates. Fruit baits of ripe bananas mixed with sugar and a little beer are the most successful, while pan traps, comprising any yellow, red or orange container filled with water, will lure and trap a variety of pollinators.

Light traps are particularly effective in attracting nocturnal flying insects. A simple version can be made by suspending a bright light, or one rich in ultraviolet radiation (such as a mercury vapour or fluorescent UV lamp), in front

Using a light trap to attract insects

of a white sheet that extends onto the ground (where insects dropping from the sheet can be seen easily). Light traps work best on warm, still, moonless summer nights and if placed at a distance from other light sources. Periods just prior to rain are particularly effective.

Pitfall traps are useful for collecting a range of ground-dwelling insects. They consist of vertical-sided containers buried with the rim flush with the ground and half-filled with water laced with a few drops of detergent, or some antifreeze. It is advisable to use small-diameter tubes, to exclude small mice and lizards that may inadvertently be trapped, and to monitor such traps daily.

KILLING AND PRESERVING INSECTS

Any specimens you wish to retain for a collection must first be humanely killed. This is best done by freezing them for at least an hour (preferably overnight), or placing them in a killing jar. To make a killing jar, place a layer of cotton wool at the bottom of a small, wide-mouthed jar, then pour over it a 10–20mm layer of sloppy plaster of Paris. Before the plaster hardens completely, make a few small holes through it with a narrow rod, so that the killing fluid can penetrate through to the cotton wool. When the plaster has set and dried, pour enough ethyl acetate over it to moisten the cotton wool and plaster thoroughly. Place some crumpled tissue into the jar to prevent the insects damaging themselves, and close the lid firmly. To kill your specimens, put them into the jar one, or a few, at a time and leave for at least 30 minutes. It is best to kill only a few insects at a time to prevent them from damaging one another. The ethyl acetate will evaporate after a few days and must be replaced.

Most dead insects can be preserved by simply air-drying them, after pinning and setting. However, soft-bodied insects, such as termites, aphids and most aquatic insects, as well as larval forms such as caterpillars, should be preserved in leak-proof vials filled with 70 per cent alcohol. Frozen or dried insects can be stored for long periods before they are set and mounted.

Killing jar charged with ethyl acetate

MOUNTING AND LABELLING INSECTS

Proper stainless steel entomological pins should be used for pinning insects, as ordinary pins are too thick and short and tend to rust. The pin is placed vertically through the thorax, slightly to the right of centre. Larger insects, especially butterflies and moths, are then 'set' with the wings and legs spread to display important identifying features. Most insects are set by arranging the limbs neatly and holding them in place with pins until they have dried. For butterflies, moths, dragonflies and lacewings, setting boards are used. These are cork or polystyrene boards with a groove along the centre, into which the insect's body is recessed. The wings can then be positioned horizontally and held in place with thin strips of plastic or paper, secured with additional pins. Insects too small to be pinned directly are either glued with alkaline wood glue onto the apex of a small triangular piece of white card ('pointing'), or attached to a small rectangle of card with a special miniature headless pin called a minuten. These cards or blocks are then mounted onto conventional pins.

How to label pinned insects

If your insect collection is to have any scientific or reference value, it is essential that each specimen is labelled with the exact location and date of collection, the name of the collector and, if possible, the identification of the specimen. Notes on habitat, behaviour, abundance, diet and the like are also useful. Labels may be neatly written by hand with a fine black ink pen, but are best generated in standardized form with a computer, printed at a small font size. The small rectangular label should be impaled on the pin below the specimen, or pencil labels placed into a vial along with an alcohol-preserved specimen.

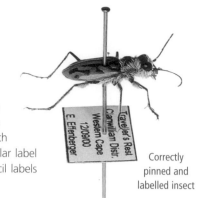

Correctly pinned and labelled insect

DISPLAYING AND CURATING

Pinned insects are fragile and prone to attack by pests, such as museum beetles. They thus need to be stored or displayed in a flat, sealed drawer or box lined with cork or high-density polystyrene and with a tight-fitting glass lid. An insect repellent such as naphthalene ('mothballs') placed in each drawer will protect the specimens against museum beetles. These beetles enter through minute spaces, so the lid needs a perfect seal. Specimens are normally arranged in rows, with related species grouped together. Modern museum collections often use unit tray systems, in which the drawers are filled with small card boxes each containing insects of a single species. This allows the collection to be expanded or rearranged without having to handle individual specimens.

Boxed insect collection

INSECT PHOTOGRAPHY

A collection of digital insect images is in many instances superior to a collection of specimens. The latter are of course essential for scientific purposes, but many successful citizen science projects rely on digital collections. Moreover, much specialized equipment is required for insect collections, whereas most people have a device capable of taking high-quality images. Insects are extremely rewarding subjects for photography, since they occur virtually everywhere, come in a spectacular variety of body forms and colours, and exhibit interesting behaviour patterns. However, as insects are small, taking good photographs of them requires the right equipment and some skill. Fortunately, camera technology has advanced dramatically in recent years and modern compact or even smartphone cameras offer close-focusing capabilities sufficient to take quality images of at least the larger insects. However, as with most technological purchases, there remain trade-offs between price, capability and convenience.

Casual photographers can take good insect images with a smartphone, this option offering relatively low cost and excellent convenience, especially for opportunistic observations. Most smartphones will focus on subjects as close as 5–10cm from the lens, good enough to capture larger insects. Inexpensive clip-on lenses can decrease this to 1.5cm, but many insects are difficult to approach that closely, and at that distance the camera may cast shadows on the subject. The fixed

Using a DSLR camera fitted with flashes and macro lens

aperture of smartphone cameras is also incapable of providing as much depth of field as is available using other options.

Compact cameras are also relatively inexpensive and convenient, and some have the additional advantages of being waterproof and shockproof. Most have excellent macro capabilities and can focus as close as 1cm, although again that working distance is uncomfortably close for skittish subjects. Some models offer additional specialized features that greatly enhance their close-up capabilities. These include a ring of LEDs around the lens, or a ring-light that redirects the LED illuminator evenly around close-up subjects (unlike flashes, these also permit shadow-free video to be filmed). Some models even offer 'focus-stacking', where the camera takes a series of images in quick succession, shifting the focus each time, then digitally combines them into a composite, in-focus image!

Although by far the most expensive and bulky option, the majority of serious insect photographers still opt for DSLR cameras, used in combination with specialized close-focus or macro lenses. The preferred macro lenses for insect photography have fairly long focal lengths (typically 90–105mm), allowing the subjects to fill the frame without the need to approach them too closely. Another advantage of DSLRs is that extension rings can be inserted between the lens and camera body to achieve even greater magnification, allowing subjects just a few millimetres long to fill the frame.

Insect photographers usually aim for images that are 'sharp' (that freeze any movement, which can originate from either the subject, or from camera shake) and have a large depth of field, and to accomplish this, one needs both a short exposure time and a small aperture. Unfortunately those camera settings, plus use of longer focal length lenses and (especially) extension tubes, all reduce the amount of light entering the camera. It thus becomes essential to boost the intensity of light falling on the subject, even in strong daylight, and this is achieved using one or more electronic flashes. Unfortunately, the position of the built-in flash, normally on the top of the camera, prevents this from being pointed at extremely close-up subjects. Thus, a free-standing flash needs to be connected to a cable (or remotely triggered using a slave unit) and positioned close to the subject and pointed directly at it. Optimally, paired flashes, or a specialized ring-flash unit, are used to eliminate or control shadows. As the duration of electronic flashes is 1/1,000 sec or less, this also effectively freezes any movement.

Most of the photographs in this book were taken using a Nikon DSLR camera with either 90 or 105mm macro lenses and an off-the-camera flash, with the aperture set at f32 or smaller. Some were taken with an Olympus Tough TG-3 compact camera fitted with a ring light and using 'microscope' and focus-stacking options.

Compact camera with light ring and focus-stacking options

INSECT RELATIVES

Insects are the only invertebrates with 3 pairs of legs, a body divided into a separate head, thorax and abdomen, and wings (although wings have not evolved in a few primitive groups and have been secondarily lost in some advanced ones). Anyone collecting insects will, inevitably, come across species in related groups. This section is intended to help the reader identify the most common of these.

ORDER ARANEAE   SPIDERS

Body divided into a fused head and thorax (cephalothorax) and a generally unsegmented abdomen, the two separated by a narrow constriction. There are no antennae, usually 8 eyes, a pair of chelicerae (fangs), a pair of pedipalps and 4 pairs of legs. The abdomen ends in a group of silk-producing spinnerets, but has no other appendages. Spiders are an abundant and diverse group, noted for their ability to construct silk webs, snares, shelters and egg sacs. All are predatory and some are venomous.

ORDER OPILIONES   HARVESTMEN

Similar to spiders, but with no constriction between the fused head and thorax and the abdomen, which is short, broad and segmented. A tubercle in the centre of the thorax has a single eye on either side. There are no antennae, and appendages consist of a pair of small, slender chelicerae (fangs), a pair of short, leg-like pedipalps and 4 pairs of very elongate slender legs. Harvestmen are abundant in moist vegetation, forest litter and caves. Most feed on small arthropods or snails, some on dead plant or animal matter.

ORDER SCORPIONES   SCORPIONS

An ancient group of relatively large species, with greatly enlarged, pincer-like pedipalps and elongate, jointed abdomen, which ends in a swollen sting. Antennae are absent and the mouthparts are formed by small pincer-like chelicerae (fangs). There is a central pair of dorsal eyes, 2–5 smaller lateral pairs of eyes and 4 pairs of walking legs. Abundant in arid areas, they emerge at night to feed on insects and other invertebrates, which they catch with their pedipalps and paralyse with their sting. The neurotoxic venom can be dangerous to humans.

'ACARI' (7 ORDERS)   TICKS & MITES

The group is best recognized by the apparent lack of any body divisions. Antennae are absent, there are 4 pairs of walking legs, and the mouthparts and pedipalps are usually inconspicuous. Ticks are external blood-sucking parasites. Mites are usually minute and occur in most habitats. Some are parasitic, others may be free-living carnivores that scavenge or pierce and suck out the contents of plant cells. Many species are of great economic and medical importance.

ORDER AMBLYPYGI — WHIPSCORPIONS

A specialized group of strange, flattened, spider-like creatures with no antennae, long, slender legs and enormously enlarged spiny pedipalps that are held flexed in front of the body. The first of the 4 pairs of walking legs is modified into elongate, whip-like feelers. Whipscorpions are secretive animals, usually found under bark or boulders in subtropical forests or arid zones. At night they move about smoothly with a crab-like gait, feeling for insect prey with their elongate first pair of legs, and capturing prey with their pedipalps. Despite their fearsome appearance they are harmless to humans.

ORDER SOLPUGIDA (SOLIFUGAE) — SUNSPIDERS

Easily recognized by the enormous chelicerae (fangs) projecting from the front of the body as a pair of vertically articulating pincers. Antennae are absent, and there are 4 pairs of walking legs. The elongate sensory pedipalps resemble a fifth pair of legs, but are not used for walking. Sunspiders are usually diurnal predators, locating insect prey with the sensory pedipalps, then capturing and tearing their prey apart with their powerful fangs. They are not dangerous, but can inflict a painful bite.

SUPERCLASS MYRIAPODA — CENTIPEDES & MILLIPEDES

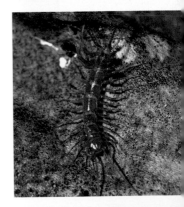

Easily recognized by their elongate body form, the absence of a thorax or abdomen and the presence of walking legs on each body segment (except the first and the last one or two). Centipedes are usually flattened and have 1 pair of legs per segment, those of the first segment being modified into poison fangs. Millipedes are cylindrical and made up of 'diplosegments', each representing 2 fused body segments and hence bearing 2 pairs of legs. Centipedes and millipedes are widespread and common. Centipedes are active predators and larger species can inflict a venomous bite, but barring allergic reactions, are not dangerous. Millipedes are harmless, slow-moving and feed on decomposing plant material.

CLASS CRUSTACEA — CRABS, SHRIMPS, WOODLICE, ETC.

Abundant and diverse in marine and freshwater systems, but relatively rare on land. Crustaceans have 2 pairs of antennae, and a variable number of legs. Woodlice (Isopoda) and landhoppers (Amphipoda) have 7 pairs of walking legs, but the bodies of isopods are flattened, while those of amphipods are compressed sideways. Woodlice and landhoppers are both confined to moist habitats such as leaf litter and compost heaps, where they scavenge on dead plant and animal material. The eggs are brooded in a marsupium under the body of the female and hatch into young that resemble small adults.

HOW TO USE THIS BOOK

This guide is designed for use in South Africa, but will be useful to readers throughout the southern African region and beyond, since many of the species featured have distribution ranges that extend into equatorial Africa.

It is impossible to include more than a representative subsample of species in a guidebook of this size and popular nature. We have concentrated, therefore, on providing information on every insect order found in the region, and as many as possible of their component families. Within these, we have featured species most likely to be encountered by readers. The species included thus tend to be those that are abundant, widespread, economically or ecologically important, conspicuous, large or unusual.

Do not be surprised if you are unable to find entries that exactly match the species you have collected. You can still learn much about their biology by identifying them to order and family level and reading the general text about these groups. The different species within a family often have much the same biology and behaviour patterns.

This guide has drawn not only on the accumulated knowledge of the authors, but has relied heavily on the knowledge of numerous specialists. Given the extraordinary number and diversity of insect species, their ubiquity across all terrestrial and freshwater habitats, and their vast importance to human life, we hope that this guide will open the fascinating world of insects to you.

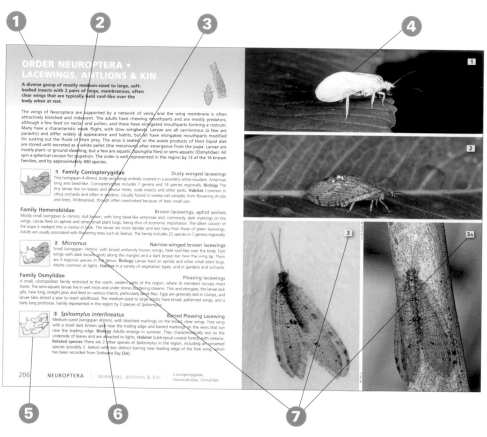

NUMBER KEY

❶ ORDER Features shared by all members of the order. Notes are included on the biology, size range, habitat and number of component species in the region.

❷ FAMILY Features that identify the family and the biology of its members. Although species may often be identified directly from the photographs, the family entry provides additional information. For some lesser-known groups, or those containing many cryptic species, identification is provided only to family level.

❸ SPECIES OR GENUS Notes on approximate size, habitat and biology of the species or genus that is depicted. An entry to genus level usually only indicates that the genus contains a number of very similar species. Bear in mind that many genera resemble one another closely, and that we have not been able to illustrate them all. 'Related species' contains information on other, often similar, members in the genus. Occasionally similar species are described that are unrelated to the species in the entry, for example mimics.

❹ PHOTOGRAPHS Wherever possible, these were taken under natural conditions and show the insect in its normal posture and typical habitat. Note that the insects are usually magnified to appear at similar sizes on the page. Small species have thus been enlarged far more than bigger ones. Since size is a key identification feature, it is important to check the actual size of species, as given in the text entries, when trying to make identifications.

❺ MAPS These give an indication of the range of each species or group featured. The distributions of many insects are poorly known, but as distribution is an important aid to identification, maps are provided that give at least a broad indication of where the insect is likely to occur. In general, the maps are not intended to convey the precise and total distribution, although most have been drawn from the existing museum and literature records for that species. Identifications should not be dismissed if they fall outside of the range indicated; the specimen may be a new record for the area and should be passed on to an entomologist or museum. It is notable that climate change has resulted in numerous recent observations of insects (and other species) well outside of what were traditionally considered to be their natural ranges. Insects, being mobile and having short life cycles, can rapidly occupy and establish themselves in suitable new areas as conditions change.

❻ SIZE The sizes given, unless otherwise stated, are wingspans or adult body lengths, excluding antennae and other appendages. A certain amount of variation around these adult lengths is to be expected. Usually, the given length is that of the specimen photographed, but a size range is given where the adult size is known to be very variable.

❼ NUMBERING SYSTEM The species or group entries on each page are numbered sequentially to match the photographs on the opposite page. Additional images of related species, or of larval or other life-history stages, are indicated by the addition of a letter to the number, e.g. 3A, 3B.

23

SUBCLASS COLLEMBOLA • SPRINGTAILS

Minute, wingless, 6-legged arthropods with soft, cylindrical or globular bodies; best recognized by the unique forked springing organ (furcula) held folded beneath the abdomen.

Collembolans, together with members of the orders Diplura (next page) and Protura (rarely encountered and not treated in this book), have now been grouped into the Class Entognatha, which are distinct from true insects (Class Insecta), but are traditionally still included in insect identification guides. Most collembolans are only 0.5–3mm long and are white or grey, although some can be brightly coloured. Mouthparts are of the biting or sucking type, but are hidden in head folds, and antennae have only 4 segments. Eyes are simple or absent. The thorax lacks any sign of wings, and the abdomen has only 6 segments, as opposed to 11 in true insects. The first abdominal segment bears a prominent 'ventral tube' – a unique organ used for adhesion and water balance. The ventral part of the fourth abdominal segment often has a springing organ (furcula) that is folded forwards under the body and clipped in place by a modified appendage (tenaculum) located ventrally on the third abdominal segment. When springtails are disturbed, blood is forced into the furcula, driving it downwards and backwards to strike the ground and hurl the animal into the air.

Springtails are among the most widespread and abundant of all terrestrial invertebrates and are particularly abundant in soil or leaf litter, floating on the surface of pools, or in other moist, sheltered habitats. Immature stages look and behave much like adults, and they moult throughout their lives. There are currently more than 8,200 species globally, with over 120 species in 17 families known from South Africa.

1 Order Poduromorpha
Plump springtails

Body plump and cylindrical, with relatively short limbs and antennae. First thoracic segment similar in size to second and third, and clearly visible from above. Body surface generally granular. Both families Onychiuridae (white) and Neanuridae (orange) are shown among compost in (**1**). The slate-coloured *Anurida maritima* (**1A**), Family Neanuridae, is the best-known species and easily identified by its marine habitat. **Biology** Most species feed on detritus or microorganisms. *Anurida maritima* scavenge on dead or dying animals, such as barnacles and beach hoppers, and retreat during high tide into intertidal crevices or the shells of dead barnacles, where a trapped air bubble affords shelter. Eggs are laid in loose aggregations in rock crevices. **Habitat** Common in soil and leaf litter and in habitats such as compost heaps, where several species may occur together. Many species also live on or near water. *Anurida maritima* commonly float in clusters on the surface of rock pools on the seashore.

2 Order Entomobryomorpha
Long springtails

Best distinguished by body shape, being relatively large, elongate and cylindrical, with longer legs and antennae than Poduromorpha (**2**), and with large furcula, best seen in side view, as per *Seira barnardi* (**2A**). First thoracic segment much reduced and often invisible from above. Body surface usually covered with hairs or scales. **Biology** Varied. Most species feed on decomposing plant and animal material and associated microorganisms. **Habitat** Diverse group found in a wide range of habitats, in or on soil, leaf litter, vegetation, caves and the nests of ants and termites.

3 Order Symphypleona
Globular springtails

Short, very round and almost globular, with little sign of segmentation, the thorax and abdomen appearing as a single unit. Antennae long and are often bent or elbowed between third and fourth segments. *Sminthurus viridis* (**3**) is typical. **Biology** Among the most athletic of collembolans, some species able to jump up to 30cm. Certain species, including *S. viridis*, which feeds on lucerne in the Western Cape, may damage crops. **Habitat** Most species live in the upper layers of leaf litter, on grasses or in other low vegetation. The pink species (from family Dicyrtomidae) shown in (**3A**) is common under rocks along stream margins and often found floating on pools of water.

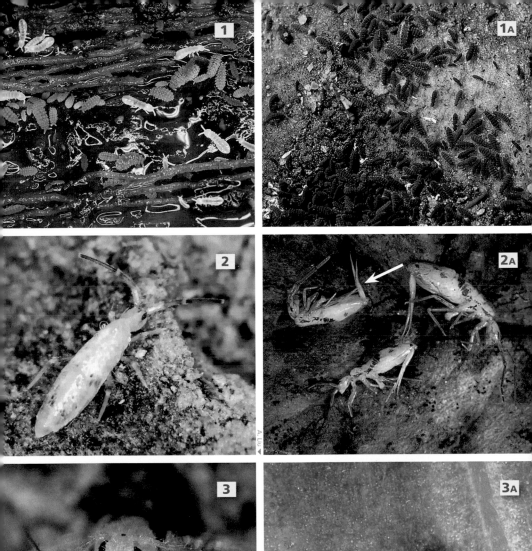

ORDER DIPLURA • DIPLURANS

Narrow-bodied, small to medium-sized, largely unpigmented, blind and wingless hexapods. The abdomen ends in a pair of appendages (cerci) that can be either long, slender and segmented, or modified into stout, pincer–like structures.

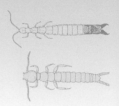

Like collembolans (previous page), diplurans, sometimes termed '2-pronged bristletails', are not true insects, but are usually included in insect identification guides, since they are easily confused with silverfish and bristletails (following pages). Most species are small, delicate, soft-bodied and pale in colour, although a few reach 50mm and have hardened and darkened cerci. Mouthparts are partially sunk into the head, which lacks both compound and simple eyes (ocelli). Antennae are elongate, mobile and made up of 10 or more conspicuous bead-like segments. Thoracic segments are clearly visible and lack any sign of wings. The abdomen is slender and has 11 usually more or less parallel-sided segments. The form of the cerci differs in the various families, being either long, tapering and segmented, or short, stout and pincer-like.

Diplurans live concealed under stones, in soil, rotting vegetation and logs or in other dark, moist places. They have biting mouthparts and may feed on dead organic matter or on a variety of live prey. Fertilization is external: males deposit stalked spermatophores, which females collect with their genital openings. Eggs are laid in soil or rotting vegetation. Development is direct, and juveniles resemble adults in both habits and appearance. About 800 species are known globally, fewer than 40 of which have been reported in the region.

1 Family Campodeidae

Small (body length 3–4mm) and pale. Characterized by the form of the cerci, which are elongate, conspicuously jointed and sometimes longer than the abdomen from which they project. About 7 genera and 27 species are present in South Africa, the most common genera in the region being *Anisocampa* and *Campodea*. The species shown here is an *Anisocampa* (body length 3mm) photographed among leaf litter on Table Mountain. **Biology** Feed on detritus and soil fungus, as well as mites, springtails and other small soil invertebrates. **Habitat** Under stones or among leaves and plant debris in damp coastal forests and mountains, particularly in the Western Cape. Common, but most species are so small they are easily overlooked.

2 Family Japygidae   Forceps tails

Relatively large (body length 10–20mm) and elongate, cerci forming stout, hardened, pincer-like structures at end of abdomen, reminiscent of the forceps of earwigs. The species shown here is a *Japyx* (body length 18mm) taken from leaf litter in the Knysna forests. Four genera and about 12 species are known from the region, most of them within the genus *Japyx*. **Biology** Carnivorous, capturing other small arthropods such as springtails, isopods and insect larvae, either with the mouthparts or by grasping them with the pincer-like cerci. Eggs are laid in soil and guarded by the female. **Habitat** In soil, damp leaves, rotting compost and logs or under stones in the moist coastal and damper eastern parts of the region.

ORDER ZYGENTOMA • SILVERFISH

Primitive, wingless insects with soft, spindle-shaped, flattened bodies, eyes well separated (or absent), and 3 long, equal, segmented, terminal appendages.

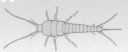

Until recently the Zygentoma were regarded as a suborder of the Thysanura and appeared under that heading in earlier editions of this book, but the group has now been given full order status. Silverfish are so called because of their undulating 'fish-like' locomotion. They have a gleaming, metallic appearance and greasy feel caused by the tiny overlapping scales that cover the bodies of most species and that rub off easily when touched. Most species are nocturnal and rely on their antennae and long terminal appendages to feel their way in the dark. They have a simple life cycle, the young resembling small adults, and are well known for their ability to absorb water vapour from the atmosphere and hence survive without drinking. The most familiar silverfish are introduced domestic species, often found trapped in bathtubs, sinks and cupboards, but there are also many indigenous species that live in leaf litter, in caves and under rocks, or in the nests of ants or termites. Some 370 species are known globally, of which about 50 have been reported in the region.

1 Family Lepismatidae

Small to fairly large (body length 4–15mm), characterized by small compound eyes and no simple eyes. Body covered with closely fitting scales. Eight genera and about 40 species in the region. **Biology** Food consists of dry organic matter, some species possessing enzymes capable of digesting cellulose. Those species that live in ant nests are assumed to scavenge from their hosts. **Habitat** Domestic species such as the pale *Ctenolepisma longicaudata* **(1)**, which was introduced to South Africa from Europe over 100 years ago, are the most commonly encountered, often among old books and papers. They feed on sugars and starches and can cause considerable damage to the bindings of books and to stored documents **(1A)**. Ant-loving (myrmecophilous) species, including *Afrolepisma* sp. **(1B)**, are restricted to the nests of ants and termites. Indigenous species, such as the widespread *C. grandipalpis* **(1C)**, are found under stones and leaves. Other than *C. longicaudata*, 2 other introduced domestic silverfish occur in the region; *C. urbana* is very similar in both distribution and appearance to *C longicaudata* (indeed these species are often confused in the literature). *Lepisma saccharina* **(1D)** requires cooler, moister conditions and is only found around Cape Town. Under the microscope the marginal setae in *Lepisma* are bare, whereas those of *Ctenolepisma* are barbed or feathered.

2 Family Nicoletiidae

Small (body length 3–5mm), completely eyeless and lacking body scales, but may be covered in fine hairs, especially on the head. Three genera and 7 species occur regionally. Most species are golden yellow and teardrop-shaped, with a compact rounded thorax and tapering abdomen. Short abdominal appendages. The few species that live outside of ant and termite nests are more slender, with elongate abdominal appendages. **Biology** Ant-loving species are assumed to scavenge from their hosts. May be parthenogenetic, since males are rarely found in collections. **Habitat** Most species restricted to ant or termite nests. Free-living species live on the forest floor or in caves.

Wingless, tapering insects closely resembling silverfish, but distinguishable by laterally flattened body, arched thorax and large compound eyes meeting in the middle of the head. Central abdominal appendage also much longer than the lateral appendages.

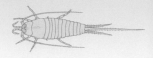

Bristletails are an ancient group of primitive insects with tapering, wingless bodies that are laterally, rather than dorsoventrally (as in Zygentoma), compressed and that are typically arched, especially over the thorax. The head bears large compound eyes that meet over the top of the head, as well as 3 ocelli. There are 3 abdominal 'tails': the central tail is much longer than the 2 lateral tails (whereas in Zygentoma they are more or less equal). When disturbed, bristletails can jump high in the air by snapping the abdomen downwards. They are less tolerant of dry conditions than silverfish, and most are found in leaf litter or under bark and rocks in moist habitats. They feed mainly on algae, lichens or decaying organic matter. Males attach spermatophores to a thread spun from the abdomen, and these are picked up by the female following a courtship dance. The eggs are laid in crevices, and the young resemble tiny adults, taking up to 2 years to reach maturity. Unlike most insects, bristletails continue to moult after reaching adulthood. Lifespan is unusually long at up to 4 years. There are about 350 known species globally, of which fewer than 20 occur in the region.

1 Family Meinertellidae

The only family of bristletails found in the region. *Machiloides* is the most diverse and common genus, the component species differing in characteristics including colour of the body scales, eye size and eye colour **(1, 1A)**. Represented by 4 regional genera and 18 species. Hardly any research has been undertaken on this group, and the last taxonomic account for the region was published in the 1950s. Almost certainly any new survey and analysis, using modern taxonomic and genetic techniques, would reveal that many more species occur in the region than are currently described. **Biology** Nocturnal, although some species crawl about actively during the day. Feed on algae, lichens and dead organic matter. Reach maturity in about 2 years and live approximately 3 years. **Habitat** Under stones or logs, especially in indigenous forests in southern and eastern parts of the region. The dark and apparently undescribed species shown in **(1B)** has unusually long antennae and tails and is restricted to the twilight zones of sandstone caves on the Cape Peninsula.

ORDER EPHEMEROPTERA • MAYFLIES

Delicate, water-associated insects with 2 or 3 long, jointed caudal filaments (cerci) extending from the tip of the abdomen in both adults and nymphs. Adults very short-lived; wings held rigidly upright when at rest; hind wings much reduced or absent.

Adult mayflies are usually found near water and most species live for only a few hours, as they lack functional mouthparts and are unable to feed. They have short antennae and large compound eyes, those of males sometimes being divided into distinct upper and lower sections. The legs are slender and weak; the first pair in males is elongated for grasping females in flight. The second pair of wings is much smaller than the first and may be absent. Uniquely among insects, they moult a second time after attaining the winged state. The dull subimago or 'dunn' emerges from the nymph and, after a few hours, moults again into a shinier, clear-winged adult or 'spinner'. Mating swarms of adults often gather over streams. Identification beyond family is difficult and based mostly on wing venation and male genitalia. Nymphs are aquatic, breathing by means of tracheal gills on the side of the abdomen. They are more commonly encountered than adults and form an important component of the invertebrate fauna of streams and rivers. Of the 3,000 species known globally, about 105 occur in South Africa.

1 Family Leptophlebiidae
Prong-gill mayflies

Bodies of nymphs slightly flattened with 3 widely spread caudal filaments of equal length. Gills characteristic in form; first gill may be absent or formed from a single flattened plate partially covering the other gills; remaining gills each split into 2 lobes, which may be long and pointed, as in **(1)** *Aprionyx*, or leaf-like, or split into filaments at their tips, depending on species. Adults have hind wings present but reduced, wings sometimes suffused with brown; abdomen ends with 3 long caudal filaments, as in **(1A)** *Aprionyx*. Eyes of males divided into a large upper section made up of large facets and a smaller lower portion made up of smaller facets. One of the more diverse families in the region, with about 20 known species. **Biology** Nymphs shred dead leaf material. **Habitat** Nymphs usually found clinging to rocks or gravel in flowing waters, more rarely in still waters.

2 Family Heptageniidae
Flat-headed mayflies

Nymphs distinctive, with very broad, flattened and oval heads, short antennae (shorter than head) and flattened, sprawling legs, e.g. *Afronurus harrisoni* **(2)**. Gills leaf-like with a tuft of filaments beneath gills 1–6, but not 7. Adults **(2A)** fairly large (body length 8–12mm), yellow to brown, with yellowish wing membrane and 2 caudal filaments. Eyes huge, dorsally contiguous in males, and not divided into separate sections. Hind wings prominent. About 15 species known from the region. **Biology** Nymphs scuttle about crab-like when disturbed and are thought to feed on fine organic matter, trapped with the mouthparts. **Habitat** Nymphs cling to rocks in fast-flowing streams. Adults emerge during summer, mostly at night.

3 Family Baetidae
Small minnow mayflies

Most diverse and commonly encountered mayfly family in the region, with over 45 known species. Nymphs **(3)** elongate and spindle-shaped, with antennae longer than head and 6 or 7 pairs of abdominal gills, each consisting of a single oval plate without basal tufts. Adults small (body length 5–14mm), yellow to dark brown, with glassy fore wings (sometimes with dark front border), very reduced (or absent) hind wings and 2 caudal filaments. Eyes of males **(3A)** turbinate, divided into 2 sections, the orange or yellow upper part raised on a stalk or column. Image **(3B)** shows female emerging from subimago. **Biology** Nymphs thought to feed on algae and detritus scraped from solid surfaces, although a few species are predatory. Swim with rapid body wriggle, resembling tiny fish, hence common name. **Habitat** Common in very wide variety of habitats, from mountain streams to low-lying rivers and still ponds.

1 Family Caenidae
Small square-gills

Nymphs **(1)** easily identified by gill structure, first pair reduced to small filaments, second pair expanded into flattened, rectangular flaps that overlap in the middle of the back and cover posterior pairs. Adults **(1A)** (*Caenis* shown here) very small to minute (body length 2–5mm), white or grey, best recognized by the fore wings, which are fringed with setae posteriorly, have reduced venation and only a single row of cross-veins. Hind wings absent. Eyes not divided into sections, but large and widely spaced, and larger in males. Abdomen with 3 caudal filaments. Six species in South Africa. **Biology** Nymphs thought to feed on fine detritus. More tolerant of pollution than other mayfly families. **Habitat** Nymphs prefer slow-flowing waters, where they colonize mud and plant debris. Widespread in region.

2 Family Polymitarcyidae
Pale burrowers

Nymphs **(2)** easily identified by prominent serrated 'tusks' projecting from front of head, and by their long, feathery gills, which are folded over the back. Gill 1 may be single or double, gills 2–7 are double. Adults **(2A)** may be large (body length up to 10mm), grey or white, and best identified by their greatly reduced legs, of which only the first pair is functional in males. Fore wings divided into small rectangular blocks by a dense network of cross-veins. Hind wings relatively large, about one-third the size of fore wings. Males have 2 caudal filaments, females 2 or 3. Only 3 species known from the region. **Biology** Nymphs use their 'tusks' and strong legs to burrow into submerged soil or wood and then feed by filtering particles from a stream of water, which they pump through the burrow by beating their enlarged gills. Adults live for only a few hours, during which they perform mass mating flights before collapsing and dying, often on the surface of rivers or ponds. Dying females expel their eggs on the water surface. **Habitat** Restricted to more tropical areas, where the nymphs burrow into riverbanks.

3 Family Prosopistomatidae
Water specs

Nymphs small (body length 3–4mm), with unique body shape **(3)** comprised of flattened, oval, shield-like carapace that covers gills, wing buds and most of the abdomen and into which the head is recessed. Adults rarely seen and not depicted, very small (body length 5mm) with 2 pairs of wings and no cross-veins on either fore or hind wings. Legs reduced, especially in females. Only 4 species known from the region. **Biology** Some nymphs are carnivorous, feeding on microscopic zooplankton. Adults emerge at dawn and live for only an hour or so. Eggs released by female as she lies on water surface after mating. **Habitat** Nymphs found under stones in fast-flowing water, usually in warmer lower reaches of rivers. The family has recently been discovered in the Olifants River in the Western Cape.

4 Family Tricorythidae
Stout crawlers

Nymphs **(4)** heavily built and flattened, with stout femora on legs; those of the most common genus *Tricorythus* with a distinctive brush-like fringe of setae projecting forwards from mouthparts (visible from above under a microscope). Adults (not depicted) small (body length 5–8mm), sooty-coloured, with black eyes in males widely spaced and not divided into sections. Hind wings usually absent, and with 3 (occasionally 2) elongate caudal filaments. There are 10 species known from the region. **Biology** Adults seldom seen since they live only a few hours. Nymphs use their brush-like mouthparts to collect fine deposited or suspended detritus. **Habitat** Nymphs found on underside of stones in areas of rapid water flow.

5 Family Teloganodidae
Spiny crawlers

Nymphs **(5)** flattened, posterior margin of thorax with distinct notch; gill 2 operculate and partially covering filamentous posterior gills; swim with characteristic dorsoventral undulations of body. Adults **(5A)** with small hind wings, males with divided eye. **Biology** Cold-adapted group restricted to southern and Western Cape. **Habitat** Nymphs in waterfalls and fast-flowing currents in mountain streams. Adults nocturnal.

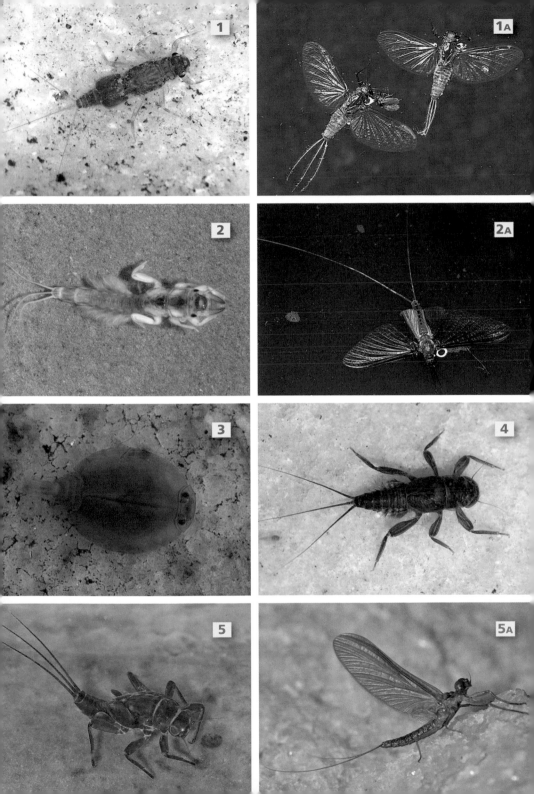

ORDER ODONATA • DAMSELFLIES & DRAGONFLIES

Medium-sized to large insects, with long, slender abdomen, 2 pairs of large, clear wings showing complex venation, very large compound eyes and short, bristle-like antennae. Dragonflies (Anisoptera) are larger and more robust, have eyes that meet above the head, broader hind than fore wings and hold their wings open and flat when at rest. Damselflies (Zygoptera) are smaller and more slender, have widely separated eyes, slender, more equal fore and hind wings and usually hold their wings together vertically above the back when at rest.

Can be confused with antlions (Neuroptera, p.206), but are distinguished by their shorter, bristle-like antennae. Both adults and nymphs are predatory. Adults hawk insects out of the air using their bristly legs, while the aquatic nymphs grab prey by shooting out their very elongated lower lip or 'labial mask'. Most species are sexually dimorphic, and their mating behaviour is unique. Males, which are usually more brightly coloured and hold territories, have a secondary reproductive opening on the second and third abdominal segments, to which they transfer sperm from the terminal genital opening. When mating, the male grasps the female behind the head using the claspers at the end of his abdomen. To obtain sperm, the female then bends her abdomen forwards to reach into the secondary reproductive organ of the male, linking the pair together in a mating position called the 'tandem wheel'. Eggs are scattered over the surface of water or injected into the stems of water plants and hatch in 1–3 weeks. Dragonfly nymphs are stout and use rectal (internal) gills to breathe; damselfly nymphs are slender and have plate-like tracheal gills projecting from the back of the abdomen. The nymph stage may last from a month to more than a year. Nearly 6,000 species occur globally, with over 160 in South Africa.

SUBORDER ZYGOPTERA   Damselflies

1 Family Calopterygidae   Demoiselles

The Glistening Demoiselle (*Phaon irridipennis*) is the only species in the region and is very large (wingspan 80mm) with exceptionally long legs and abdomen and a coppery or green sheen to the body. The broad wings are iridescent and clear brown, with many cross-veins. Nymphs (**1A**) are slender, very long-legged and easily identified by the exceptionally enlarged and elongated first segment of the antennae (see arrow). **Biology** Always associated with running water, and prefer to fly in the shade of trees. Perform a fluttering courtship dance. Often gregarious. **Habitat** Rocky streams in subtropical forest, but also margins of larger rivers in subtropical bushveld. Range extends into equatorial Africa.

2 Family Chlorocyphidae   Jewels

Largish (wingspan 50mm), stocky, with long wings folded above the broad body when at rest. Brilliant blue or red abdomen in males. Three species in region. The Dancing Jewel (*Platycypha caligata*) (**2**) has a wingspan of 45mm. Dorsal area of broad abdomen bright sky blue in males, with thin blue dorsal stripe along brownish-red thorax. Tibiae in males very broad and flat, with red outer and white inner surfaces. Male Boulder Jewel (*P. fitzsimonsi*) (**2A**) has only the last 4 segments of the abdomen blue and no blue thoracic stripe, while female (**2B**) is yellowish, with black stripes and markings. *Chlorocypha consensua* (**2C**) male has an entirely red abdomen and is found in KwaZulu-Natal (but rarely) and more commonly in Zimbabwe. Nymphs are unique in that they have only 2, not 3, caudal gills. **Biology** *P. caligata* males court females by displaying the red outer surface of the expanded tibiae, then vibrating the legs and displaying the white inner surface, while flying around the female with a pendulum-like movement. Adults present October–February. **Habitat** Widespread and common in rocky streams and pools, usually at higher altitudes, but also in the lowveld. Often seen perched on wood stumps or rocks in forest streams.

1 Family Lestidae
Spreadwings

Small to large (wingspan 40–75mm), generally green or brown, with long transparent wings, long wing-spot, long, slender abdomen and large forceps-like appendages at end of the abdomen in males. Like Malachites (below) characterized by their habit of perching with wings fully spread, but malachites are metallic green in colour. Labial mask of nymphs **(1A)** very narrow and elongate, broadening abruptly towards the end, like a spoon with a long handle. Seven species in South Africa. **Biology** Most abundant in warmer areas and typically found perching on emergent vegetation close to the banks of ponds or slow-flowing rivers. Eggs laid in slits in reeds or grass stems and nymphs reach maturity in a few weeks. **Habitat** Margins of quiet pools and streams in open country, adults usually close to water. Nymphs found in slow-flowing streams and pools.

2 Family Synlestidae
Malachites

Fairly large (wingspan 50–85mm), generally with metallic green bodies, thorax with yellow lateral thoracic stripes. Males sometimes with distinctive chalk white and brown bands across wings. Characteristically rest with wings spread. Nine species in South Africa, 8 of them endemic, some with restricted ranges. The White Malachite (*Chlorolestes umbratus*) **(2)** is the smallest (wingspan 48mm) representative and adult males are easily recognized by the unusual all-white ('pruinose') back and strongly banded wings. Unbanded and non-pruinose males also occur and are difficult to identify to species. Occurs in forests and fynbos in the southern Cape. The Mountain Malachite (*C. fasciatus*) **(2A)** (wingspan 55mm) is the most widespread species and has a bright metallic green body and bi-coloured wing-spots; wings may be banded or not (latter shown). Nymphs **(2B)** are large (body length 35mm) and slender with long legs; body mottled light and dark brown or cream; abdomen with 3 paddle-like gills at the end, each with a rounded tip and a dark band running along its centre. **Biology** *C. umbratus* occurs as an adult during most of the summer, *C. fasciatus* is common in eastern parts of the region, October–May. **Habitat** Forested, sunlit, stony mountain streams and pools. Nymphs occur on upper surface of rocks and on roots.

3 Family Platycnemidae
Stream damsels

Now incorporates the threadtails (classified as Family Protoneuridae in earlier editions of this book). Diverse family of small to medium-sized damselfies with 6 South African species; all with long, dense spines on the legs. Some robust, but threadtails (genus *Ellatoneura*) characterized by elongate, slender abdomens. Nymphs **(3)** with thickened, sausage-like gills.

4 *Ellatoneura glauca*
Common Threadtail

Smallish (wingspan 40mm), males have pale blue thorax with black lateral stripe. Top of very slender abdomen is black, undersurface white, tip pale blue. Females **(4A)** and immature tan, ringed at each segment with a narrow whitish band. **Biology** Commonly found in pairs during summer, perching on rocks or plant stems. **Habitat** Locally common and found in numbers near shady streams and riverbanks. Range extends to equatorial Africa.

5 *Allocnemis leucosticta*
Goldtail

Easily recognized, medium-sized (wingspan 50mm), with blue face and underside to eyes, abdomen long, slender and dark, with narrow white band at each segment and distinctive gold tip. Wings smoky brown with wide white wing-spot. Both sexes similar, although tip of abdomen in female may be pale bluish. **Biology** After mating, the male continues to clasp the female behind the head **(5A)**, allowing her to hang vertically with most of her abdomen below the water surface and to lay her eggs in rotten submerged twigs. **Habitat** Near streams in shady wooded ravines in higher-rainfall areas, perching on plants near the water. Nymphs occur under stones in the backwaters of these streams.

1 Family Coenagrionidae
Citrils, sprites, bluetails, bluets

Very large and diverse family of small to very small damselflies including over 1,200 species worldwide, with 38 reported in South Africa. Includes many of the commonest African damselflies. Nymphs **(1)** are slender, with a short mask that only reaches back to the first pair of legs when folded, and 3 wide leaf-like gills.

2 *Ceriagrion glabrum*
Common Citril

Medium-sized (wingspan 46mm); male easily recognized by combination of orange abdomen, yellowish-sided thorax, and green eyes. Females olive brown becoming orange with age. **Biology** Common and widespread. Adults make very short weak flights from low perches just above the water surface. **Habitat** Around streams and pools that have extensive reedbeds. Widespread throughout Africa.

3 *Pseudagrion hageni*
Hagen's Sprite

Medium-sized (wingspan 50mm), male thorax varies geographically – being orange and black in the Cape and green and black further north. Head black and red, abdomen dull bronzy green with characteristic violet tip. Female brown. **Biology** Flies low over water, settling on twigs or lily pads just breaking the surface of the water. **Habitat** Widespread. Common in summer in streams and pools at various altitudes.

4 *Pseudagrion massaicum*
Masai Sprite

Smallish (wingspan 39mm), male with red head and eyes, thorax orange-red dorsally with black longitudinal stripes, blue ventrally, abdomen dark above, greenish below, with paler rings, last 2 segments bright blue. Female greenish. **Biology** Usually seen perching on floating vegetation such as lily pads. **Habitat** Still waters and slow-flowing rivers. Widespread into East and Central Africa. The most common of 4 similar red sprites in the region.

5 *Pseudagrion sublacteum*
Cherry-eye Sprite

Medium-sized (wingspan 46mm), front half of eye and head bright red, rear of eye dark. Thorax dark blue, abdomen dull bronze with last 2 segments bright blue. Dull blue bloom on thorax in mature males. Female brown. **Biology** Flies for long periods low over water and settles on floating vegetation or rocks emerging from water. **Habitat** Fast-flowing rivers and broad streams. Distribution patchy. Range extends to equatorial Africa.

6 *Pseudagrion kersteni*
Kersten's Sprite

Medium-sized (wingspan 46mm), male uniformly blue-grey except for black lines along thorax and paler tip to abdomen. Eyes dark brown above and green below, distinctive white powdery patch between eyes. Female greenish or yellowish-brown with dark markings. **Biology** Flight low and accurate, close to water surface. **Habitat** Found near running water, but absent from vleis and pans. One of the most common damselflies in the region and occurs in most parts of Africa.

7 *Ischnura senegalensis*
Common Bluetail

Small (wingspan 33mm). Male similar to *Africallagma* (p.42), but underside of abdomen is tan, not blue. Thorax dark dorsally and brilliant blue ventrally. Wing-spot is divided into blue and white triangles. Abdomen with blue tip. Most females **(7A)** have a pale green or orange thorax, and a greenish-black abdomen. Immature females can be orange and are difficult to identify. Nymphs occur in standing water, even when polluted. **Biology** Adults occur October–June, often gregariously. **Habitat** Found in almost any stagnant or slow-flowing body of water, often the only damselfly present. Rare in mountain streams and seems to tolerate saline water. Arguably the most common and widespread damselfly in the region. Range extends to equatorial Africa as well as Japan and India.

1 *Africallagma glaucum* Swamp Bluet

Small (wingspan 34mm). Mature males sky blue with black dorsal stripes on thorax and abdomen. Underside of abdomen of males uniformly sky blue (compare *Ischnura senegalensis* on p.40); females brownish or yellowish. Nymphs pale green or orange to brown, slender and delicate, with 3 very large, flattened translucent caudal gills with pointed tips. The commonest of several similar blue *Africalligma* species in region. **Biology** Flies weakly, close to water surface, and settles on twigs, reeds and grass. Lays eggs in submerged plants. **Habitat** Vleis, pools, streams, rivers, even brackish pans, in a variety of vegetation types. Widespread and, with *I. senegalensis*, probably the most common damselfly in the region. Range extends to equatorial Africa.

SUBORDER ANISOPTERA Dragonflies

2 Family Gomphidae Clubtails

Easily recognized, medium-sized to large (wingspan 50–97mm), thorax and abdomen of both sexes usually dramatically marked in black and green or black and yellow stripes, end of abdomen typically with broad plate-like expansions. The only dragonfly family with eyes widely separated. Hunt from perches. Nymphs have 4-segmented, short and stubby antennae, the last segment often very small. They vary from broad **(2)** to slender and are bottom-dwellers, often living buried within mud. There are 17 species in 10 genera known from South Africa.

3 *Ictinogomphus ferox* Common Tigertail

Largest (wingspan 90mm) gomphid in the region, robust, with downward-facing oval flaps near end of abdomen (larger in males). Eyes blue, thorax yellowish to green with black stripes, abdomen strikingly banded black and yellow. **Biology** Fast and powerful fliers, usually seen perching on reeds or branches alongside or over water. Males territorial. Females lay eggs in shallow pools at stream or pond margins. Adults occur October–April. **Habitat** Prefers wide rivers, pools and streams with marginal reedbeds or overhanging trees. Widespread in warmer parts of the region. Range extends to equatorial Africa.

4 *Ceratogomphus pictus* Common Thorntail

Large (wingspan 65mm). Sexes similar in colour, eyes green, thorax yellow with wide pale green to blue and thin black oblique bands; abdomen slender, yellow dorsally, with black lateral markings. Flanges at end of abdomen in males yellow, edged with black, much reduced in females. **Biology** Rests on mud or stones, taking brief and low flights over the water. **Habitat** Common near streams with extensive grassy or rock cover and on ponds and dams. Widespread. **Related species** *C. triceraticus* (Western Cape) is larger with more black markings. Also, *Paragomphus cognatus* **(4A)**, which is common in the eastern parts of the region and recognized by the distinctive hooked upper claspers and outward-curling lower claspers at the tip of the male's abdomen.

Family Macromiidae Cruisers

Large, strikingly patterned dragonflies with very long legs and long, slender abdomens that end in a club-like expansion. Small family – only 1 genus with 3 species in South Africa.

5 *Phyllomacromia contumax* Two-banded Cruiser

Very large (wingspan 97mm), sexes alike. Easily recognized by plain brown thorax and 2 conspicuous broad yellow bands at base and before swollen tip of abdomen. Eyes bright green. **Biology** Usually seen in flight, patrolling strongly back and forth. Perch vertically. **Habitat** In woodland areas in warmer parts. Range extends across Africa.

Family Aeshnidae

Hawkers

Large to very large (wingspan 80–140mm), brightly coloured dragonflies, with very large eyes. Powerful fliers, spending long periods on the wing, patrolling large territories, often along a river or road, and feeding on large insects, including other dragonflies. Mating occurs on a perch, not in flight. Females have well-developed ovipositors used to insert eggs into aquatic plants. Nymphs elongate, with a flat and long labial mask. Eleven species in 5 genera in the region.

1 *Zosteraeshna miniscula*

Friendly Hawker

Large (wingspan 80mm). In males, base of hind wings incised, thorax with conspicuous, oblique brown and yellow bands. Blue band across second abdominal segment in mature specimens, rest of abdomen banded in pale blue and brown. Females with yellower thorax and golden yellow tinted wings. Large (body length 40mm), robust, cylindrical brown nymphs **(1A)** with labial mask reaching base of first pair of legs. Lateral spines on last 3 abdominal segments. **Biology** Flies low over water or grass, patrolling slow-flowing rivers or ponds. **Habitat** Widespread, favouring still waters in open hilly country. **Related species** The Stream Hawker (*Pinheyschna subpupillata*), which lacks the blue 'saddle' on the first abdominal segments.

2 *Anax imperator*

Blue Emperor

Very large (wingspan 105mm), males with green thorax and sky blue abdomen. Females **(2A)** similar, but with mottled brown coloration on abdomen. Nymphs **(2B)** very large (body length 55mm), with flattened head, eyes nearly meeting in the midline, labial mask reaching the base of the last pair of legs when folded, and lateral spines on last 3 abdominal segments. Early-stage nymphs have very broad contrasting bands of cream and black. **Biology** Prefers standing water, where females settle on water plants and curl the ovipositor underneath to pierce plant tissue and insert eggs. Frequently preys on other large dragonflies such as *Pantala flavescens*. Nymphs also attack large prey, including fish and tadpoles. Migratory adults occur October–June. **Habitat** Common near pools, streams and rivers, including brackish waters and organically enriched stagnant bodies of water. Widespread, extending to Europe and Western Asia. **Related species** There are 3 other local species in the genus. *A. tristis* (wingspan 130mm) has a yellow-and-black abdomen and is the largest African dragonfly, *A. ephippiger* has a blue abdominal saddle, and *A. speratus* is uniformly orange-red.

Family Libellulidae

Skimmers, dropwings, widows, etc.

The most diverse and commonly encountered family of dragonflies. Mostly medium-sized (wingspan 34–90mm), typically red or blue. Males differ in colour from females and only attain full coloration when mature, resulting in identification problems. In blue species, a waxy bloom develops gradually in adult males. Mating occurs on the wing, and males are often seen leading females over the water, where the females whip their abdomens over the surface to lay eggs. Nymphs have a cupped labial mask that covers much of the front of the head. There are 59 species in 26 genera in the region.

3 *Notiothemis jonesi*

Jones' Forestwatcher

Medium-sized (wingspan 52mm), male with bright blue-green eyes, diagonal green stripes on thorax, and slender abdomen expanded towards end. Bright pale green ring on abdominal segment 7 is diagnostic. **Biology** Shy and alert woodland species. **Habitat** Bush or forest streams and pools. Range extends into East Africa.

4 *Hemistigma albipuncta*

African Piedspot

Medium-sized (wingspan 53mm), slender. Blue bloom on body of males. Black streak along front of fore wings and long black-and-white wing-spot characteristic. Female banded yellow-brown and black. **Biology** Flies over small pools and rivers, settling frequently. **Habitat** A savanna species, locally common along the grass and sedge verges of pans and rivers in the subtropical region, range extending to equatorial Africa.

1 *Trithemis arteriosa* — Red-veined Dropwing

Medium-sized (wingspan 55mm). Adult males red-bodied with red leading wing veins and small yellow patch at base of hind wing; abdomen slender-waisted, with black lateral marks and distinctive black tip. Immature males and females yellow or dull orange. Nymphs are oval (body length 20mm) with dark brown mottling, spines on top of abdomen and no body hair. **Biology** Settles with wings downwards and forwards on reeds and grasses from which short forays are made for food or, in the case of males, to pursue intruders. **Habitat** Abundant in a wide variety of vegetation types in most aquatic habitats, including brackish pans in arid regions. One of the most common dragonflies in the region, with a very wide range that includes all of Africa and the Middle East. **Related species** *T. kirbyi*, the Orange-winged Dropwing **(1A)**, also common across the whole region, has red wing veins too, but a shorter, wider abdomen with less black, and much larger orange patches on the bases of both wings.

2 *Trithemis furva* — Dark Dropwing

Medium-sized (wingspan 62mm). Mature male uniform cobalt blue. Female yellow and black. **Biology** Slow but agile fliers. **Habitat** Often seen perching low over water along streams and large rivers. Common and widespread across whole region, range extending to equatorial Africa. **Related species** There are several other, hard to distinguish, uniformly dark blue species in the genus, but they aren't as widespread.

3 *Trithemis stictica* — Jaunty Dropwing

Medium-sized (wingspan 59mm), combination of pale blue thorax and yellow- and black-striped abdomen in adult males distinctive. Females banded black and yellow. Both sexes have yellow staining at base of hind wings. **Biology** Perch on plant stems overhanging water with wings held forward. **Habitat** Locally common in well-vegetated forest streams and a variety of open waterbodies. Range extends to equatorial Africa.

4 *Trithemis annulata* — Violet Dropwing

Medium-sized (wingspan 60mm) and distinctively coloured. Male with purplish thorax merging into red abdomen. Wing veins red. Female coloration consists of drab browns and yellows. Small amber patch on base of hind wings in both sexes. **Biology** Males prefer to settle on rocks in the sun, moving to trees as soon as the sun is obscured by clouds. **Habitat** Found along many different forms of waterbody, in a range of vegetation types, including those in semi-arid regions. Wide-ranging across Africa, Europe and Asia.

5 *Crocothemis erythraea* — Broad Scarlet

Medium-sized (wingspan 66mm). Males are a uniform vivid red with a blue marking at rear of the eye, yellow wing-spots, amber patch at base of hind wing, abdomen distinctively broad. Female mustard yellow with pale line between wing bases. **Biology** Males territorial and perch on rocks and stems around streams and ponds. **Habitat** Common and widespread across entire region and through Africa into Europe and Asia.

6 *Crocothemis sanguinolenta* — Small Scarlet

Medium-sized (wingspan 60mm), uniformly red but smaller with somewhat narrower abdomen than *C. erythraea* (above), and with redder wing-spot. Females yellowish. Both sexes with amber patches at base of hind wings. Nymphs squat and robust, with labial mask cupped in profile, no body markings and no spines along top of abdomen, which has a tuft of hair at the tip. **Biology** Common all through summer, often with *C. erythraea*. **Habitat** Along the banks of pools and streams, often perching on rocks. Range extends across Africa into Asia.

7 *Urothemis assignata* — Red Basker

Large (wingspan 75mm), male with entirely red body and characteristic opaque dark red patch at base of hind wings (*Crocothemis* have paler, transparent patches). Female with paler orange thorax, but still red abdomen and wing patches. **Biology** Flies low and strongly, returning persistently to the tip of the same reed. Females lay eggs among waterlily pads. **Habitat** Stagnant bodies of water overgrown with reeds, grasses and waterlilies; also in running water. Range extends to Somalia and Senegal.

1 *Acisoma inflatum* — Stout Pintail

Small (wingspan 48mm). Anterior of abdomen characteristically swollen, the posterior portion tapering abruptly. Male largely blue, thorax and abdomen marked with variegated black pattern. Female (depicted) paler blue, with yellow and black markings. **Biology** Adults occur sporadically through summer on the margins of pans and slow-flowing rivers. **Habitat** Subtropical pools or the banks of quiet streams with thick vegetation. Only recently recognized as distinct from the Asian *A. panorpoides*, under which name it appears in earlier editions of this book.

2 *Diplacodes lefebvrii* — Black Percher

Small (wingspan 49mm). Male recognized by its entirely blue-black coloration. Wings clear with long dark wing-spot. Females boldy marked in pale yellow and black with yellow wing-spot outlined in black. **Biology** Perches on emergent plants near water. **Habitat** Common and widespread around pans and slow-flowing rivers across eastern half of region.

3 *Diplacodes luminans* — Barbet Percher

Medium-sized (wingspan 63mm), spectacular species. Male with face, eyes, thorax and first part of abdomen scarlet, posterior section of abdomen strikingly banded in black and yellow, wings with amber patches at base and yellow wing-spot with dark centre. Female yellow and black with similar wing colouring. **Biology** Locally common in summer in northern tropical parts of the country. **Habitat** Found along margins of still water, including seasonal pools.

4 *Orthetrum julia* — Julia Skimmer, Cape Skimmer

Medium-sized (wingspan 60mm), with 2 subspecies, the eastern Julia Skimmer *O. j. falsum* (**4**) and the western Cape Skimmer *O. j. capicola* (**4A**). Males have a pale blue abdomen, but eastern form has dark thorax and wing-spot, while western form has paler thorax and yellow wing-spot. Western females (**4B**) attractively banded with pink and yellow stripes. Nymphs (**4C**) with very small eyes, entire body covered in hairs. **Biology** Rest with wings well forward and downwards. Eggs are scattered just under the surface of water. **Habitat** Common along streams, rivers and pools in bush, woodland or forest. **Related species** *O. caffrum* is another very common blue species found across the region. Both sexes have 2 distinct cream stripes along the thorax between the wing bases.

5 *Nesciothemis farinosa* — Eastern Blacktail

Large (wingspan 70mm). Male distinctive; eyes black dorsally becoming blue below, thorax powder blue, abdomen bi-coloured, blue anteriorly, black posteriorly. Female yellow and black with yellow wing-spot and slightly smoky tips to wings. **Biology** Perches conspicuously close to water during summer. **Habitat** Found along margins of still or running water. Widespread in region, but most common in the northeast, ranging across Africa to the Middle East.

6 *Brachythemis leucosticta* — Banded Groundling

Medium-sized (wingspan 53mm). Adult males with dark blue-black head and body, short abdomen and broad, dark brown bands across wings, wing-spot bi-coloured. Immature males banded in yellow and black, and females (**6A**) lack wing bands. **Biology** Gregarious and confiding, with the curious habit of following a person and settling nearby. Flies low over muddy edges of pools. Cryptic when settled, but conspicuous when disturbed. **Habitat** Locally common on mud or on bare ground beneath very large shady trees in the warmer parts of the region. Also found in equatorial Africa.

1 *Palpopleura lucia*　　　　　　　　　　　　Lucia Widow

Smallish (wingspan 49mm), male **(1)** with unmistakable broad black wing markings and bi-coloured wing-spot. Body colour highly variable, some with red eyes and yellow-and-black abdomen (depicted), others with black eyes and blue abdomen. Female **(1A)** with narrower black wing markings with pale yellow shadowing behind. **Biology** Common summer resident in warmer eastern parts. **Habitat** Common in a range of rivers, floodplains and marshes. **Related species** The closely related Portia Widow (*P. portia*) has a similar range but narrower and more scalloped black markings on the wings, and the female's hind wings lack a yellow 'shadow'.

2 *Palpopleura jucunda*　　　　　　　　Yellow-veined Widow

Very small (wingspan 39mm) and stout, with yellow legs, sides of thorax and abdomen. Distinctively patterned wings with extensive yellow and black markings, yellow wing-spot and yellow or orange shading. Abdomen of adult males becomes blue. **Biology** Flies for brief periods, settling frequently on grasses and reeds. **Habitat** Vleis, small pools and streams well vegetated with reeds and grasses. Range extends to equatorial Africa.

3 *Pantala flavescens*　　　　　　　　　Wandering Glider

Medium-sized to large (wingspan 80mm), face bright yellow, wings long with orange wing-spots, hind wings exceptionally broad with small yellow patches at tips. Male with orange back, female (shown) attractively banded in yellow and orange. Nymphs **(3A)** large and hairless, with oval abdomen bearing large and characteristic curved spines at end. **Biology** Frequently migrates, flying erratically in front of advancing storms with characteristic gliding flight. Patrols long flight paths in search of a mate or in the quest for food. Females attracted to temporary pools in which to lay eggs. Voracious nymphs feed on almost anything, including their own kind, and develop very rapidly. **Habitat** Common in most waterbodies in all parts of Africa. Extremely widespread. Range extends to Asia, the Americas and Australia.

4 *Tramea basilaris*　　　　　　　　　　　Keyhole Glider

Large (wingspan 88mm). Both sexes easily recognized by distinctive dark brown patches set within paler shading at base of hind wings. Abdomen of male orange, that of female pale yellow; abdomen black-tipped in both sexes. **Biology** Nomadic and unpredictable, but can be common. **Habitat** Around margins of wetlands in the eastern part of the region. Ranges across Africa, Asia and Indian Ocean islands.

5 *Rhyothemis semihyalina*　　　　　　　Phantom Flutterer

Medium-sized (wingspan 64mm) and unmistakable. Adults of both sexes have purple bodies and very wide hind wings with broad metallic purple panels at their bases. **Biology** Has a distinctive butterfly-like flight, hence its common name. **Habitat** Around the margins of marshes and rivers in savanna regions. Ranges across Africa.

ORDER BLATTODEA • COCKROACHES & TERMITES

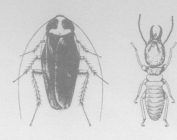

Recent studies have shown that cockroaches and termites (previously order Isoptera) are very closely related and in fact belong to a single order, Blattodea, within which there are a number of groups. One of these is the superfamily Blattoidea, which contains all termites and a group of termite-like, wood-inhabiting cockroaches (family Cryptocercidae) that diverged from their cockroach relatives about 170 million years ago.

Although cockroaches and termites may not resemble one another in appearance, both use symbiotic microbes in their guts to aid in the digestion of cellulose; primitive termites and cockroaches produce an egg case (ootheca); and certain cockroaches live in small colonies within wood. They also share a preference for dark, humid places, and both feed their young with faecal pellets. Both cockroaches and termites have biting mouthparts and are soft-bodied insects. Cockroaches may be winged in both sexes, or have winged males only. A few are totally wingless. In termites, wings are present only on reproductives.

1 COCKROACHES

Cockroaches are oval and flattened, medium-sized to large insects, with the head concealed beneath the shield-like front of the thorax (pronotum). Many are social, and live in aggregations, with some exhibiting parental care **(1)**. Cockroaches are largely nocturnal, and have become secondarily wingless in some species. Where present, the fore wings form leathery 'tegmina'. These protect the hind wings, which have an enlarged pleated fold at the base (the anal fold) **(1A)**. The legs are similar in size and are adapted for running, with large oval coxae. Scent glands are often present on the top or bottom of the body, and a pair of short structures (cerci) project from the end of the abdomen. Eggs are produced in a 2-tiered egg case (ootheca) **(1B)** that may contain 5–60 eggs. These purse-like structures may be deposited immediately or carried on, or in, the body. Nymphs occur in the same habitat as adults and may be confused with wingless adults. Cockroaches feed on a range of foodstuffs, largely of vegetable origin. A few species are associated with humans and have become household pests. All 6 known families occur in the region and include 20 endemic genera that occur mostly in the southern and eastern parts. There are 35 endemic species in the southwestern parts of the Cape, and most of the genera of this region (*Aptera*, *Temnopteryx* and *Blepharodera*) are endemic. The arid parts of the Western Cape and Northern Cape have the smallest number of endemic species (20 species in 3 endemic genera).

Family Blattidae   Domestic cockroaches

Fairly large family of medium-sized insects of diverse appearance, identified microscopically by a symmetrical plate under the male genitalia, the same plate being made up of 2 parts in the female. Females carry the egg case for 1–2 days before depositing it. About 50 species are known from the region.

2 *Cartoblatta pulchra*   Gregarious Spotted Cockroach

Medium-sized (body length 21–25mm), with shiny black nymphs marked with rows of white and orange dots across the body. Tegmina of adult females short or rudimentary; those of males extend beyond the abdomen. **Biology** They advertise their conspicuous warning coloration by aggregating in exposed positions on tree trunks by day, heads facing inwards. Aggregations of up to 150 nymphs are thought to be initiated by the secretion of pheromones. Abdomen of nymphs covered in a sticky, protein-rich secretion, thought to deter ants. Egg case contains more eggs (95) than that of any other species of Blattidae. **Habitat** Subtropical bushveld and coastal forest. Range extends to Kenya. **Related species** There is 1 other species of this genus in the region. Additional species occur in tropical Africa and Asia.

1 *Periplaneta americana*   American Cockroach, Common Cockroach

Large (body length 27–40mm), familiar, shiny reddish-brown, with long legs and spiny tibiae. Both sexes winged and fly readily, males with longer tegmina. **Biology** Aggregates in mixed groups of nymphs and adults. Survives in temperate regions in association with humans. Egg cases are deposited immediately, and nymphs take 6–12 months to mature. **Habitat** Dwellings, sewers, ships and other areas associated with human habitation. Cosmopolitan, appearing to favour coastal areas of high humidity (ancestral distribution probably tropical West Africa). **Related species** There are 4 other *Periplaneta* species in the region; the genus is easily confused with *Pseudoderopeltis*.

2 *Deropeltis erythrocephala*   Redhead Black Velvet Cockroach

Large (body length 27–37mm), black, with long-winged males **(2)**, and rotund wingless females **(2A)**. Head and legs typically red-yellow, rarely black. **Biology** Gregarious and slow-moving, often stand with hunched body. **Habitat** Widespread and very common in a range of vegetation types, living under stones that lie on larger rocks, or under dry tree trunks. Abundant in arid parts of the Cape, occasionally becoming established in homes. **Related species** There are 14 other species in the region, all with a large pad on the hind metatarsus.

3 *Pseudoderopeltis albilatera*   Orange-shouldered Cockroach

Medium-sized (body length 14–27mm), with a yellow to orange band on either side of the prothorax. Wings reduced or absent in females **(3)**; long, shiny, black and fully developed in males **(3A)**. **Biology** Males secrete a substance from tufted glands on the abdomen, on which females feed before mating. **Habitat** The genus is widespread in the region, in a range of habitats including forest and stony mountain tops. **Related species** There are about 24 other species in the region, apparently the centre of origin of the genus.

Family Ectobiidae (formerly Blattellidae)   Domestic cockroaches, wood cockroaches

Very diverse family of small (body length 5–20mm), delicately built and fast-moving cockroaches, with uniform spination along the undersides of the mid and hind femora. Females carry the egg case, sometimes until the nymphs hatch. Females of subfamily Blattellinae are able to rotate the egg case so that the keel faces 90° to the left prior to deposition. Males of many species have glands on their backs that produce a secretion to attract females. Before mating, males of some species offer females a sperm package coated with uric acid. Both sexes are normally winged and fly to lights. There are approximately 90 species in the region.

4 *Blattella germanica*   German Cockroach

Small (body length 11–15mm), yellowish-brown, with 2 dark stripes running along prothorax. **Biology** Aggregates in mixed groups of adults and nymphs. Depends on central heating and other artificial sources of heat to extend its range, and feeds on almost any kind of food. Takes 3 months to reach adulthood. Flies short distances. Females attract males from afar by emitting a sex pheromone, to which some people develop severe allergies. Eggs begin developing inside a large ootheca carried by the female and hatch into 3mm-long black nymphs within about 24 hours of having detached from the female's body. **Habitat** Human habitation, especially kitchens and drains. The most common inland domestic roach. Cosmopolitan, probably originating in South Asia. **Related species** There is 1 other regional species in this genus.

5 *Temnopteryx larvalis*   Cape Zebra Cockroach

Smallish (body length 20mm), flattened, with circular, reduced tegmina and hind wings in males; females wingless. Abdomen striped black and yellow. **Biology** Often found in small groups under flat, exfoliated sheets of rock. Females may have an egg case protruding from the tip of the abdomen **(5A)**. **Habitat** Usually associated with rocks overlying boulders in the mountainous parts of the fynbos biome. Common in the Western Cape. **Related species** There are 6 other species of this endemic genus, all with black and yellow stripes on the abdomen.

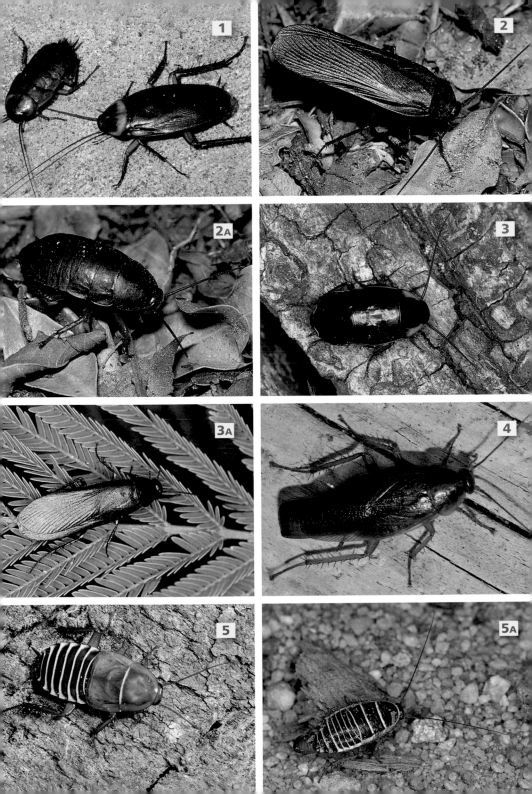

1 *Saltoblattella montistabularis* — Table Mountain Leaproach

Small (body length 6–10mm), unique, grasshopper-like body, with enlarged and elongated hind legs, very long, thin antennae, large bulging eyes, and wings reduced to tiny non-functional pads. Males (1) slimmer than females (1A). **Biology** World's first known jumping cockroach; a very proficient jumper covering distances of 30cm and landing accurately on grass culms. Apparently a case of convergence with grasshoppers in terms of body design and habits. Females lap at tergal secretions produced by the male during mating, and carry an egg case. **Habitat** Restio plains and open areas of low fynbos vegetation. **Related species** Since the discovery of *S. montistabularis* numerous additional species of leaproach have been discovered in the Western Cape.

2 *Supella dimidiata* — Slender Black-spot Cockroach

Small (body length 17mm), slender, golden yellow, with black markings on thorax and tegmina. Small triangular head. Wings fully developed in the male, extending beyond the end of the abdomen. In the flightless female the tegmina are short and hind wings rudimentary. **Biology** Agile and fast-moving. Females secrete a sex pheromone from an abdominal gland, attracting males from some distance. The egg case is not rotated, so the keel always faces upwards. Clusters of egg cases deposited into cracks on tree trunks. **Habitat** Found in trees and shrubs in a range of vegetation types. Also occurs in East Africa. **Related species** Of the 6 Afrotropical species in this genus, only 1 other occurs in the region. Also related is the Brown-banded Cockroach (*S. longipalpa*), a ubiquitous household pest in North America.

Family Blaberidae — Giant cockroaches

A large family of cockroaches of variable appearance, all small to large (body length 12–50mm), sluggish and heavily built, with the body carried close to the ground. The egg case is retained in the body, and the young, in many cases, are born live. In some species, egg cases are deposited at a later stage of development. Many species have wingless females that burrow in the soil. Can be distinguished from Ectobiidae by unspined ventral surfaces of the mid and hind femora, and the varying size and shape of spines on fore and hind margins of the femora. Males of some species offer a sperm package coated in uric acid as a nuptial gift to the female. Mostly subtropical, a few species being cosmopolitan pests. About 20 genera and 90 species known from the region.

3 *Aptera fusca* — Table Mountain Cockroach, Giant Cockroach

Large (body length of females 30–40mm, of males 29mm), bulky, with reddish head and brown to black body segments edged in yellow. Dark brown wings in males (3); females (3A) are wingless. Femora heavily spined, hind tibiae having 2 very wide ridges with only 2 rows of spurs. **Biology** Adopts a tail-up defensive stance when alarmed and squeaks loudly. The brightly coloured body and release of noxious fluids when molested probably protect it from bird predation. About 18–24 young are born live and protected by the female for a while under her body. Heavily infested with protozoan gut parasites. **Habitat** Widespread on low vegetation in open areas in the fynbos biome, where it often basks on top of plants. **Related species** The other species in the genus, *A. munda*, recorded from the Richtersveld, is rare.

4 *Bantua* — Barrel cockroaches

Medium-sized (body length 24mm), elongate, shiny black or reddish-brown, with wings in males only. Pronotum is covered in small granules and ends in a projection on each side, both sides strongly keeled into ridges. Of the 6 known species, 4 occur in the region. **Biology** Sluggish. Generally found in small groups under tree bark. Cylindrical body probably an adaptation for burrowing into decaying wood. **Habitat** Subtropical bushveld and thornveld.

5 *Derocalymma* — Trilobite cockroaches

Medium-sized (body length 20mm), very flattened oval body, generally a matt dark brown with lighter markings. Well-defined ridge on sides of triangular pronotum, body of pronotum distinct from side flanges. Males (5) brown with a rough surface to wings and body, wings often far longer than body. Females (5A) wingless. Of the 16 known African species, 7 occur in the region. **Biology** Granules on body surface are sensory. Nymphs do not aggregate like other Blaberidae. **Habitat** Under the bark of both dead and living trees in all types of forest and open woodland.

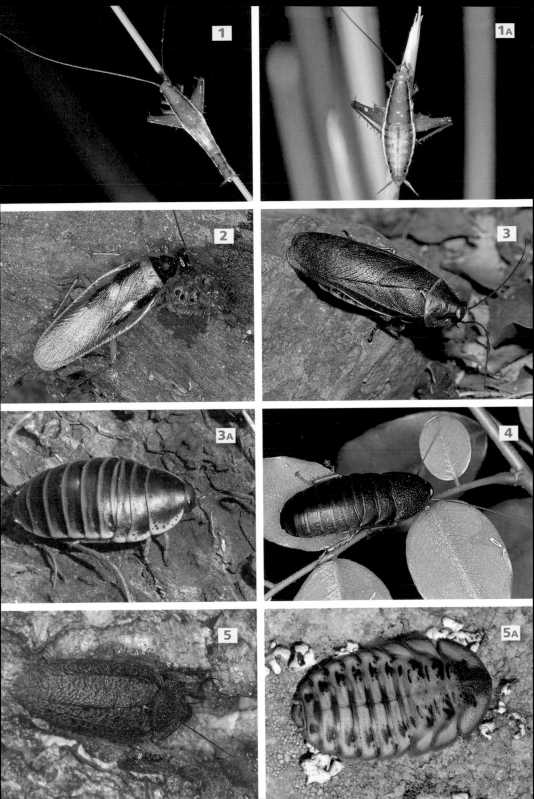

1 *Zuluia deplanata* Rock Cockroach

Medium-sized (body length 20mm), flattened, tan-coloured, with numerous speckles and dots providing superb camouflage to match the rocks on which it occurs. Very large, spade-like pronotal shield. Females and nymphs **(1)** oval and wingless; males **(1A)** almost spherical with granulated and camouflaged tegmina. **Biology** Lives in small groups of nymphs, females and males. **Habitat** The surface of lichen-covered granite rock. **Related species** There are 4 other species in the genus, all restricted to the eastern parts of southern Africa.

2 *Blepharodera discoidalis* Burrowing Cockroach

Medium-sized (body length 20–22mm), stout and oval. Body convex, shiny and black with a cream border. No terminal tooth at corners of pronotal shield. Wings in males only. Fringe of long setae along sides of body in females; restricted to pronotum in males. **Biology** Burrows a few centimetres below the surface of organically enriched soil beneath karroid bushes. **Habitat** Semi-arid areas with low vegetation. **Related species** There is 1 other species in the genus, which is also restricted to the arid southwestern parts of the country.

3 *Oxyhaloa deusta* Red-head Cockroach

Medium-sized (body length 12–20mm), with black legs, dark brown hind wings, and blackish-red pronotum covered in erect hairs. Both sexes winged. **Biology** Males produce a long-distance sex pheromone to attract females. Popular as pets with cockroach hobbyists. **Habitat** Under stones in a variety of (mostly subtropical) vegetation types, occurring in most parts south of the Sahara. **Related species** Of the 10 Afrotropical species known, only 1 other, *O. buprestoides*, occurs in the region; it is also widespread in tropical Africa.

4 *Gyna caffrorum* Ghost Porcelain Cockroach

Medium-sized to large (body length 27mm), distinctive and striking. Cream pronotum with black butterfly-shaped central marking. Numerous cream spots on the sides of the dark brown tegmina. Both sexes winged. **Biology** Nymphs of Afrotropical species live in ground litter, tree holes and ant and termite nests; adults live in the treetops and never at ground level. Other *Gyna* species feed on bat guano in caves. Popular pets with cockroach hobbyists. **Habitat** Mostly subtropical bushveld and forest. Vast numbers found in bat caves in Mpumalanga. **Related species** One other species occurs in the region; as with *G. caffrorum*, both sexes are winged.

5 *Perisphaerus*

Medium-sized to small (body length 23mm), oval, shiny black, with rounded edge to corners of pronotum. Thorax has granulated surface. Males with long wings, females wingless or with very short tegmina. Convex on top and flat below, allowing rolling up of the body into a protective ball. Endemic to southern Africa, with 24 species in the genus. **Biology** Commonly seen resting under flat stones or on large rocks on mountain tops. Other species in the genus apparently brood their young. **Habitat** Found in ground litter, under rocks or bark in a variety of vegetation types.

6 *Hostilia*

Smallish (body length 15mm), stout, light brown and mottled. Hind tibiae very flattened, with the 2 rows of spurs poorly developed. Pronotum similar to that of *Bantua*, but the edges are bent downwards, and the central disc is distinct from the lateral flanges. Males **(6)** are winged; females **(6A)** are convex and wingless. *Hostilia carinata* **(6B)** is pale with serrated projections along the sides of the abdomen. There are 3 endemic species in the region. **Biology** Sluggish. Live under tree bark. **Habitat** Bushveld and forest.

Family Corydiidae
Sand cockroaches

Yellow or maroon roaches, mostly occupying arid areas. Males are winged, and anal fold of hind wings is simply folded, not pleated fan-like, when at rest. Females are wingless. Both sexes emit pheromones that are long-distance attractants. The egg case is carried around for a brief period before it is deposited. About 10 species are known from the region.

1 *Tivia termes*
Desert Cockroach

Small (body length 12mm), oval, pale reddish-yellow. Wings in males extend beyond end of abdomen. **Biology** Lives under bark of living and of dead trees, especially in very dry woodlands. Some species of *Tivia* have been collected from ants' nests, others from bat guano in caves. **Habitat** Arid open bushveld. Widespread in Africa, range extending to Western Australia. **Related species** There are 2 other species in the genus; they are found in Mpumalanga and KwaZulu-Natal.

2 TERMITES

Termites are now regarded as cockroaches that have evolved complex social systems, typically occupying distinctive nests (termitaria) built of soil, although some live in wood. A colony is made up of 4 different forms (castes). Very large primary reproductives (queens and kings) usually retain vestiges of wing bases, and are darkly pigmented. Secondary reproductives are less pigmented and may have wing buds. Soldiers are sterile males and females with specialized and hardened (sclerotized) heads. Workers are sterile males and females with unmodified heads and faint sclerotization. Termite reproductives ('flying ants') have 2 pairs of similar wings with reduced venation **(2)**; these are shed, detaching along a line of preformed weakness, after a brief nuptial flight. Mature queens are hugely swollen and enlarged, can lay up to 25,000 eggs a day and are attended to and fed by nymphs **(2A)**. Soldiers lack eyes, and their heads are often bigger than their bodies. Two types of soldier can exist in a single colony: small minor soldiers (with small heads and minute mandibles) and large major soldiers (with massive mandibles). Workers are the most numerous caste and are usually pale and eyeless. They build and maintain the nest, obtain food and feed all other members of the colony. Parent colonies release eyed reproductives at dusk, often after rain. A pheromone released by the female attracts the male, and the pair burrow into a suitable substrate, where eggs are laid. Reproductives are produced only after a few years. Colonies exist for decades, possibly centuries in some of the species with large colonies and termitaria. All termites feed on dead or living vegetable matter and (except for Termitidae) digest cellulose using single-celled organisms and bacteria in their guts – the latter capable of harvesting atmospheric nitrogen. Nests are very complex and structure is species-specific. Many other species of insect are associated with the nests as scavengers or predators. Some termite species are of economic significance because they damage wood, crops and fodder. Soldiers (or reproductives) are mostly used for identification purposes.

Family Termopsidae
Damp-wood termites

A small family of primitive termites, represented in South Africa by 2 rare genera (each with a single species).These termites are found in small nests in wood that is in contact with the soil. There are no worker castes, and in soldiers the pronotum is narrower than the head capsule.

3 *Porotermes planiceps*
Southern Log Termite

Medium-sized (body length of soldier 13mm), head of soldier large, reddish-brown, rectangular and flattened, with slight constriction halfway along its length. Toothed mandibles less than one-third the length of the head. **Biology** Excavate flat tunnels and galleries in dead wood only, usually protea stumps embedded in the soil, but also other dead trees such as pine and keurboom (*Virgilia oroboides*). Galleries extend into the roots. **Habitat** Open mountain fynbos vegetation with proteas. **Related species** No other species of *Porotermes* in southern Africa, but some species do occur in Chile, Tasmania and Australia, where they may be destructive pests of dead and living eucalypts.

Family Kalotermitidae
Dry-wood termites

Small colonies of primitive termites that live entirely in dry wood and have no worker caste. They infest furniture and other wood in homes; some species are able to survive only in wood that is exposed to precipitation. Capable of reabsorbing water from their hindgut, allowing them to tolerate dry conditions and live in wood unconnected to the soil. Restricted largely to coastal and high-rainfall areas, with 12 species known from the region.

1 *Bifiditermes durbanensis*
Dry-wood Termite

Small (body length of soldier 5.5–7mm). Soldiers have very stout bases to their mandibles, which are parallel-sided and toothed on the inner margin. Reproductives with dark brown bodies and wings (**1A**). **Biology** Lives in dead branches and branch stubs of a range of living trees; also in standing and fallen dead trees. Feeds only on hard, barely rotted wood, opening it for later fungal attack. Galleries are always above ground. **Habitat** Coastal forest and in oaks and other trees in urban areas, especially Cape Town. **Related species** There is 1 other regional species in the genus.

2 *Cryptotermes brevis*
House Termite

Easily identified by the small (body length 5mm) soldiers, which have characteristic flattened and rounded (phragmotic) foreheads, with rough edges. **Biology** Soldiers use their heads to block tunnels from ant attacks. Mature colonies of *C. brevis* are very small, with only about 300 individuals. A serious cosmopolitan pest of household wood, the introduced *Cryptotermes brevis* originates from the West Indies, and its poppy-seed-sized droppings indicate an infestation. Zinc-tipped mandibles allow it to bore into and feed on dry hardwoods. **Habitat** Dry wood, of all kinds, including timber, furniture and dead limbs of indigenous forest trees. Now known from South and Central America, Hawaii, Australia, Africa and Madagascar, as well as Asia.

Family Hodotermitidae
Harvester termites

Large termites, recognized by functional compound eyes in all castes, mandibles with several inner teeth and the presence of pigmented workers. Commonly seen in winter running on the soil surface, collecting and cutting grass and twigs and piling cut vegetation before carrying it underground. Nests partially or totally underground. Two species of this primitive family occur in the region.

3 *Hodotermes mossambicus*
Northern Harvester Termite, Rysmier

Large (body length of soldier 15mm, worker 8–13mm, queen 25mm), with eyes in all castes. Workers have very dark brown heads, and bodies striped in brown and cream. Jaws of soldiers curve strongly inwards, and are pale red and robust, with the tips and teeth on the inside edge black. Reproductives ('flying ants') with 2 stumpy abdominal appendages above the cerci. **Biology** Pigmented, eyed workers (**3A**) forage in bright sunshine (especially in autumn and early winter), cutting fresh and frost-killed grass and other plant matter, and dragging this into piles around foraging ports. Diffuse nest system 1.5–7m underground, with tunnels leading from piles of cut grass and numerous turreted soil dumps (**3B**) to temporary subterranean grass-storage shelves. All tunnels finally descend to spherical hives (up to 20), each around 60cm in diameter, surrounded by flat grass-storage chambers, made of fragile, horizontal black shelves of chewed vegetable matter (carton). They are filled with white nymphs (hence the Afrikaans name, which means 'rice ant') and house the queen and king. Heavy infestations can deplete the veld of browse for cattle (leading to soil erosion) and can damage lawns. Estimated to remove 1–3 metric tons of forage per hectare. Swarms of winged reproductives occur before rain in summer. A vast number of animal species feed extensively on this termite. **Habitat** Grassland and most other habitats, especially open areas. Favours disturbed and overgrazed areas. Absent from forest. **Related species** Other members of this genus occur throughout Africa. Workers closely resemble *Microhodotermes viator* (p.64), but the latter has indistinct striping on a more uniformly brown abdomen.

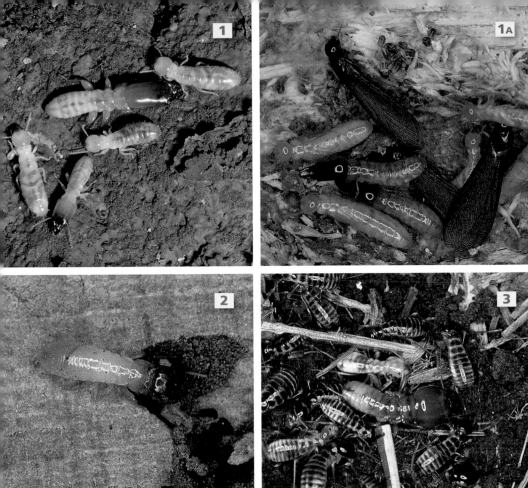

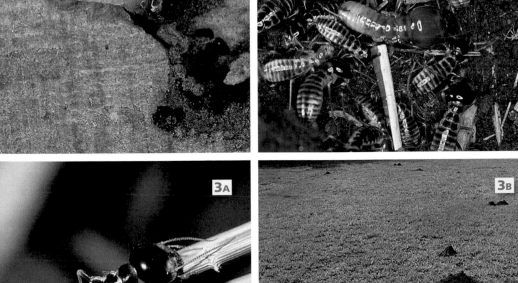

1 *Microhodotermes viator*   Southern Harvester Termite

Large (body length of soldier 7–13mm, worker 6–8mm), soldier mandibles black with reddish bases. Reproductives ('flying ants') with 2 abdominal appendages above and of equal length to cerci. **Biology** Ubiquitous component of Karoo fauna, impacting equally on flora and fauna. Produces sharp conical termitaria in soils with a high clay content. At the base of the mound is the large spherical hive with its delicate horizontal layers of chewed vegetable matter (carton) in which queen, king and nymphs live. Soldiers emerge to guard the release of winged reproductives **(1)**. Mounds generally become covered in sand, forming massive, long-lived structures, called *heuweltjies* (up to 20m in diameter and 2m in height) that are evenly spaced, like all termitaria, across the landscape. Owing to the large amounts of frass (pelleted faeces) that pour out on the surface, the mounds are richer in nutrients than the surrounding soils. They are colonized by plants that favour disturbed sites, including colourful annuals, creating remarkable patterns in the landscape **(1A)**. Workers forage in large numbers by day, preferring woody plants **(1B)**, especially *Pteronia* and vygie species. Green and dead twigs are cut and dragged to temporary storage areas around foraging ports. Reproductives have black head capsules and wing bases. **Habitat** Found in a range of vegetation types, but prefers open veld and avoids fynbos on sandstone. **Related species** *Hodotermes mossambicus* (p.62) looks very similar, but the jaws of *M. viator* soldiers are thinner and longer and curve inwards only at the tips. Other species in the genus occur in North Africa.

Family Rhinotermitidae   Subterranean termites, damp-wood termites

Psammotermes allocerus is the most common, and is abundant in Kalahari sands. Frontal gland (fontanelle) of soldier caste opens from a short, cone-shaped projection on the head, and is used to secrete a defensive sticky drop when disturbed. Nests are dark, papery structures up to 35cm in diameter, often located about 30cm below the sand surface at the base of a grass clump. The termites feed on a range of dead plant matter (including timber), and construct sand-covered runways to reach food sources. Seven species are known from the region, some of which are introduced.

2 *Psammotermes allocerus*   Desert Termite

Small (body length of soldier 7mm; worker 5mm). Nests largely subterranean, with an inconspicuous grey surface dome located in loose sand at the base of a grass clump or dead shrub **(2A)**. Underground portion of nest 20–70cm in diameter, and 20–90cm in height **(2B)**. May contain more than 1 small vaulted royal chamber housing queen and king. Possibly consists of a complex of closely related species. **Biology** Feeds largely underground on grass roots and subterranean portions of shrub branches and trunks, ultimately killing the plants and forming a bare oval disc ('fairy circle') around the nest. Can cause considerable damage to wooden buildings located in arid regions. Also destroys wooden fence posts and, in the northern parts of its range, crops. **Habitat** Both winter- and summer-rainfall regions in a range of arid vegetation types. KwaZulu-Natal population exists under high-rainfall conditions.

3 *Coptotermes formosanus*   Formosan Subterranean Termite

Large (body length of soldier 12–15mm), body pale brown, head with short tube (fontanelle) that opens at front of head just before mandibles. Mandibles of soldiers slender, curve inwards and lack smaller 'teeth' on inner margin. **Biology** Forms very large colonies in old dead wood **(3A)**, filling excavated chambers with carton (macerated wood and saliva). **Habitat** Attacks the heartwood of both living and dead trees and timber, and capable of foraging for distances of 100m underground. Introduced to South Africa in 1924, and a serious economic pest globally. Restricted to Simon's Town and Komatipoort. **Related species** Four other species of *Coptotermes* occur in Africa, and around 21 species are native to the tropical parts of the world.

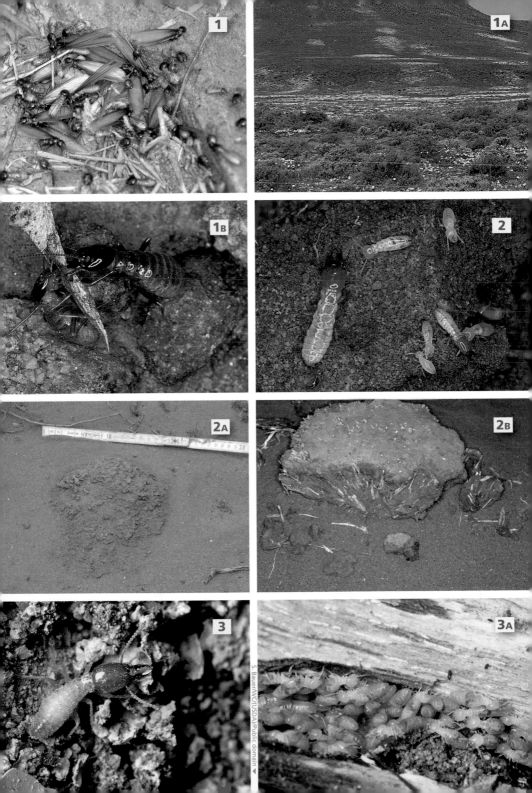

Family Termitidae
Higher termites

A large family that includes 80 per cent of the region's termite species. Some species lack soldiers. They feed on a range of vegetable matter, and to aid their digestion of cellulose, their guts contain only bacteria, lacking the single-celled organisms found together with bacteria in the guts of other termite families. Nests are not always obvious, and some species move into the vacated nests of others. The first part of the thorax (pronotum) is saddle-shaped and has lobes that project forward. The family includes fungus-growing species as well as the 'nasute' termites, in which soldiers have a long cone on the head, bearing the opening of the frontal gland. An economically significant family with 190 species known from the region.

1 *Macrotermes natalensis*
Large Fungus-growing Termite

Very large (body length of large soldiers 18mm, workers 5–8mm). Small and large soldiers, both with very large orange head capsule and slender untoothed jaws that cross when closed. Nests **(1A)** are very large (on average 75cm, but up to 200cm tall) conical or domed mounds and may be topped irregularly with pinnacles. Numerous wide 'ventilation tunnels' run into the nest, the nest cavity (diameter 60–90cm) usually just below or (in waterlogged soils) above the ground. **Biology** Fungus gardens **(1B)** for which the genus is known are grown on delicate horizontal shelves, on moulded faecal pellets. The colony cares for and feeds on the developing fungus. Estimated to consume around 20kg of plant matter per hectare annually. **Habitat** The most common species in the genus, especially in KwaZulu-Natal and Limpopo, where it occurs in open bushveld and grassland. Occurs widely across Africa and Madagascar. **Related species** *Macrotermes* includes 5 other species in the region. *M. mossambicus*, which is dominant in Limpopo province, constructs mounds over 4m tall, with mud runways from nest to food source. Workers are known to forage in bright daylight on dead wood, dead grass, dung, leaf litter and, to a lesser extent, live vegetation. The forage is plastered with a sand cement before it is eaten. Winged females alight head-down on grass stems after long nuptial flights, and emit a pheromone that attracts males. The enormous (100mm long) queen is imprisoned along with the smaller (25mm long) king in a royal cell below the fungus garden.

2 *Odontotermes badius*
Common Fungus-growing Termite

Large (body length of soldier 9 mm), similar to *Macrotermes*, but with a single soldier caste, members of which have 1 tooth on the inner edge of mandibles. Winged reproductives **(2A)** have dark heads and wings. Underground nests, detectable by subtle soil elevation or low grass-covered hump with soil dumps above the hive. Shrubs and trees may arise from the high mounds (up to 1m tall and 6m in diameter) of large colonies. **Biology** Probably the most destructive of all termites in wooden homes. Forages on wood, dry dung, leaf litter, tree bark and dead grass. Food source (including tree trunks) is plastered into nest cement **(2B)**. The queen is imprisoned in a spherical clay cell in the central hive within which lies the fungus garden. This, in turn, is protected by a very thick clay layer. **Habitat** Widespread and abundant in a wide range of habitats, including the foundations of buildings. **Related species** *Odontotermes* includes 21 other species in the region, including the Lesser Fungus-growing Termite (*O. latericius*), which builds clay chimneys **(2C)** that descend to points near the hive and probably function as air shafts.

3 *Microcerotermes*
Carton nest termites

Small (body length of soldier 5.5mm), with long, parallel-sided, rectangular-shaped heads. Mandibles of soldiers with fine, even serrations on inner margin. Small nests (diameter 13cm) made of concentric layers of yellowish chewed vegetable matter (carton) are situated underground or against or in the nests of *Cubitermes*, *Macrotermes* or *Trinervitermes* species. There are 24 species in the region. **Biology** Forage on fallen and standing timber, dry dung, and the woody stubble at the base of grasses. Occasionally attack wooden fence posts or house timber. **Habitat** Abundant in open woodland.

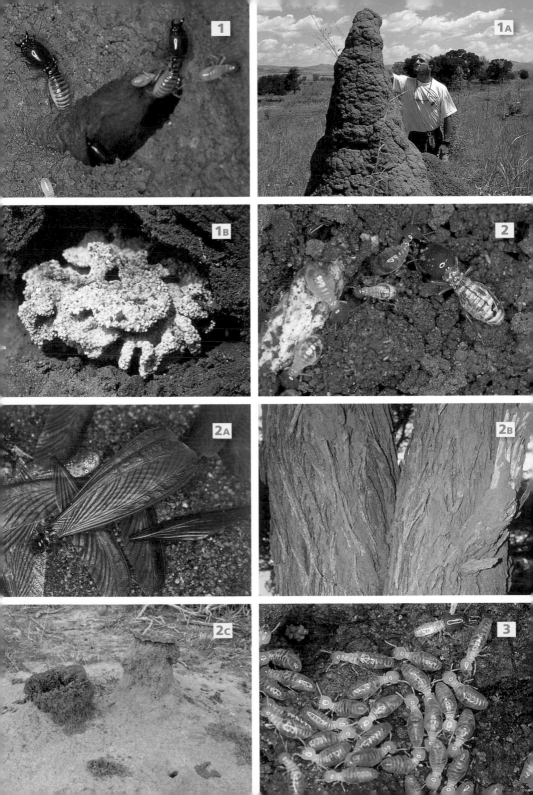

1 *Trinervitermes* — Snouted harvester termites

Small (body length of soldiers 5.5mm, of workers 5mm), recognized by pear-shaped head without jaws in major and minor soldiers, ending in a long, thin snout (possible confusion with related genera, in absence of microscopic examination of soldier head capsule). Nests small to medium-sized (35–100cm tall), smooth, domed mounds (**1A**), which may be fairly flat, with a thin but hard outer shell of cemented sand and an inner structure of open cells and galleries. There are 12 species known from the region. **Biology** Grass-feeding, foraging at night, using long underground tunnels that radiate from the nest. Vast amounts of cut grass are stored in the nest, and dense populations may compete with cattle for fodder. In de-vegetated areas, sparsely populated nests are constructed as granaries that connect with the primary nest. (Extensive fields of such nests with many dead colonies occur in the Free State and elsewhere.) If attacked, soldiers squirt a sticky fluid from the head, which entangles ants and contains repellent turpenes. This is insufficient, however, to repel their major predator, the aardvark. Colonies live for a few decades. **Habitat** Prefer wetter grassland parts of the region. Absent from the northwestern Cape although widespread and very common in some parts of the Cape.

2 *Amitermes hastatus* — Black-mound Termite

Small (body length of soldiers 4.5mm, of workers 4mm), all castes having a bloated abdomen packed with grey material. Mandibles curve strongly inwards, forming an arch when closed. Each mandible has a strong tooth that ends in a 90° turn about halfway along its length. Nests are characteristic small (35cm tall), domed, grey to black mounds (**2A**) with a nodular or honeycombed surface and tunnels that radiate out towards food sources. **Biology** Prefers to feed on decaying wood and humus, but will attack dead wood. Black digested faecal paste is used to construct the mound. Bodies of old queens are stained brown from the saliva of workers. Most active after spring rains, when winged reproductives are released through holes in the top of the mound. Colony growth is very slow, taking about a decade to reach maturity. If the queen is killed, the colony can continue to survive through the production of secondary reproductives. Soldiers release a drop of irritant fluid onto attackers through the frontal gland opening (fontanelle) on the head. **Habitat** Common in a variety of habitats. **Related species** *A. messinae* and *A. unidentatus*, which are tropical species that occur in the far north of Limpopo.

ORDER MANTODEA • MANTIDS

Relatively large, elongated and predatory insects, easily recognized by the mobile triangular head, large eyes, heavily spined prehensile fore legs, and characteristic 'praying' posture.

Mantids have extraordinarily mobile heads, with large compound eyes set high on their upper corners. The first thoracic segment is elongate and bears characteristic, well-developed, heavily spined fore legs, usually held folded below the head, as if in prayer. The remaining legs are sometimes decorated with ornamental lobes, but are otherwise unmodified. The wings, which are often reduced or absent in females and ground-dwelling species, are folded over the elongate, 11-segmented abdomen. The relatively narrow, tough fore wings are used mainly to cover and protect the broad, membranous hind wings.

All mantids are alert predators, using their specially modified fore legs to ambush and grasp live prey. Eggs are laid in a frothy mass that hardens as it dries to form a characteristic egg case (ootheca). Nymphs are wingless and often differ in appearance from the adults, some mimicking ants during their early developmental stages. Of the 2,400 or so species known globally, about 185 occur in South Africa.

Family Hymenopodidae   Flower mantids
Spectacular mantids that often mimic flowers. A raised process is usually present on the middle of the head, and the first thoracic segment has lateral lobes. The inner margins of the front femora have alternating long and short spines; those on the tibiae are closely spaced and lie at an angle. The legs bear ornamental lobes, and the fore wings are often decorated with bands or spiral markings. Some 14 species and 9 genera occur in South Africa.

1 *Phyllocrania paradoxa*   Ghost Mantid
Medium-sized (body length 44mm). Sexes differ in appearance. Males **(1)** slender and mottled brown with dark shoulders and a darker cross on the hind wings. Females **(1A)** larger and heavier, with triangular lateral flanges on the prothorax and abdomen and no cross on fore wings. Head with characteristic long, erect, wavy dorsal projection. Nymph has similar leaf-like expansions to adult; abdomen tightly curled over back, enhancing leaf mimicry. Legs ornamented with prominent leaf-like lobes. Egg case **(1B)** long, flattened and narrow. **Biology** A superb mimic of dead leaves, remaining motionless while waiting for prey to come within grasp. May make swaying movements to mimic windblown vegetation. **Habitat** Subtropical vegetation and forest margins. Ranges across Africa and southern Europe.

2 *Pseudocreobotra wahlbergi*   Spiny Flower Mantid
Medium-sized (body length 42mm). Attractively mottled in variable pinks, browns, reds and greens with a prominent, eye-like, spiral marking on each fore wing. Femora of mid and hind legs ornamented with rounded lobes. Large lateral extensions on underside of abdomen. Wingless nymphs **(2A)** carry the abdomen curled above the body and are spectacularly ornamented and striped in pink and green. **Biology** Perches on and mimics flowers, and ambushes visiting insects. Spreads fore wings to display eyespot in threat display to frighten off attackers. When threatened, nymphs can expand the raised abdomen to reveal a single dorsal eyespot. **Habitat** On flowers and in vegetation in warmer regions.

3 *Harpagomantis tricolor*   Flower Mantid
Small to medium-sized (body length 31mm). Resembles *Pseudocreobotra wahlbergi* (above), except for the absence of fore wing eyespot. A double spine projects forward from middle of the head. Eyes pink with white spots, and produced into forward-pointing spines, giving the head a W-shape when viewed from above. Fore wings attractively banded in mauve and green. Colour variable, with some specimens having vibrant yellow and green markings. **Biology** Stands immobile and unseen on flowering plants, waiting to ambush visiting insects. **Habitat** In association with flowers with which its colours blend. Widespread and fairly common throughout the region.

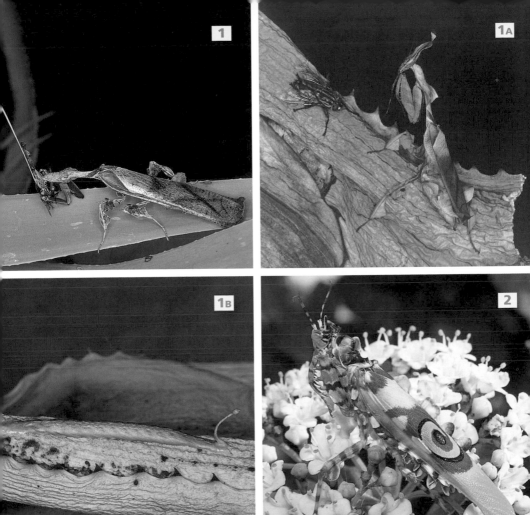

1 Galinthias amoena
Cone-eyed Mantid

Small (body length 25mm). The enormous eyes project strongly forward into points, giving front of the head a distinct U-shape when viewed from above. First thoracic segment greatly elongated, with a rounded expansion above the origin of the fore legs. Fore wings greenish-yellow and not patterned except for net-like venation. Hind wings red anteriorly, becoming brown behind. Body and limbs banded in pale pink and green. **Biology** Stands immobile and unseen on flowering plants, waiting to ambush insects. **Habitat** In weedy habitats and in association with flowers. Range extends into Central and West Africa.

2 Oxypiloidea tridens
Bark Mantid

Small (body length 24mm), flattened, mottled and extremely cryptic, with powerful grasping limbs. Identifiable by upright teeth on the prothorax, a small pair close together in front and a larger, separated pair behind. **Biology** One of several similar-looking species that scuttle about on the bark of trees. Capable of rapid movement. Antennae vibrate continuously. **Habitat** The trunks of trees in arid areas.

3 Oxypilus
Boxer mantids

Small (body length 30mm), body dark brown to black, wings lighter brown patterned with white lines. Fore legs robust, with white spines on coxa, femora (grasping 'hand') broad and white externally. Nymphs **(3A)** rotund and patterned in brown and cream, fore legs massively expanded, femora white with very large spines. Females wingless, with rotund abdomens. Three similar species in the region. **Biology** Nymphs resemble bird droppings. Conspicuous fore legs are opened and extended in unusual, repeated movements, revealing the brightly coloured inside parts, possibly as an aggressive display. **Habitat** Adults and nymphs found on bark, where they are very cryptic.

Family Mantidae
Common mantids

Most typical green and brown mantids belong to this family, as do many species that mimic bark. Inner margin of the grasping legs has a row of alternating short and long spines; those on the margins of the tibiae are erect and well separated. By far the largest family of mantids, with 47 genera and 140 diverse species in South Africa.

4 Dystacta alticeps
Brown Leaf Mantid

Large (50mm). Head oval, eyes not projecting; male (depicted) has elongate thorax, constricted behind origin of the fore legs, wings flattened, broadening posteriorly, light mottled brown, darkening posteriorly, criss-crossed with pale lines. Legs elongate, raptorial limbs fairly slender. Female pink, with much broader, rounded and flattened abdomen and abbreviated wings. **Biology** Cryptic and strongly sexually dimorphic, with wingless females. Moves fore limbs as does *Oxypilus* (above). **Habitat** Vegetation; distribution extends into East and Central Africa.

5 Sphodromantis gastrica
Common Green Mantid

Large (body length 55mm), robust and bright green, usually with a white spot near the anterior corner of each fore wing. Sides of abdomen may be mauve and yellow. Females much fatter than males. Nymphs lack wings and curl the abdomen up over the body. Familiar boat-shaped foam-like egg cases **(5A)** laid on walls, fences, etc, and comprise a series of vertical chambers along either side of a central keel. Each chamber contains a single egg, and the nymphs hatch after about a month. **Biology** Unusual in that its diet consists mainly of caterpillars. **Habitat** On the foliage of trees and shrubs in domestic gardens and a variety of types of undisturbed vegetation. Often attracted to lights. The most common urban mantid in the region.

6 Omomantis zebrata
Zebra Mantid

An attractive large (body length 52mm), slender green mantid. Characterized by coloration of the fore wings, each decorated with diagonal dark brown stripes and a pair of yellow spots surrounded by black. Legs long, thin and unornamented. **Biology** Often captures flying prey. Attracted to lights, and sexes not aggressive to one another. **Habitat** Bushes and trees in savanna, ranges from Cape to Kenya.

1 *Miomantis* — Delicate mantids

Small to large (body length 25–60mm), delicate, translucent, pale green or light brown mantids, with broad heads and few other distinguishing features. Females short-winged and fatter than males. *Miomantis* includes 20 easily confused regional species. **Biology** Young nymphs of some West African species are convincing ant mimics. *M. caffra* has been introduced from South Africa to New Zealand and become invasive. **Habitat** Common in a variety of vegetation types.

2 *Polyspilota aeruginosa* — Marbled Mantid

Very large (female body length over 80mm, male smaller) and spectacular. Male relatively slender, green ventrally, becoming darker brown on dorsal surface. Wings usually dark mottled brown, with a white flash midway along margin. Inner surface of fore legs diagnostically bright blue and yellow, with a conspicuous dark spot at base of femur. Female **(2A)** fatter and more muted brown, but still with paler spot on wings and blue inside fore leg. **Biology** Heavy-bodied and formidable, and may rear up aggressively when threatened, revealing the coloured areas on inner sides of fore legs. Can draw blood when grasped, using the fierce spines on its fore legs in attack. Often attracted to lights. **Habitat** Trees in bushveld and savanna.

3 *Ligariella* — Ground mantids

Very small (body length 11mm) with enlarged heads, short, squat, mottled, flesh-coloured bodies, and banded legs. Males winged, females wingless. Best identified by unusual ground-dwelling habits. There are 5 species in *Ligariella*. **Biology** Mimic stones and scurry around on the ground in search of insect prey. **Habitat** Arid regions. Found on the quartz plains of the West Coast and Karoo. **Related species** There are several similar species in the related genus *Ligaria*.

4 *Episcopomantis chalybea* — Grass Mantid

Large (body length 50mm), pale, uniformly straw-coloured. Eyes and corners of head greatly extended into elongate points, giving head a striking V-shape when viewed from above. Hind wings with patch of black mottling. **Biology** Well camouflaged in dry grasses, where it lies pressed to the stem waiting for insect prey. **Habitat** Grassy vegetation in drier regions.

5 *Popa spurca* — African Twig Mantid

Female **(5)** large (body length up to 80mm), robust, brown, with densely knobbed and ridged surface. Distinct ridge down centre of prothorax, from the margins of which project a row of rounded knobs. Limbs ornamented with spikes and ridges. Keel on femur of fore legs and series of toothed lobes on posterior limbs are good identifying features. Male **(5A)** smaller (up to 70mm), slender and dark, and characteristically holds fore legs stretched out. The only *Popa* species in the region. Previously known as *P. undata*. **Biology** Nymphs ground-dwelling, adults tree-living. **Habitat** Dry vegetation. **Related species** *Gonypetella deletrix* **(5B)**, which is a small, very dark, cockroach-like species with a very short pronotum.

Family Tarachodidae

Previously a subfamily of Mantidae, but recently recognized as a distinct family. Characterized by large head, somewhat flattened body and shortened middle and hind legs. Males usually fully winged, but females have shortened wings.

6 *Tarachodes* — Bark mantids

Medium-sized (body length 37mm). Mottled grey-brown with reticulated silvery wings and often flattened from top to bottom; closely resembles bark and lichen. Wings attractively reticulated, with dark blotches. Eleven similar-looking species in the region. **Biology** Move about on the trunks of trees in search of caterpillars and other prey. Some species **(6A)** have brushes of setae along the margins of the thorax and limbs, which help them to blend into their surroundings. Several African species show maternal care, the females guarding the egg case for up to 70 days until the nymphs hatch. **Habitat** Usually on tree trunks in the warmer parts of the region.

1 *Pyrgomantis rhodesica* Grass Mantid

Large (body length 70mm), unmistakable, with long, cylindrical, light brown body and short legs, the first pair folding neatly into the sides of the body. Head elongate and drawn out into a long pointed cone. Pink on inner surface of fore legs. Shortened wings in some individuals. **Biology** Lies motionless along the length of a grass stem, perfectly disguised, with limbs folded, waiting for prey. Species in the genus exhibit fire melanism, black individuals predominating after a bush fire. **Habitat** Grassland.

Family Thespidae Thespid mantids

A small family of lightly built mantids with a slender, elongate prothorax. Tibia or 'finger' of the fore limbs is less than half the length of the femur on which it closes. Females are wingless.

2 *Hoplocoryphella grandis* Stick Mantid

Large to very large (body length 60–80mm), very slender and delicate, pale brown, with spindly, elongate limbs, remarkably widened 'hammer-shaped' head and dark bands across the eyes. Beautiful, translucent, bubble-like egg case **(2A)**, with a sculptured ridge down 1 side, is usually conspicuously attached to a grass stem. **Biology** Primarily a species of arid regions. **Habitat** Dry grasslands.

3 *Hoplocorypha macra*

Medium-sized (body length 40mm). Similar to *H. grandis* but smaller and lighter. Body mottled grey and brown, with slender limbs and spine-like projections on the posterior margin of each abdominal segment. Females relatively robust and wingless, males slender-bodied and winged. Egg case a transparent yellow ball. **Biology** An ambush predator that seeks its prey in vegetation. **Habitat** Vegetation in drier areas; range extends into East Africa.

Family Amorphoscelidae Bark mantids

Small, sometimes delicately built mantids, which lack spines on the fore tibia. Males are typically winged while females are wingless.

4 *Compsothespis*

Small (body length 32mm), extremely slender and delicate, with such minute raptorial limbs it could be mistaken for a stick insect. Uniformly brown except for a pale yellow line along each fore wing and orange base to hind wings. Short-winged and flightless. **Biology** Weak fore limbs suggest it can take only very small prey. **Habitat** Low, often herbaceous, vegetation.

Family Sibyllidae Sibyllid mantids

Fragile mantids with a forked and laterally toothed projection on top of the head. The long, slender prothorax is armed with lateral lobes and a sharp dorsal projection. Only 1 species is known from the region.

5 *Sibylla pretiosa* Cryptic Mantid

Medium-sized (body length 45mm), distinctive, delicate and ornate. Head with large, forked dorsal spike armed with a pair of lateral teeth. Prothorax very elongate, slender and cylindrical, with sharp lateral lobes. Long, thin legs decorated with small femoral lobes. Fore wings mottled brown. **Biology** Cryptic, moving around on the bark of trees and feeding on other insects that share this specialized habitat. Commonly eats flying prey. **Habitat** On tree trunks in woodlands.

Family Empusidae
Cone-headed mantids

Large, usually slender and elongate mantids, often with quaint or bizarre body forms or adorned with leaf-like lobes, especially along the walking legs. Conical projection from top of head, divided at the tip. Antennae of males are elongate, with comb-like projections along both sides. Spines alternate along the inner surface of the femur on the fore legs, with 1 long spine following every 2–4 short ones.

1 *Empusa guttula* — Cone-headed Mantid

Very large (body length 74mm), overall colour mottled pink and green; top of head produced dorsally into a large, elongate cone. Antennae with comb-like lateral projections in male (shown); thread-like in female. Prothorax greatly elongated, laterally expanded above the origin of the fore legs, the expansion semicircular and bearing teeth anteriorly. **Biology** Alert and voracious predator. Often attracted to lights. **Habitat** Various vegetation types.

2 *Idolomorpha dentifrons* — Ornate Cone-headed Mantid

Very large (body length 80mm) and impressive, similar in colour and shape to *Empusa*, but with longer antennae and an even more elongate prothorax that is only slightly expanded above the fore limbs, the expansion rounded, not toothed, anteriorly. Inner surface of fore legs attractively patterned, coxa red with white dots near base becoming blue distally, femur orange. Males attracted by odours released by females. Nymphs slender and green with an upturned abdomen and very long, lobed legs; fore legs with similar distinctive coloration on medial surface as in the adults. **Biology** A powerful species capable of taking large prey. **Habitat** Vegetation in tropical areas.

3 *Hemiempusa capensis* — Giant Cone-headed Mantid

One of the largest mantids in the region (body length up to 85mm). Looks similar to others in the family, with a projection on top of the head, comb-like antennae and very elongate prothorax, but lateral expansions of the prothorax are rounded lobes. Legs with small lobes above junction of tibia and femur. Marked lateral lobes along sides of abdomen. Nymphs **(3A)** mottled green and mauve, very slender, with long, lobed legs and upturned abdomen. **Biology** An uncommon voracious predator. **Habitat** Low vegetation.

Long, flattened insects, easily identified by the conspicuous curved forceps at the tip of the elongated abdomen.

Long, predominantly brown or black insects, with short, square, leathery fore wings (tegmina), which protect the broad, fan-shaped, elaborately folded hind wings; rarely wingless. Mostly nocturnal and hide in damp, confined spaces by day. Most are omnivorous, but some capture live insects with the forceps, which can also be used for display and for folding and unfolding the hind wings. Eggs are laid in a burrow and tended by the female, which may continue to guard the early-instar nymphs. Nymphs look and behave like miniature adults. Harmless but may damage garden crops. Of the 2,000 global species, about 50 occur in the region.

Family Labiduridae — Long-horned earwigs

Contains a wide variety of winged and wingless, large to small, dull to brightly coloured species. The second tarsal segment is cylindrical, and not extended beneath the third. Antennae with 16–30 segments.

1 *Labidura riparia* — Shore Earwig

Large (body length 30mm) and pale, with dark markings on leathery fore wings and across abdomen. Huge curved forceps in male (shown); those of female closer together and straighter. **Biology** Aggressive, raising forceps over the head when threatened. Attacks smaller insects with the forceps before consuming them. Large white eggs are laid under logs or stones and guarded by the female **(1A)**. **Habitat** Under wood or stones on beaches and riverbanks. Naturally cosmopolitan.

Family Forficulidae — Common earwigs

Identifiable by second tarsal segment, which is expanded laterally and extends around the third. The antennae have 10–15 segments.

2 *Forficula senegalensis* — Common earwig

Small (body length 8mm). Thorax and leathery fore wings golden brown, with a dark central band; abdominal segments black. Size and shape of forceps unusually variable, being shorter and straighter in female **(2)** and often very long, curved and ornate in male **(2A)**. **Biology** Scavenges rotting fruit, organic debris and dead insects. **Habitat** Crevices, leaf litter, arum lilies; also, under stones or on trees. Very common.

Family Pygidicranidae

A primitive family of stout, flat earwigs, with a broad head and textured body surface, and often covered in short, stiff bristles. All species are winged. Forceps are simple and symmetrical.

3 *Echinosoma wahlbergi*

Medium-sized (body length 13mm), unmistakable due to rough surface texture. Black, with paler hind wings that project beyond the leathery fore wings. Forceps strongly curved and knobbed on inner surface. Legs banded brown or black. **Biology** Details not known. **Habitat** Under bark and beneath logs. Uncommon.

Family Anisolabididae

Forceps usually asymmetrical – the right more curved than the left. Most species are secondarily wingless. Antennae have 14–24 segments, and the tegmina (when present) are small and rounded, not meeting at their base.

4 *Euborellia annulipes* — African Earwig, Ring-legged Earwig

Medium-sized (body length up to 20mm), completely wingless, dark brown with paler 'neck'. Antennae with fewer than 20 segments. Legs pale with dark band across the femur, hence common name. Forceps stout and often asymmetrical, those of male more curved than in female **(4)** (shown, with nymph). **Biology** Reaches considerable densities in damp habitats such as compost heaps. **Habitat** Cosmopolitan and possibly the most widespread earwig in the world, introduced from southern Europe to South Africa about 1912, common in suburban gardens.

ORDER EMBIOPTERA (EMBIIDINA) • WEB-SPINNERS

Small, elongate, soft-bodied and often wingless insects living in silken galleries under bark, logs or stones. Best identified by the greatly swollen first tarsal joints of the front legs and the kidney-shaped eyes.

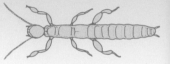

Web-spinners, or embiids, are a small and little-known order with elongate, cylindrical bodies and short legs. They use the greatly swollen first tarsal segments of the front pair of legs, which are packed with ball-like silk glands, to weave silken tunnels and galleries on and under tree bark, under stones, or in soil and leaf litter. The femora of the hind legs are enlarged to propel the insect rapidly backwards – one of several adaptations to life in narrow tunnels. Females and nymphs are always wingless, but in some species adult males develop 2 pairs of similar, widely spaced, dark wings. These are extremely flexible, so as to bend easily in the tunnel. The abdomen in males terminates in 2 short tails (cerci), the shape of which are important for family identification.

Most web-spinners live in small colonies, seldom emerging from their galleries, which are gradually expanded by the colony. They feed on bark, lichens, mosses and dead plant material. Adult males leave the colony in search of mates, and in winged species they may be attracted to lights, where they are sometimes mistaken for small reproductive termites. Eggs are laid in clusters within the galleries and attended by the females. Juveniles resemble adults and moult 4 times during their development. Although they live in family groups, they are not truly social. There are only about 360 described species globally and an estimated 37 species in South Africa, from families Embiidae, Teratembiidae and Oligotomidae, the majority of which remain undescribed.

Family Embiidae

Most South African species are included in this widespread family of relatively large, robust species. Adult males are characterized by 1 or 2 lobes on the inner surface of the left cercus. The cerci are covered with short, peg-like setae. Many species from this family are thought to occur in the region, but to date these remain poorly known and are undescribed.

1 *Apterembia*

Medium-sized (body length 11mm), both sexes wingless. **Biology** Feed on leaf litter. Construct galleries where leaf litter and debris collect against margins of stones. **Habitat** The only named genus of embiid in the Western Cape, occurring beneath mats of vegetation growing against rocks; also spins webbing between fallen leaves. The specimen shown here was collected on the slopes of Table Mountain.

Family Oligotomidae

This family originates in Asia, but several species have been widely dispersed as a result of commercial traffic in logs, plant material and fruit. The left cercus in adult males lacks inner lobes and is divided into 2 segments, separated by a membranous joint.

2 *Oligotoma saundersii* — Saunders' Embiid

Small (body length 7mm). Adult male winged, female wingless. Best recognized by sickle-shaped hook on underside of the genitalia in adult male. The only member of the family in the region. **Biology** Spins flat webs radiating from fissures in the bark of shade trees. Probably feeds on lichens and mosses on the trunks of host trees. Females remain within the tunnel webs throughout their lives. The winged males **(2A)** have to pump up their flexible wings with blood before they can fly, and are often attracted to lights. **Habitat** Indigenous to India, but has been widely distributed by commercial activity, and is now found in East Africa, Madagascar, Australia and other moist tropical and subtropical regions around the world. Common in gardens and parks in Durban, where its webbing can be detected under flakes of bark on tree trunks. Probably more widespread, but poorly sampled due to its cryptic habits.

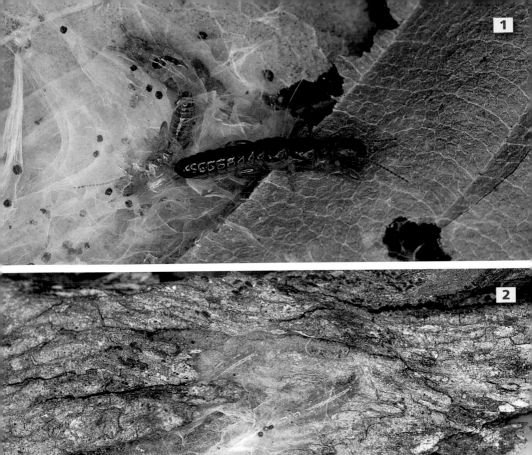

ORDER PLECOPTERA • STONEFLIES

Small to medium-sized, soft-bodied insects associated with flowing water. Adults are elongate and rarely fly, but are fast runners. The membranous wings are folded flat onto, or furled around, the abdomen. Large folded anal fan present on hind wings.

Adult stoneflies have large compound eyes, weak mandibles, long thin antennae, and long conspicuous cerci (reduced to 1-segmented stumps in Family Notonemouridae). The aquatic nymphs are recognized by their long paired cerci, a body shape similar to that of adults and (in Family Perlidae) tufts of gills at the base of the legs. The larval stage lasts from a few months to years. Mature nymphs (recognized by black wing pads) climb up a rock and then moult, leaving behind the empty 'shuck'. Adults appear on the surface of rocks on cold overcast days, but otherwise hide in cool shadowed rock cracks or beneath rocks in, or very close to, rivers. They also occur on vegetation bordering streams. Adults fly rarely, and have a weak, fluttering, termite-like flight. Males are smaller than females. Notonemouridae are most abundant in the southwestern Cape, and Perlidae in larger, more slow-flowing rivers in the warmer, summer-rainfall parts of the region. Nymphs are vulnerable to predation by introduced trout and bass.

1 Family Perlidae
True stoneflies

Large (body length 25mm), yellow to tan, with large compound eyes, wings that fold flat over the abdomen, and a pair of long cerci that project beyond the wings. At least 5 superficially similar *Neoperla* species occur in the region. **Biology** In summer adults are attracted to lights close to fast-flowing rivers and streams. Mating takes place by day, the female developing a conspicuous egg ball at the tip of the abdomen. Eggs have elaborate sculpturing, useful in separating species. *Neoperla* nymphs **(1A)** have well-developed bushy white gills at the base of the legs and on the abdomen **(1B)** (see arrows), apparently enabling them to occupy less-oxygenated streams than Notonemouridae, and are voracious predators on other aquatic insects, such as mayfly and blackfly larvae. Presence of larval skins (shucks) on rocks **(1C)** indicates emergence periods. **Habitat** Large, often turbid, rivers, as well as smaller forest streams with rocky beds.

Family Notonemouridae
Southern stoneflies

Small (body length 5–8mm) with thin lead grey or dark brown bodies and wings tightly furled around the abdomen, giving the appearance of pencil lead. Antennae are long. Cerci long and conspicuous in nymphs, but inconspicuous and reduced to small, single-jointed stumps in adults. Short-winged or wingless individuals may occur. Adults mate during the day, and can be numerous in winter in the Western Cape. They feed on plant matter and live for only a few days. Nymphs are important shredders of dead leaves in mountain streams. They resemble wingless adults and occur in leaf packs or under stones, usually in streams with a canopy of vegetation. Most species occur in the Western Cape, a few in the southern Cape, Lesotho, KwaZulu-Natal and the Mpumalanga Drakensberg. Restricted to cold, unpolluted mountain streams. Six genera and about 30 species occur in the region.

2 *Desmonemoura*
Porcupine stoneflies

Adults small (body length 5mm) and readily identifiable, with wings conspicuously banded in tan and dark brown. Nymphs relatively hairless, with a few bristles on the abdomen, identifiable when mature by banded wing pads. There are 2 species in the genus. **Biology** Adults emerge in mid-summer. **Habitat** Nymphs found under stones in fast-flowing mountain streams, adults in shaded rock crevices in or near streams.

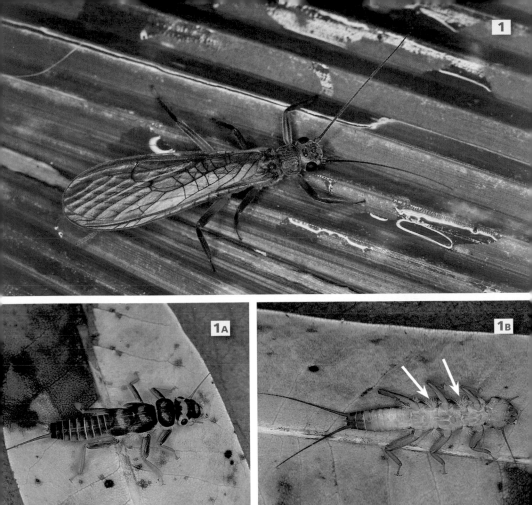

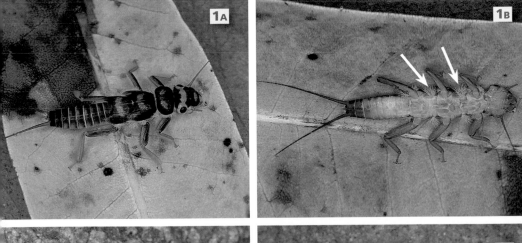

1 *Aphanicerca* — Cape stoneflies

Small (body length 5–10mm) (but among the largest southern stoneflies), slender and elongate. Body and wings dark brown, fore wings with a clear patch in the middle of the wing. Nymphs **(1A)** elongate and smooth, with several characteristic short tufts of hair along the inner side of each antennal base **(1A inset)**. There are 8 described and many more undescribed species in the genus. **Biology** Males tap ('drum') the substrate during courtship. Adults emerge in wet mid-winter and spring. **Habitat** Good-quality shaded mountain and forest streams. Nymphs abundant in leaf packs and under stones, adults resting in crevices on or under rocks. Restricted to the Cape Fold Mountains in the Western Cape.

2 *Aphanicercella* — Hardy stoneflies

Adults (not depicted) very small (body length 4–6mm long) and thin, with grey-black wings. Nymphs **(2)** have incomplete whorls of hair on the abdomen, apart from the last few segments where they are complete; wing pads become black. *Aphanicercella* includes 11 species, most of them found in the Western Cape. **Biology** Most widespread of all of the genera, often in large numbers on rocks in streams when overcast. Adults emerge in wet mid-winter and spring. **Habitat** Both open and closed-canopy streams, on rocks but also adjacent vegetation.

3 *Balinskycercella* — Drakensberg stoneflies

Small (body length 4–9mm), dark grey with a clear fore wing patch, as for *Aphanicerca*. *B. tugelae* shown **(3)**. Males with small recurved hooks at end of the abdomen. Nymphs **(3A)** hairy, with complete whorls of hair on each abdominal segment, and wing pads also covered in long bristles. There are 3 species in the genus. **Biology** Adults occur in summer, November–January. **Habitat** Mountain streams in Lesotho and the KwaZulu-Natal Drakensberg.

4 *Afronemoura* — Tufted-nymph stoneflies

Small stoneflies (body length 4.5–8.5mm), wings grey-brown with clear patches, legs banded grey and tan. Distinctive nymphs **(4A)** have a conspicuous tuft of short hairs one-third of the way from the tip of the antennae **(4A inset)**. There are 3 species in the genus. **Biology** Adults occur throughout the year, but most abundant May–December. **Habitat** Rocky forest streams in Hogsback, Amatola and KwaZulu-Natal Drakensberg mountains.

5 *Aphanicercopsis* — Denticulate stoneflies

Adults small (body length 3–6mm), dark grey to black, resemble *Aphanicercella* but with slightly thicker mesothorax. Nymphs **(5)** with oval prothorax, small compound eyes, and body largely free of bristles. There are 4 species in the genus. **Biology** Adults occur in wet mid-winter and spring. Short-winged and totally wingless forms occur, usually much smaller than winged forms. **Habitat** Open and forested rocky mountain streams in the Cape Fold Mountains.

ORDER ORTHOPTERA • CRICKETS, KATYDIDS, GRASSHOPPERS & LOCUSTS

Medium-sized to large insects with bulky body, broad blunt head, hind legs usually modified for jumping and, typically, massive development of the saddle-like first part of the thorax (pronotum). The fore wings are leathery, forming tegmina; the hind wings are membranous and fan-like. One or both pairs may be absent in adults.

Orthopterans have chewing mouthparts and, as with most primitive insects, short tactile projections (cerci) at the end of the abdomen. Juveniles resemble miniature adults, but lack wings. There are 2 relatively distinct subgroups. In the largely nocturnal 'long-horned' crickets and katydids (suborder Ensifera) the antennae exceed the body length; males attract females with sounds produced by rubbing the fore wings against one another, and this is detected by 'ears' **1** on their fore legs; and females have long ovipositors. In the mainly diurnal 'short-horned' grasshoppers and locusts (suborder Caelifera) the antennae are less than half the body length; sounds are produced by rubbing the hind legs against the fore wings or abdomen and detected by hearing organs on the sides of the body; and females have short ovipositors. Orthoptera are generally good fliers, but few (with the notable exception of migratory locusts) are capable of sustained flight. Short-horned grasshoppers lay many eggs in a pod in the soil **2**, while katydids lay single eggs in or on plants. The highest concentration of endemic species occurs in the arid western parts of the region, with most of the summer-rainfall species occurring widely across the Afrotropical region. Over 800 species have been recorded in South Africa.

SUBORDER ENSIFERA
Katydids, crickets

More primitive than the grasshoppers and their kin (suborder Caelifera), their name ('sword-bearer') refers to the large blade-like ovipositor of the females. Hind legs are only somewhat enlarged, but often very long. Among the most vocal of all insects, the song being used to attract and court females, and to challenge rival males. Sound is produced by rubbing specialized areas of the fore wings against one another. 'Ears' are located on the thorax or fore tibiae.

Family Anostostomatidae
King crickets, Parktown prawns

Includes king crickets – large, brown or reddish insects (formerly included in family Stenopelmatidae) that live in underground burrows by day and emerge at night to feed on small animals and plant matter. Males of many species have very enlarged palps and mandibles, used aggressively in fights with other males. They stridulate by rubbing the legs against the abdomen and, like the 4 families that follow, have 4-segmented tarsi. Life cycles are long, the eggs taking up to 18 months to hatch, and adults requiring 1–3 years to reach maturity. Adults live for about a year. Males may keep harems, and females brood eggs and young in a specially excavated chamber. At least 51 species are known from the region.

3 *Borborothis*
Medium-sized (body length 30–35mm), wingless and uniformly dark brown. Hind femur armed with 2 rows of teeth along its top length. Small eardrums on both sides of fore tibiae. *Borborothis* comprises 2 species. **Biology** Shallow burrows are constructed under logs in moist forest. Among the most primitive of king crickets. **Habitat** Under logs in coastal and montane forest.

4 *Onosandrus*
Medium-sized (body length 28mm), with a single spine on the inner side of upper fore tibiae (in addition to spines near the apex) and unmodified mandibles of the same size in male and females (see also *Libanasidus*, p.90). Black markings create barred pattern on the cream body. Hind femora thickened basally, tapering towards apex. *Onosandrus* comprises about 10 southern African species. **Biology** Nocturnal, roaming about at night to feed. **Habitat** In wet soil (e.g. under sodden moss) and under or in rotting logs.

1 Nasidius

Large (body length 28mm), black with reddish head and legs. Inner side of upper fore tibiae with 2 spines (in addition to spines near apex). Blunt cone on forehead of males (less blunt in females). Mandibles enlarged ventrally in males, not towards the front as they are in *Libanasidus* below. About 14 species in the region. **Biology** Emerge from burrows at night, especially after rain. Presumed carnivorous. **Habitat** Under logs in indigenous forest, or in soil burrows under stones.

2 Henicus brevimucronatus

Platypus King Cricket

Medium-sized (body length 20mm), bizarre. Males with grossly enlarged head and mouthparts, with tusk-like process just above mandibles. Upper lip (labrum) enormous and spoon-shaped **(2A)**. Both sexes have 2 spines on inner side of upper fore tibiae (besides spines near apex). **Biology** Adults reside in pairs in burrows. **Habitat** Under and in rotten logs in Afromontane forest. **Related species** There are 9 other species in the genus. *H. monstrosus* (occurring in mountains of the Western Cape) is similar and even more bizarre, with a very enlarged head and antler-like mandibles in the male. It uses its enlarged upper lip to block its burrow and reputedly produces rasping sounds by gnashing its jaws.

3 Libanasa

Medium-sized (body length 23mm), with a single spine on inner side of upper fore tibiae (in addition to spines near apex). Cream-coloured, with black marbling. May be confused with *Onosandrus* (p.88), but hind tibiae are very thick at base, and do not taper towards apex. About 7 species in the region. **Biology** Nocturnal, emerging from burrow at night. **Habitat** Under stones, rotting logs and leaf litter in subtropical forests.

4 Libanasidus vittatus

Parktown Prawn, King Cricket

Very large (body length 60–70mm), familiar, with head and thorax reddish, abdomen with orange and black bands, and orange legs. Enlarged mandibles in male bear sharp horn-like processes (tusks) that overlap. Female (shown) with sword-shaped ovipositor. Single spine on inner side of upper fore tibiae (in addition to spines near apex) in both sexes. **Biology** Nocturnal, emerging from burrow in the early evening. Varied diet includes snails, insect grubs, earthworms, seedlings, fallen fruit and dog droppings. Adults can consume up to 4 snails a night. When disturbed, vocalizes by rubbing its abdomen against the hind legs and ejects foul-smelling faeces. **Habitat** Originally restricted to forest litter in indigenous forests, but has spread to well-watered, rich soils in city gardens. Common in the northern suburbs of Johannesburg.

Family Stenopelmatidae

Jerusalem crickets, sand crickets

Large, very rotund, wingless, cricket-like insects, with long antennae, short, strong burrowing limbs and a swollen abdomen. They have a large head bearing robust jaws, and feed on dead plant matter, as well as worms and insects. The only genus in the region is *Sia*, which is endemic, with 4 fairly common regional species.

5 Sia pallidus

Cape Sand Cricket

Large (body length 30mm), cream, with black-banded abdomen, honey-coloured head, thorax and legs, and 4-segmented tarsi. **Biology** As with other members of *Sia*, it constructs burrows under stones. Has a severe bite, probably carnivorous. Reportedly rubs its hind legs against the body to generate sounds. Emerges at night and climbs vegetation. **Habitat** Fairly dry sandy soils, where it constructs burrows under rocks and logs.

Family Gryllacrididae
Leaf-rolling crickets

Small, rounded crickets, fully winged (*Stictogryllacris*) or wingless (*Glomeremus, Ametroides*); 4-segmented tarsi. Antennae disproportionately long.

1 *Glomeremus*
Wingless leaf-rolling crickets

Medium-sized (body length 10mm), body pale tan; 3–4 spines on mid and fore tibiae (in addition to spines near apex). Tip of ovipositor pointed (but rounded in *G. obtusus*). There are 5 species of *Glomeremus*. **Biology** Mute, probably carnivorous. **Habitat** Western Cape *Glomeremus* are known to hide between very young leaves on buds of the Silver Tree (*Leucadendron argenteum*), which they bind with an oral secretion from the labial glands. **Related species** There are 2 species of (tropical) *Stictogryllacris* (body length about 15mm) and 1 of *Ametroides* (**1A**) from Mpumalanga, the latter recognized by 2 spines on either side of the fore and mid tibia, excluding apical spines.

2 Family Schizodactylidae
Dune crickets

Medium-sized (body length 11mm), unusual-looking, pale. Tarsi highly modified, 4-segmented and with long lobed structures. Often have knobs on cerci. A few species are winged, with spirally coiled wing tips. *Comicus*, with 8 wingless species, is the only African genus. **Biology** Inhabit very sandy areas in arid western parts, where they spend the day in burrows and emerge at night to feed on other insects. The enlarged tarsi enhance digging ability and act like snow shoes when they walk over sand dunes, resting in depressions with outstretched antennae, presumably waiting for prey. The large lobed structures on the tarsi help them to leap several metres. Males produce calls by rubbing the femora of the hind legs against the abdomen. **Habitat** Prefer consolidated sand to very loose, shifting sand dunes. Most southerly records from red sands of Vanrhynsdorp and Vredendal.

Family Rhaphidophoridae
Camel crickets, cave crickets

Wingless, humpbacked crickets covered in fine golden hairs, with very long legs, palps, and antennae; tarsi 4-segmented. Females with long ovipositors. Only the subfamily Macropathinae occurs in Africa, with the 2 known species restricted to caves in the Western Cape.

3 *Speleiacris tabulae*
Table Mountain Cave Cricket

Medium-sized (body length 10mm), humpbacked, with uniform brown body. Head and ends of legs reddish-orange. Eyes small and dark. **Biology** Scavenges on various types of decaying matter in caves, especially bat guano and dead bats. Runs and jumps well and is generally found on moist cave walls and rocks. Needs very high humidity levels to survive. **Habitat** In sandstone caves and grottoes on Table Mountain; also in crevices among large boulder aggregations in moist forest on the southern slopes of the Cape Peninsula. Also recorded from a small cave in the Cape Town suburb of Oranjezicht. **Related species** *S. monslamiensis*, which has recently been described from caves in the Hex River mountains, near Ceres.

Family Tettigoniidae
Katydids, bush crickets, long-horned grasshoppers

A very large family of predominantly green species, mostly concentrated in the warmer parts of the region. Identified by the combination of very long, thin antennae, 4-segmented tarsi and a pair of 'ears' on fore tibiae. Males rub a toothed file on 1 fore wing against an oval disc on the other, producing a buzzing mating call or an aggressive (probably territorial) call. Typically, the ovipositor is very long and sword-like. Most live in vegetation and are often excellent leaf mimics. Includes herbivores, scavengers and carnivores. Females of arboreal forms lay eggs in leaf sheaths, or use a serrated ovipositor to penetrate plant tissue and lay eggs in it. About 160 species are known from the region, approximately two-thirds of which are endemic.

Subfamily Hetrodinae
Armoured ground crickets, koringkrieke

Large, flightless and often spiny crickets, with the antennae originating from below the lower margin of the eye. Males retain small circular wings (hidden under the spiny pronotum). Primarily herbivorous, but will scavenge and can turn cannibalistic, devouring roadkilled kin. Eaten by only a few animals and large birds, including the Bat-eared Fox, jackals and the Kori Bustard. Males produce loud, buzzing mating calls, usually at night. Abundant in arid and semi-arid zones. About 30 species in 5 genera are known from the region. Most genera and all species are regional endemics.

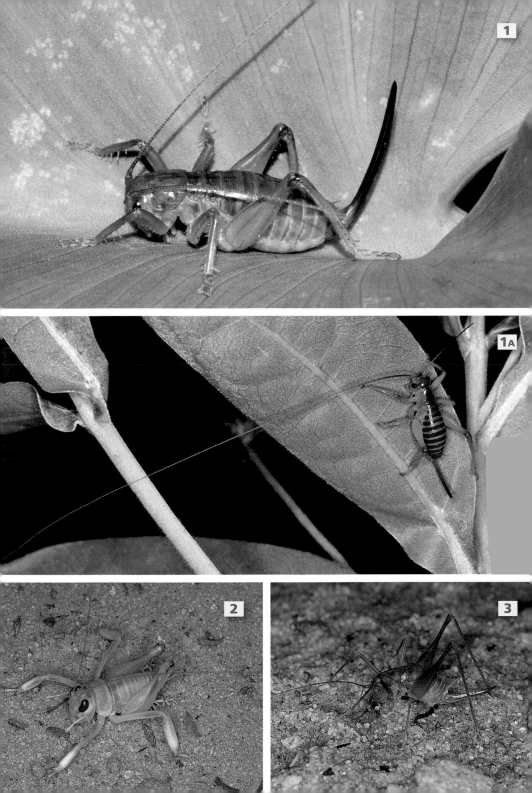

1 *Acanthoplus armiventris*   Green Ground Katydid, Koringkriek

Medium-sized (body length 21mm) with yellow pronotum with black spines, including 2 short ones in front; brown, yellow and green abdomen, red antennae. **Biology** Feeds on *Senegalia* leaves, but its catholic tastes include stoneplants (*Lithops*), citrus fruits and quelea chicks. Drops from vegetation when alarmed, the male producing a series of short alarm chirps. Squirts yellow blood from the thorax when handled. Males attract females with a high-pitched vibrating call, but also tap their abdomen against the substrate. Eggs are laid in shaded soil in May, in packages of 3–11, and hatch the following March, 2 days after the first rains. Eggs can lie dormant for another year if summer rains fail, resulting in large swings in population size in different years. **Habitat** Low thorn trees, scrub and subtropical bushveld. **Related species** There are 5 other species in the genus, all with an unspined dorsal surface on the hind femora and straight spines on the pronotum. *A. discoidalis* **(1A)**, which is very common across the entire region except for the west coast, also lacks spines on the underside of the hind femur and the male has short, pointed, inward-curving cerci.

2 *Acanthoproctus cervinus*   Antlered Ground Katydid

Large (body length 39mm), greyish-brown. Abdomen with dark brown top, pronotum with black, branched, antler-like spines projecting from front corners and simple spines laterally; large yellow patch on back. Lateral spines on abdominal segments. Female wingless. **Biology** As in all members of the family, reduced wings of males have a row of file-like teeth that produce the characteristic continuous, high-pitched call when rubbed together. Present for most of the year. **Habitat** Low, semi-arid, scrubby vegetation and open thornveld. **Related species** There are 2 other large species in the genus, both with 2 side-facing projections arising from the pronotum; one of these, *A. vittatus*, occurs in the Free State and Eastern Cape.

3 *Enyaliopsis*   Horned ground katydids, koringkrieke

Large (body length 30–50mm), brown, with a pair of blunt outward-projecting horns on each side of the pronotum. Back edge of pronotum light yellow. There are 13 African species, but only the Northern Armoured Katydid (*E. transvaalensis*) occurs in the region (an unnamed species from Zimbabwe shown here). **Biology** Largely terrestrial, often found moving slowly on soil and rocks. When handled, they squirt bright green blood up to 25cm. Call from bushes at night. **Habitat** Open woodland and grassland.

4 *Hetrodes pupus*   Common Ground Katydid, Koringkriek

Large (body length 36mm), grey or brown, abdomen with black spines and tan stripes. The 4 short spines in the middle of the pronotum form a chatacteristic rectangular pattern. **Biology** Active at night, when males produce a continuous piercing buzzing call. Adults occur from September to June. Females lay about 14 very large eggs. **Habitat** Very common in low and open vegetation, ranging from fynbos to succulent vygie veld. **Related species** No other species in the genus, but there are 4 subspecies.

Other subfamilies

The following Tettigoniidae belong to subfamilies other than Hetrodinae.

5 *Eurycorypha*   Leaf katydids

Medium-sized (body length 22mm), with elongated oval eyes. Open 'ears' on fore tibiae in males. Tegmina bend sharply, lending a humpbacked appearance. Usually has a yellow stripe through the eyes, pronotum and tegmina. A large genus with many similar African species, approximately 7 occurring in the region. **Biology** Eggs are inserted into leaves and hatch in 3 months. Nymphs of some species mimic ants **(5A)**, being black or dark brown, with ant-like behaviour and shape (and have been found in the company of *Polyrachis* ants). The penultimate instar is green, resembling the adult. Nymphs reach adulthood in 7–11 weeks; adults live for 3–7 months. Adults are excellent broadleaf mimics, and even rock from side to side (wind mimicry). **Habitat** Subtropical forest and scrub.

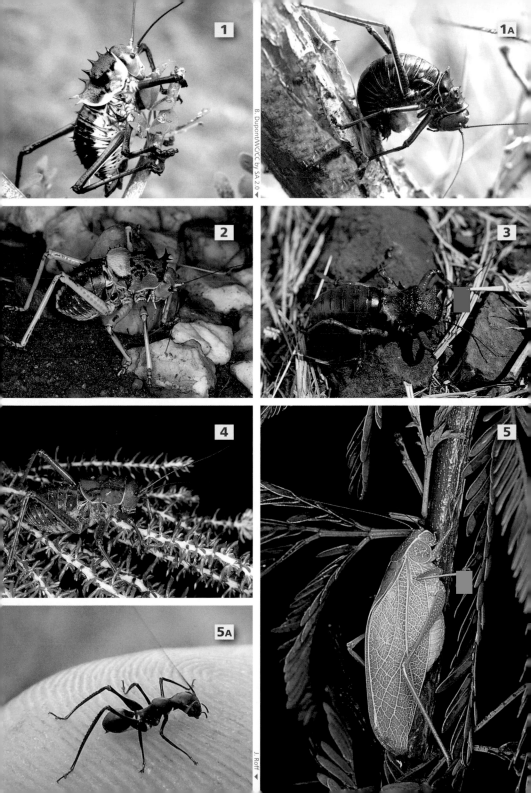

1 *Melidia brunneri* — Yellow-bodied Leaf Katydid

Medium-sized (body length 35mm), lemon green with a yellow abdomen, and red markings on antennae and abdomen. Cells of hind margin of wings green (not brown). More robust than *Phaneroptera* (below) and with broader wings. Sound-producing organ on left fore wing in male has brown markings. **Biology** Like other katydids, feeds on buds, leaves and fruits. Thought to produce a short, intermittent, rasping call at night. **Habitat** Subtropical bushveld as well as succulent karoo. A small genus that also occurs in Central and East Africa. **Related species** Three other African species; only *M. brunneri* reported in the region.

2 *Phaneroptera* — Leaf katydids

Medium-sized (body length 30mm), green, with round eyes and unstriped green abdomen. Ends of hind tibiae usually have 3 apical spurs on each side. Fore wings opaque green, at least in the last half; hind wings have a tan dorsal stripe and extend well beyond fore wings. *P. sparsa* is the most common of 12 species. May be confused with *Tylopsis* (below). **Biology** Eggs are laid at the edge of the leaves between the upper and lower epidermis; female stabilizes leaf between legs before slitting the leaf open to oviposit. Feed on a range of plant parts, including insect galls containing other insects' eggs and larvae. Minor pests of tea, cotton and vegetable crops. **Habitat** A range of vegetation types. Widespread in most of the Old World.

3 *Tylopsis* — Grass katydids

Medium-sized to large (body length 40mm), uniformly green, resembling *Phaneroptera*, but with slit-like 'ears' on the fore legs. A stripe, yellow in males (3), white in females (3A), runs along the bottom of the abdomen. *T. continua* and *T. rubrescens* have a brown or reddish pronotum. The third regional species, *T. bilineolata*, is very common in the wetter parts of the region. Genus widespread in Africa, extending into Europe. **Biology** Feed on plant parts, and also on aphids and caterpillars. Appear to mimic grass, not leaves. **Habitat** Bushveld and open grassland.

4 *Conocephalus caudalis* — Long-tailed Meadow Katydid

Small (body length 20mm, excluding 30mm ovipositor), fairly slender and delicate, green with brown dorsal stripe, long hind legs and ovipositor. Females wingless. **Biology** Feeds on grasses, grass seed and moths' eggs. **Habitat** Common in lush grassland, especially near water. **Related species** There are 14 other regional species in this common genus, which has a global distribution, including *C. conocephalus* (4A), which is fairly widespread, tan with a brown dorsal stripe and wings in both sexes.

5 *Megalotheca longiceps* — Wingless Meadow Katydid

Medium-sized (body length of female 29mm), slender and elongated, with short legs and a very long ovipositor. Either tan (5) or green (5A), with pinkish back and antennae, and a white stripe along the side of the body. Female wingless, but male may have short tegmina. **Biology** A restio mimic, resting with legs held close against the stems of restios, on which it probably feeds. **Habitat** Restio veld and grassland. **Related species** There are 3 other species in the genus. Two occur in KwaZulu-Natal, where they feed on grasses.

6 *Terpnistria zebrata* — Acacia Katydid

Medium-sized to large (body length 44mm), with broad thorn-like spines on legs and a small crest on rear of pronotum. Entire body and wings barred in cream and green. **Biology** Leg spines resemble *Senegalia* leaflets and coloration matches *Senegalia* foliage. **Habitat** A typically southern African genus of bushveld and (often arid) thornveld, although 1 species extends to East Africa. Found on *Senegalia* and close relatives. **Related species** *T. lobulata* (6A), from KwaZulu-Natal, is the only other species in the genus; it has a smaller crest on the pronotum and lacks leafy outgrowths on the hind tibiae.

1 *Ruspolia* — Cone-headed katydids

Large (body length 60mm), elongate, uniformly green or brown. Jaws asymmetrical with a yellow base. Purple-striped colour forms (**1A**) also occur. Approximately 9 regional species. **Biology** Active at night, feeding on grasses and grass seed, manipulated and then cracked with the very powerful asymmetrical mandibles. Prefer *Digitaria* and *Chloris* grasses, but also feed on (and damage) 'milk' seed of sorghum, millet and maize. Males produce a very loud, continuous (up to 5-minute long) hissing call. One of the few katydid genera that forms large locust-like swarms. Eggs are laid in the ground at the base of a grass clump, the nymphs hatching after 1–2 months and reaching adult maturity in 2–3 months. Often attracted to lights, where the fatty adults are caught by members of certain tribes and boiled, roasted and eaten. **Habitat** Grassland and open bushveld.

2 *Cedarbergeniana imperfecta* — Cederberg Rock Katydid

Large (body length 35mm, hind legs 70mm long) but slender, with pale green and tan body. Legs very thin and long, banded in black, pale green and orange, with black barring on hind femora. Ovipositor long and broad. Adult males not known, but probably have very small wing pads. **Biology** This primitive katydid lives in groups in small, constantly cold caves within cracks in rock faces. At night, exits to feed on vegetation at the cave entrance. Mating call of male is a short click. **Habitat** Restricted to Wolfberg Cracks in the Cederberg Mountains.

3 *Aprosphylus olszanowskii* — Olszanowski's Black-kneed Katydid

Medium-sized (body length 21–28mm), body pale tan, ends of hind femora black; wings shorter than length of hind femora. Fore wings banded in black and white, hind wings finely marked with black patches. One of 3 species in the genus that are in the same primitive subfamily as *Cedarbergeniana* (above) and that are restricted to arid ecosystems. **Biology** Shelters during the day under piles of rocks, rapidly jumping and taking flight when disturbed. **Habitat** Occurs under rocks in the arid western parts of the region. **Related species** Four large (up to 45mm long) species, which live in deep rock crevices in arid parts of Namibia and the Northern Cape, including *Pseudosaga maculata* (**3A**), which has finely mottled wings and barred legs.

4 *Zitsikama tessellata* — Knysna Forest Ground Katydid

Medium-sized (body length 25mm), dark brown with wavy cream and black longitudinal stripes. Antennae and legs very long, the latter banded in black, orange and tan, with spines on apexes of fore tibiae. Male smaller than female; tegimina very short and square in male, reduced to wing scales in female. Ovipositor short, upcurved. The only species in the genus. **Biology** Nocturnal. Males sing using the reduced tegmina. **Habitat** Leaf litter on wet indigenous forest floor.

5 *Alfredectes semiaenea* — Alfred's Shieldback

Medium-sized (body length 21mm), copper-pink body evenly speckled with small black dots. Parts of body with a metallic tinge. Base of hind tarsi with 2 elongated pads, presumably to assist in jumping. Large elongated tan pronotum, with bright green and black markings, and hind femora marked in black. **Biology** Male use rudimentary wings concealed beneath the large pronotum to call. Mostly carnivorous, feeding on smaller insects. **Habitat** Low bush, or on the ground in clearings; also in secondary growth in forest areas. **Related species** *Thoracistus* species (**5A**) (body length 45mm, including 20mm ovipositor), which have a very inflated olive pronotum with green and black shading. *Thoracistus* is a subtropical genus that includes 8 regional endemics, which occur on the forest floor in Limpopo, the Eastern Cape and KwaZulu-Natal.

6 *Clonia melanoptera* — Short-winged Predatory Katydid

Huge (body length 103mm, including 40mm ovipositor), squat and robust. Body olive green with cream, grey and purplish markings. Tegmina and hind wings very reduced, anal area of hind wings dark brownish-black. **Biology** Nymphs present in winter. Adults carnivorous, appearing in summer. Male's buzzing mating call very loud, carrying 1.5km. Flashes black hind wings when threatened. **Habitat** Low scrubby vegetation, often in strandveld and fynbos. Adults live in low thorny bushes. **Related species** *Leptoclonia* species, which are small (about 50mm long), wingless or short-winged, and widespread in the drier central parts of the region.

1 *Clonia wahlbergi* — Winged Predatory Katydid

Large to very large (body length 40–65mm), slender, body apple green with silvery markings. Wings fully developed or slightly shortened, as in all members of the genus. Anal area of hind wings whitish with concentric brown bars. *Clonia* comprises 14 regional species. **Biology** A voracious predator of other large insects **(1A)**. Can bite severely if mishandled. Active by night, male producing a low buzzing call. Eggs are laid in the ground at the base of a bush. **Habitat** Widespread in bushveld, forest margins and grassland.

2 *Zabalius aridus* — Blue-legged Leaf Katydid

Large (body length 60mm), bright green, with a few peculiar raised orange tubercles on the pronotum, and pale blue, pink and tan hind legs. Nymphs **(2A)** resemble adults, but have a yellow midline stripe and green hind legs. **Biology** A broadleaf mimic. Feeds on leaves of trees and shrubs. Normally slow-moving, but kicks out defensively with the colourful spined hind legs when disturbed. **Habitat** Lush forest vegetation, often in gardens. **Related species** *Z. ophthalmicus* from the forests of KwaZulu-Natal is the only other *Zabalius* species in the region. The closely related Afrotropical species *Z. apicalis* feeds exclusively on young *Ficus* leaves and lays eggs in rows on the tender shoots, its life cycle from egg to adult taking about 100 days.

3 *Cymatomera denticollis* — Bark Katydid

Large (body length 55mm), with 3 ridges along light grey prothorax. Abdomen barred in red, orange, yellow and black. Fore wings intricately marked in grey, brown and orange. Broad grey hind wings banded in light brown. **Biology** Magnificent lichen or bark mimic. Like other leaf katydids, active at night. If disturbed, raises wings in a threat posture **(3A)**, revealing colourful abdomen, and is capable of spraying a defensive chemical from an abdominal gland. **Habitat** Dry subtropical bushveld.

4 *Brinckiella* — Flightless spring katydids

Medium-sized (body length 15–20mm), humpbacked, green to yellow, wingless or with very short wings. Legs and antennae very long and slender. *Brinckiella* comprises 8 regional species. **Biology** Males call by rubbing the hind legs against the body. Adults occur in spring. **Habitat** Semi-arid fynbos and succulent karoo.

Family Gryllidae — Crickets

Familiar and very common insects, generally in shades of brown and black, with 3-segmented tarsi, long, thin antennae, hearing organs on the fore legs, and sound-producing organs on the fore wings (when present). Fore wings are folded horizontally over the body and downwards on each side, forming a saddle-like box over the body. Each fore wing has its own file and mirror (cf. Tettigoniidae, p.92), sound production being complex and very well developed. Most species are nocturnal and omnivorous. After courtship, the male transfers the sperm package to the female, which has a very long, thin, straight ovipositor. About 70 species are known from the region, but the fauna is undoubtedly far richer.

5 *Acanthogryllus fortipes* — Brown Cricket

Large (body length 25mm), stocky, with a very large, broad pronotum and head. Generally dark brown with tan areas behind the head, on the sides of the prothorax and on the elytral margins. Fore tibiae with long apical spurs; hind tibiae heavily spined with about 8 very pronounced long spurs on both inner and outer sides. Both sexes winged, elytra reaching the end of the abdomen in males. The only species in the genus. **Biology** Crops grasses and stores the clippings around the 2 burrow entrances. Adults present from November to January, the chirping call of males heard from burrow entrances in spring and summer. A serious pest of lawns, cricket pitches and young seedlings. **Habitat** Common in short-cropped grass.

1 *Gryllus bimaculatus* — Common Garden Cricket

Large (body length 25mm), male shiny black, female, shown on left in **(1)**, dark brown; both sexes with 2 yellow shoulder patches. *Gryllus* includes 1 other regional species. **Biology** When approached, male changes chirping song to a softer, lower call. Highly territorial, males very aggressive towards one another. Never constructs burrows. **Habitat** Back yards, alleys and drains in urban areas; rarely in undisturbed grassy areas. Very widespread, occurring all over Africa and into Europe and Asia, typically in association with human habitation and cities.

2 *Teleogryllus wernerianus*

Small (body length 15–25mm), without head markings, both sexes with white palps. Hind wings very long. **Biology** The only member of the genus that does not produce the familiar 2-part call day and night. Never constructs burrows, instead using cracks in soil or dry leaves for concealment. Adults present from December to March, migrating to wetter regions in winter (mid-year). **Habitat** Mainly in grassland and hedges. **Related species** There are 5 other regional species. The genus occurs across Africa and Asia.

3 *Cophogryllus* — Mute crickets

Small (body length 10mm), both sexes wingless. Body tan, with dark brown markings. Ovipositor longer than hind tibiae. Comprises 7 similar species. **Biology** Adults hide under stones and logs by day, moving about freely at night. Males are mute. **Habitat** Open, exposed (often arid) areas in grassland or woodland.

4 *Grylloderes*

Medium-sized (body length about 19mm), uniformly dark brown, with a pale stripe along edge of fore wings. Hind tibiae heavily spined. *Grylloderes* comprises 5 regional species. **Biology** Mainly surface-dwellers, although some live in cracks, and 2 shiny species with small hind legs dig their own burrows. Emerge to forage and mate at night (mating call a chirp). **Habitat** Abundant in cultivated fields; also seasonally swampy grassland and bushveld. *G. primiformis*, the most common species, occupies a range of habitats from coastal dunes to montane grassland.

5 *Neogryllopsis* — Gaudy crickets

Medium-sized (body length 17–23mm), short-winged. Head and prothorax orange-brown, with ivory-coloured markings. Abdomen dark brown, with light ivory stripes. Males have only fore wings; females wingless. Comprises approximately 22 very similar regional species. **Biology** Active by night, when they leave their burrows and males **(5)** call from the ground or in bushes. **Habitat** Open sandy woodland.

6 *Brachytrupes membranaceus* — Tobacco/Giant Burrowing Cricket

Massive (body length 40mm), heavily built, uniformly pale yellowish-brown, with dark markings on head and hind legs, very broad head and prothorax, and very strongly spined hind legs. **Biology** Lives in tunnels 150–250mm deep, with a large chamber at the end. Reputedly the largest and loudest cricket in the world, with a powerful, high-intensity, continuous buzzing call that can be heard over 1.5km away. Males call from outside their burrows. A pest of crops, including young tobacco plants. Hunted by *Chlorion maxillosum* (p.488) as food for its young. **Habitat** Arid sandy areas. Extends across Africa. **Related species** There are 2 other regional species in the genus.

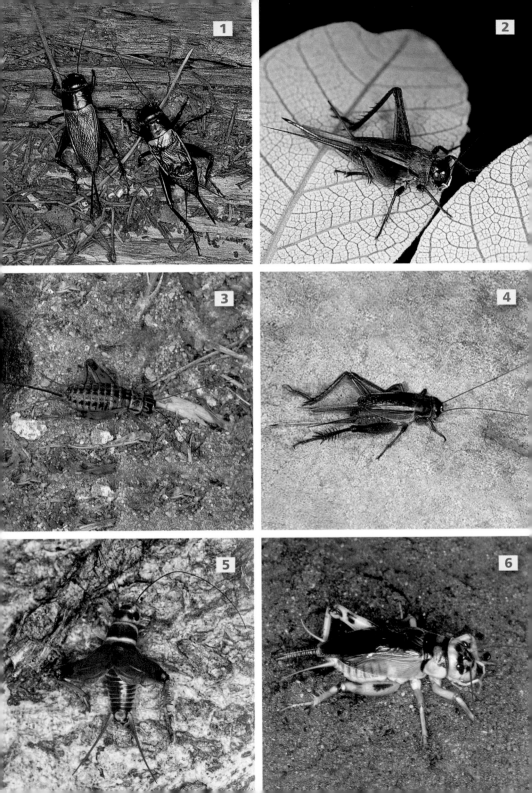

1 *Oecanthus* — Tree crickets, thermometer crickets

Small (body length 16mm), fragile, pale green or straw-coloured, elongate, with translucent wings. All species winged and capable of flight. There are 10 regional species. *O. capensis* is very common in the Western Cape: male **(1A)** has black abdominal stripes and the short hind wings don't extend beyond the tegmina. **Biology** Chirp rate is temperature dependent and can be used to identify species. For *O. capensis* the temperature in degrees Celsius equals the number of chirps in 3 seconds + 11; for *O. karschi*, the number of chirps in 6 seconds + 12. Leaves are used as cones to amplify the mating chirp, 1 species calling from just behind a small hole that it gnaws in a leaf. Eggs are laid in a row in deep slits in young shoots, often causing wilting and death of peach tree shoots. Food includes aphids and young caterpillars. **Habitat** Trees, shrubs, grasses, bushes and weeds, often in gardens. Favour the garden tree *Ficus nitida*.

Family Gryllotalpidae — Mole crickets

Cylindrical, light brown and furry, with fore legs massively developed for burrowing, much like those of true moles; hind legs not enlarged. Some species are wingless, but those in the region have short fore wings with very large, fan-shaped hind wings folded beneath them, the tips sticking out like 2 spikes. There are files on each fore wing in males. A small family, with 4 species in the region, 1 of which is endemic.

2 *Gryllotalpa africana* — Common African Mole Cricket

Medium-sized (body length 25–30mm), light brown, with shortened fore wings and a large dome-shaped pronotum covering the head. **Biology** Male produces a characteristic uninterrupted deep buzzing sound (hard to locate), by rubbing together the files on the fore wings. Does not jump, but often flies to lights. Feeds on plant roots, and may be a pest of (wet) lawns or bowling greens. Feeds on potatoes and strawberries at night. Eggs are laid in a nest chamber in the burrow, and the 10 larval stages take 2 years to reach maturity. Constructs a permanent burrow up to 1m in depth, with 2 openings. Males call from a fork near the mouth of the burrow, which amplifies the sound. **Habitat** Very moist or waterlogged soils in gardens or near vleis, riverbanks and streams. **Related species** There are 3 other regional *Gryllotalpa* species, but they are less widespread than *G. africana*.

3 Family Trigonidiidae — Sword-tailed crickets

Fore wings short and hardened, without stridulatory apparatus. Hind tibiae without smaller spines between long spines. Ovipositor flattened and upturned. Second tarsal segment short, widened and flattened. *Trigonidium erythrocephalum* **(3)** (body length 5mm) has a cherry red head, black body with grooved beetle-like tegmina, and yellowish-green legs. Family includes miniature, subtropical and tropical crickets. **Biology** Perch in exposed positions on leaves during the day, and very agile when disturbed. More common in the warmer eastern parts of the region. **Habitat** Trees and shrubs in areas with high summer rainfall.

4 Family Mogoplistidae — Scaly crickets

Slender-bodied and flattened, covered in translucent scales. Wings very short or absent. Hind tibiae lack long spines (but have apical spurs). Hind femora are stout. The species in the region are small (body length 8mm), cylindrical and brown, with long cerci held parallel to the ovipositor. *Arachnocephalus* has an almost cosmopolitan distribution in the tropics. *A. bricki* recorded from the region; probably many more regional species to be described. **Biology** Apparently nocturnal, hiding during the day between leaf bases. **Habitat** Shrubs and trees in bushveld and on forest edges.

SUBORDER CAELIFERA
Grasshoppers, groundhoppers, pygmy mole crickets

A familiar group, with antennae shorter than the length of the body, hind legs powerfully enlarged for jumping, and no 'ears' on the thorax. The hearing organ, when present, is on the first abdominal segment. Sound produced by rubbing hind legs against wings. Females have a very short ovipositor used to excavate soil for oviposition.

1 Family Tridactylidae — Pygmy mole crickets

Very small (body length less than 5mm), shiny black or sand-coloured, with shortened fore wings and long hind wings (but may nevertheless be flightless). Fore legs well-developed, hind legs with characteristic tibial spur longer than tarsi, used for digging. Hind femora may be bigger than abdomen. There are 4 genera in the region: *Tridactylus* (1 species); *Xya* (8 species); *Afrotridactylus* (2 species), and *Dentridactylus* (1 species). **Biology** Live in small tunnels just below the surface of the soil, and emerge to feed on damp soil, ingesting the algae associated with it. Able to walk on and even jump off the surface of water and to swim. **Habitat** Moist soil associated with the sandy shores of pans, rivers or the sea. The sandy coloured *A. pallidus* lives in dense colonies near the high-water mark of marine beaches in the Eastern Cape, and constructs shallow mole-like burrows in the sand, cleaning unicellular algae off the sand grains as it does so.

Family Tetrigidae — Groundhoppers, grouse locusts

Like small grasshoppers, but easily distinguished by a massive extension of the hood-like pronotum, which usually covers the entire top of the body. Fore wings reduced to small scales, hind wings large and fan-like. Diurnal, but attracted to lights. Cryptic, but may be numerous, and can be netted in large numbers from moist open ground. Many swim well. Adults feed on small plants and algae and breed in spring; unlike grasshoppers their eggs are not laid in a pod. Widespread in damp areas, often in seepages or on the margins of streams and rivers. Tetrigidae has 12 genera and about 30 species in the region.

2 *Tettiella*

Small (body length 7mm), stocky, brown, with a large, broad, curved pronotal hood. The genus occurs in many colour forms, some with light hood markings; all wingless. There are 3 regional species. **Biology** Slow-moving, but jump when disturbed. **Habitat** Leaf litter.

3 *Ascetotettix capensis*

Small (body length 10mm), dark brown, with yellow-striped pronotal hood extending above and in front of head. **Biology** Flightless. The sole representative of the genus. **Habitat** Among fallen leaves in moist Knysna forests.

Family Euschmidtiidae — Bush hoppers

Medium-sized grasshoppers (body length 20–30mm), body flattened from side to side, antennae short. Many species short-winged or wingless. When magnified, inner end of hind tibiae shows a single spine. Eight regional species.

4 *Amatonga*

Medium-sized (body length 24mm), green or brown, with characteristic cylindrical body tapering at both ends, very short antennae, and a yellow cheek stripe. Sit with hind legs positioned at 90° to the body and flat on substrate. Two species in region. **Biology** Foodplants varied, including ferns and conifers. **Habitat** Normally found on shrubs (rarely on grasses) in tropical grassland, bushveld and forest margins.

Family Thericleidae — Forbhoppers

Small grasshoppers, very similar to Euschmidtiids, but distinguished under magnification by 2 well-developed spurs at inner ends of hind tibiae. Occur on shrubs and small trees rather than grasses. There are 78 species known from across the region.

5 *Lophothericles*

Small (body length of males 10mm), stocky, with short rod-like antennae, very large eyes and abdomen upturned at the end. Wingless, with white markings on brown body. Legs green. There are approximately 15 regional species. **Biology** Found on the toxic plant Gifblaar (*Dichapetalum cymosum*). **Habitat** Shrubs in grassland and bushveld.

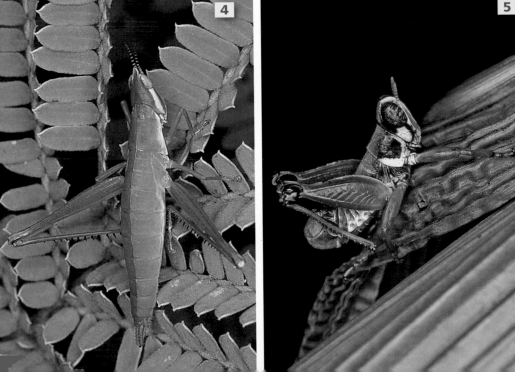

1 *Schulthessiella minuta*

Small (body length 10mm), stocky, light tan, pink or plum-coloured, with very short antennae and large eyes. Wingless. Hind femora green, very broad, with toothed margin and very large brown knee. **Biology** Feeds only on young shoots of a few species of shrubs. **Habitat** Subtropical bushveld and grassland.

Family Pneumoridae   Bladder grasshoppers

Unmistakable, large, usually bright green, arboreal grasshoppers, with generally very large and inflated pronotum, disproportionately so in nymphs. Males have very large and broad fore wings, and shorter hind wings without the large folding (anal) lobe. Females usually have reduced wings that are almost hidden beneath the pronotum. Hind legs are not enlarged for jumping. In males, the abdomen is hollow and inflated, acting as a resonating chamber for sound production; the hind femora have ridges that are rubbed against ridges on the abdomen. Mating call of males very loud and far-reaching. Restricted feeding habits, with usually only a few host plants. Feeding generally takes place in the day, mating at night. Some live in forests, others in the drier western parts of the region. Virtually endemic, with 17 species known from the region (2 forest species of *Physophorina* extending into East Africa).

2 *Pneumora inanis*   Bladder Grasshopper

Very large (body length 62–86mm), characterized by a weakly inflated pronotum. Emerald green or brownish-red, with white edge to pronotum and 2 white spots on tegmina. Females have silvery spots on body and on the short wings. The only species in the genus. **Biology** Male's call is a long screech followed by a repeated 'khonia' sound. Nymphs disperse soon after hatching and develop through 6 instars from March to November. **Habitat** Subtropical forest.

3 *Bullacris intermedia*   Bladder Grasshopper, Hekiejee, Gonna

Medium-sized (body length 44–49mm), male **(3)** green with brown-and-yellow bar along metathorax, female **(3A)** green with 3 rows of silvery markings on the abdomen. Pronotum always smoothly arched along its length in both sexes. The 4–6 nymphal instars have a very large pronotum and silvery markings. **Biology** Adults appear in early spring and, when approached, hide behind branches. Males produce a loud frog-like groan in the early morning, which carries up to 2km; they are territorial and fight one another. Small wingless males, without the inflated abdomen, employ 'sneak' tactics, locating and mating with females without making a mating call **(3B)**. Cup-shaped egg pods deposited in a shallow hole under the foodplant; hatching initiated by rain. The genus is fairly host-specific, usually on Asteraceae, e.g. *Metalasia* (*B. membracioides* feeds on *Helichrysum* and the thistle *Berkheya*). **Habitat** Patchily distributed in semi-arid vegetation, typically succulent karoo and renosterveld. **Related species** *Bullacris* includes 6 other similar-looking species, including *B. serrata* (body length 50–55mm), in which the pronotum has an irregular outline and a red line along the top and the female **(3C)** is more evenly marked with white patches than the male **(3D)**.

4 *Physemacris variolosa*   Silver-spotted Bladder Grasshopper

Medium-sized (body length 39–50mm), olive green, both sexes heavily and evenly marked with silver spots. Pronotum has a few large tooth-like ridges. Male usually winged, female wingless and resembles nymph. Certain males are wingless and small and engage in 'sneaky' matings, short-circuiting the normal courtship of winged males and females. **Biology** Like most pneumorids, males fly at night. They spend much time flying around and calling from low bushes late at night (from 22h00), the high, screeching call sounding like 'hatchigeeee'. The male flies to the female after she responds to his mating call with her softer call. Known to feed on Renosterbos (*Dicerothamnus rhinocerotis*). **Habitat** Open fynbos and renosterveld.

Family Charilaidae   Cone-head grasshoppers

Medium-sized grasshoppers, easily recognized by a unique pair of ridges (carina) running along the length of the pronotum. Although a few species have fully developed wings, these are vestigial in others. A small, primitive family of 5 species in 3 genera, restricted to southern Africa, except for a single species in North Africa.

1 *Paracharilaus curvicollis*

Medium-sized (body length 32mm), cone-shaped head, with thin furrow in front; relatively long greyish-white antennae, and unbroken cream lines running along brown areas on head and prothorax. **Biology** Sluggish, some charilaids produce sounds by their rubbing hind wings against the fore wings. **Habitat** The only charilaid in South Africa. Occurs in open plains with sparse grass cover.

Family Lithidiidae   Stone grasshoppers

The smallest grasshoppers in the region. Robust and cryptically coloured with short limbs and large bulging eyes; most wingless. Occupy extremely hot, arid environments. The 3 genera in the region are concentrated in the Richtersveld.

2 *Lithidiopsis carinatus*   Richtersveld Stone Grasshopper

Small (body length 8mm), body robust, surface granulated, dark brown; eyes large. Legs with 2 black bars, and a black mark on the inner surface. **Biology** Survive on very hot soils but seek refuge in the shade of large stones. **Habitat** Areas of consolidated sand covered in small stones. **Related species** There is 1 other regional species in the genus.

Family Pamphagidae   Rain locusts

Large (body length up to 70mm), primitive, heavily built grasshoppers, normally cryptically coloured in dull earthy shades, or superb stone mimics. When viewed from above, the snout region of the head has a short furrow running towards the eyes (the fastigial furrow). Antennae often very broad, sword-shaped, triangular (not round) in cross section. Very rough body surface, often bearing tubercles and spines. Pronotum has a raised keel-like crest, which may be punctured by a series of small holes. Many species have winged males and wingless females. Most males are able to stridulate. Of the 71 species known from the region, about half are endemic.

3 *Hoplolopha*   Saw-backed locusts

Fairly large (body length 38–43mm), typically with wingless females (3) and winged males (3A). Females various shades of brownish- and greenish-grey, males steel grey with a cream cheek and prothoracic stripe, and pale yellow hind wings. Antennal segments somewhat flattened. There are 6 regional species. **Biology** Sluggish. Females produce an egg case with a swollen base. **Habitat** Arid, shrubby areas with low vegetation. Occur in arid northern and western parts of the country.

4 *Puncticornia puncticornis*

Large (body length 70mm), female (shown) wingless; top of pronotum and first abdominal segments with reflective, wart-like black markings over yellowish or flesh-coloured body, legs mottled brown; head with bluish patch behind the eyes. Males light tan, with large brown wings. **Biology** Slow-moving, spending time on the ground, not on vegetation. **Habitat** Arid, sandy areas strewn with quartz pebbles.

5 *Lamarckiana*   Rain locusts

Large (body length 60–100 mm), generally uniformly grey, with very flattened antennae, and cream cheek and prothoracic stripe. Males winged (5), with smoky black hind wings; females (5A) wingless. *L. cucullata* is the largest grasshopper in the region. There are 4 regional species in the genus. **Biology** Nocturnal, sluggish, males calling at night from trees, producing a very loud mechanical noise by rubbing the hind legs against a broad grooved area on leading edge of the tegmina; females rest on the ground. Avoid grasses. When flushed, males fly strongly for some distance. **Habitat** Various low vegetation types, usually in semi-arid areas with bare ground. Mostly subtropical.

6 *Stolliana angusticornis*

Large (body length 77mm), grey, with raised pronotum with a thick cream stripe. Thin cylindrical antennae. Male (shown) has dark brown hind wings and a black bar on the tegmina. Female wingless. **Biology** Males fly strongly when flushed. **Habitat** Rocky mountain slopes in arid areas. **Related species** There are 3 other regional species in the genus.

1 *Batrachotetrix stolli* — Toad Grasshopper

Medium-sized (body length 20mm), stout, wingless, with broad, flat pronotum and broad hind tibiae (common to genus). Various colour forms, colour and texture perfectly matching grey flintstone substrate. Male winged and smaller than female (shown). **Biology** Folds antennae against the head when approached and remains immobile. **Habitat** Open, well-drained, pebbly patches between low shrubs in semi-arid regions. **Related species** There are 14 other species in the genus. Flesh-coloured species (body length 24mm) are found on sand, and both sexes are winged. Most species occur in the Northern Cape and Namaqualand.

2 *Trachypetrella anderssonii* — Quartz Grasshopper

Large (body length 71mm), very squat and toad-like. A remarkable stone mimic that matches not only the colour of the semitranslucent pinkish-white quartz stones among which it lives, but also the texture, including the chipped fracture planes of stones **(2A)**. Wings vestigial. Antennae thin and can be withdrawn into sockets. Legs have a border of hairs that smooth the body contour when withdrawn, eliminating tell-tale shadows. **Biology** When molested, raises hind legs and lashes out with great strength and speed while producing a grating sound. **Habitat** Restricted to arid quartz plains with very little vegetation.

3 *Lobosceliana*

Large (body length 43–68mm), brown, with tan pronotum edged with white stripe and with 2 brown wedge-shaped markings. Antennae broad and flattened. Males winged, with brown hind wings. Females wingless. Nymphs uniformly light brown, with large pronotum. Three regional species in genus. **Biology** Not known. **Habitat** Dry leaf litter in deciduous woodland.

Family Pyrgomorphidae — Foam grasshoppers, lubber grasshoppers

Small to large grasshoppers distinguished by a combination of bright colours, conical head, fastigial furrow (see Pamphagidae, p.110), and a pair of warty crescents on either side of the fastigial furrow. All stages of wing reduction occur in various species. Many have warning coloration and can produce a foamy defensive secretion; a few can produce sound. Most feed on herbs or shrubs, but rarely on grasses. Many species are gregarious at all stages. Approximately 39 species are known from the region.

4 *Phymateus morbillosus* — Milkweed Locust

Large (body length 60–72mm), bulky, with red thorax or red knobs on green prothorax. Fore wings dark bluish-green with yellowish veins, hind wings mostly red. Red legs with purplish bases. **Biology** In late summer adults lay eggs that hatch the following spring. In common with other large species in the genus, nymphs take 2 years to reach maturity. Adult populations peak in alternate years. Prefers the milkweed *Asclepias fruticosus*, as do most species in the genus, but also recorded from *Solanum panduriforme*, and will feed on citrus and a variety of garden plants. Nymphs form bands, which migrate to new foodplants. When molested, gravid females give a threat display by raising the wings. **Habitat** Open veld, in groups on isolated milkweed bushes. **Related species** There are 5 other regional species in genus.

5 *Phymateus viridipes* — Green Stinkweed Locust

Large (body length 70mm), body and fore wings green, hind wings red and blue. Pronotum with raised serrated ridge tipped with red. Nymphs **(5A)** spotted black and yellow. **Biology** Like most *Phymateus* species, raises and rustles its wings when alarmed and produces an evil-smelling foam from the thoracic joints. Flies strongly at some height in late summer, forming swarms of many thousands and migrating long distances. **Habitat** Grassland. **Related species** *P. leprosus* **(5B)**, which has gregarious, shiny, black-and-yellow-green nymphs **(5C)** and is an occasional pest of young citrus trees in the Eastern Cape. Human fatalities from ingestion of this species are known.

1 *Dictyophorus spumans* — Foam Grasshopper, Rooibaadjie

Large (body length 64mm), bulky, familiar, with red-and-black body and legs, and abdomen with black and white bands. Colour variable, some forms redder and with less black **(1A)**. Flightless, with short to very short reddish tegmina and no hind wings. **Biology** Like many in the family, extracts and stores heart poisons (cardiac glycosides) from the milkweeds on which it feeds, and exudes these when molested **(1B)**. Known to be fatal to dogs if eaten. **Habitat** Open, often rocky, areas with low vegetation. Frequently encountered on mountain tops.

2 *Zonocerus elegans* — Elegant Grasshopper

Medium-sized to large (body length 28–50mm), head mainly black, antennae black with orange rings. Body attractively banded in blacks, reds and oranges. Tegmina long or short, pinkish-black or greenish at the base. The hoppers are black and white and gregarious. **Biology** A pest of irrigated smallholdings, damaging vegetables and fruit trees in the Northern Cape and Free State. In tropical areas, a pest of cotton, cocoa, coffee and many other crops. In nature, prefers Milkweed (*Asclepias fruticosus*), *Senecio* and *Solanum panduriforme* and, rarely, grasses. Can emit a nauseating yellow fluid when disturbed. Eggs are laid communally in autumn, hatching in spring and reaching maturity in 75 days. One generation per year. Livestock avoid eating bushes infested with this species. **Habitat** Ubiquitous on low shrubs in subtropical areas. **Related species** The other (more tropical) species in the genus, *Z. variegatus*, has a mainly yellow head with reduced black patterning, 1–2 orange bands on the antennae, and large plain green tegmina. It has been consumed by humans for centuries in both Africa and Asia.

3 *Ochrophlebia cafra*

Medium-sized (body length 32mm), lightly built. Body variegated in cream, yellow and black. Wings streaked with black and yellow, hind wings pinkish-purple. Colour variable, some forms drabber **(3A)**. **Biology** Adults are slow-moving and are found on low vegetation or on the ground. **Habitat** Semi-arid, open, succulent karoo vegetation.

4 *Taphronota*

Medium-sized to large (body length 52mm). Mostly uniform grass green, with very broad tegmina; hind wings uniform orange. Prothorax covered in bumps. There are 2 regional species in the genus. **Biology** Produce a noxious foam when molested. *T. stali* feeds on small woody plants, but other related species are considered agricultural pests in tropical Africa, where they feed on potato and amaranths. **Habitat** Grassland and open bushveld.

5 *Maura rubroornata*

Medium-sized (body length 40–50mm), body black with variable degree of patterning with large red or white dots; abdomen black with variable red or white bands; antennae black with 2 orange bands. Short- or fully-winged, tegmina black. Hind femora with 2 orange or white bands. **Biology** Often seen sitting on rocks or on aloe inflorescences. **Habitat** Rocky grassland.

6 *Stenoscepa gracilis*

Small (body length 20mm), wingless, bright green with a pink central stripe bordered by yellow stripes running along the length of the body; antennae with pink bases. **Biology** Feeds on leaves of the thistle *Berkheya coddii* and on *Senecio* species, its body accumulating nickel in high concentrations from these leaves. **Habitat** Serpentine grasslands. **Related species** The other regional species in the genus, *S. pictipes* **(6A)**, is similar, but with a row of small black spots running along the length of its abdomen.

Family Lentulidae

Endemic to eastern and southern Africa. South Africa, and particularly the Cape Floristic Region, are the centre of diversity for the family. Almost 70% of the known species are South African endemics. Includes a few large species, but mostly small, typically sluggish grasshoppers, always lacking wings, sound-producing organs and fastigial furrow (see Pamphagidae, p.110, which also have this furrow). Eyes, head and thorax very large. A few species occur in leaf litter, but the majority are tree or shrub dwellers, often found in groups. Frequently restricted to specific host species. About 60 species (in 20 genera) are known from the region, most from subtropical parts.

1 *Lentula*

Smallish (body length 19mm), cylindrical, wingless, large head and thorax giving the body a conical appearance. Prominent white stripe through cheek and thorax in some species. There are 4 regional species in the genus. **Biology** Sluggish, basking on the upper surface of leaves, but can dodge rapidly behind them. Most feed on a restricted range of plants, often including wild thistle (*Berkheya*). In the Western Cape *L. obtusifrons* feeds on Renosterbos (*Dicerothamnus rhinocerotis*) and *L. callani* on daisy bushes (*Euryops*). Heavy infestations of these species can kill plants. **Habitat** Woody shrubs and herbs.

2 *Paralentula prasinata*

Medium-sized females (body length 26mm) and much smaller males, as for all species in the genus. Very similar to *Lentula*, but with a single black stripe above the white lateral stripe on cheek and thorax. Body brown with orange speckling, hind femora olive green. **Biology** Can defoliate *Gymnosporia* and other bushes. **Habitat** Common in subtropical bushveld. **Related species** There are 3 other regional species in the genus.

3 *Devylderia*

Medium-sized (body length 35mm), wingless, convincing twig mimics. Grey, faintly mottled with pale orange. Fairly long, serrated antennae. Males less than half the size of females. There are 4 regional species in the genus. **Biology** Very common in summer, males usually carried on the backs of females, even when not copulating. **Habitat** Only in thick bushes in fynbos (especially on top of Table Mountain), renosterveld or succulent karoo.

4 *Betiscoides meridionalis*   Slender Restio Grasshopper

Medium-sized (body length 36mm), superb restio-stem mimic; body smooth, with 3 longitudinal stripes of green, grey and brown. **Biology** Probably feeds on restios. Adults present January–May. **Habitat** Restricted to restio stands on mountain slopes and plateaus. **Related species** There are 2 other common species in this genus, both smaller than *B. meridionalis*. Also, *Bacteracris* species, which are similar, but have a roughened body.

Family Acrididae   Short-horned grasshoppers, locusts

The largest and most familiar grasshopper family, recognized by their short, stout antennae, wings in adults of most species, and a hearing organ (tympanum) on each side of the abdomen. The egg-laying tube of the female is short and usually concealed, as are the male genitalia, and the abdomen in males always turns up at the end. Herbivorous, favouring a range of grasses and often consuming more even than antelope and cattle (up to 130kg of grass yearly per hectare). Nevertheless, most species are habitat-specific. Eggs are laid in groups in soil and mixed with a frothy substance that hardens to form the egg pod. Hind legs usually have pegs, which, when rubbed against a strengthened vein in the fore wing, make the characteristic chirping call. The 356 species known from the region are divided among 13 subfamilies.

5 *Truxalis burtti*

Large (body length 70mm), slender, green or brown, with grey stripes. Wavy white lines on tegmina separate green or brown longitudinal stripes. Hind wings with characteristic mottled black pattern, caused by darkened cross-veins that are denser towards the wing base (**5A**). Stridulatory serrations on inner side of hind femora and top of abdomen. **Biology** Rests on the ground between tufts of long grass. **Habitat** Subtropical grassland.

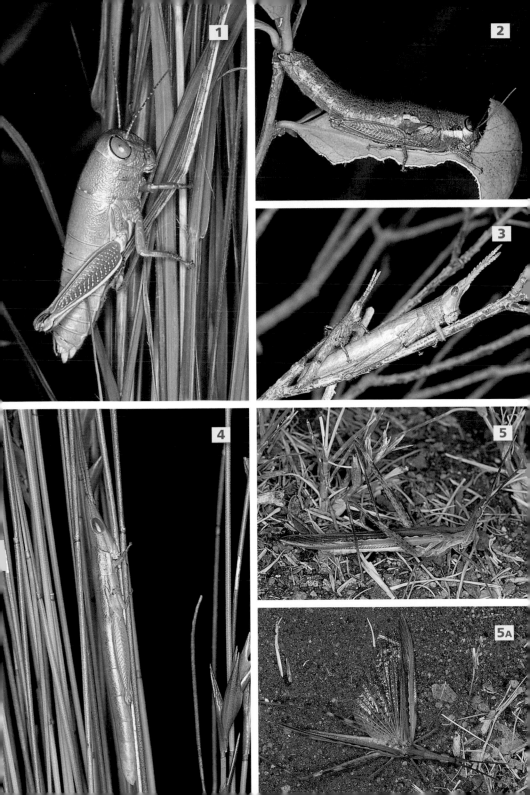

1 *Acrida acuminata*   Common Stick Grasshopper

Large (body length 66mm), elongate, green or straw-coloured (occasionally striped), with yellow or purplish hind wings lacking dark patches. Slender nymphs resemble small mantids. **Biology** Life cycle similar to that of *Cyrtacanthacris* (below), with adults present for a few summer months only. Female lays eggs in sandy areas, producing a very long egg pod with a hard wall. Feeds on various long grasses, both tough and soft species. Flies readily, making a loud crackling noise (louder in males) by rubbing the wings together. **Habitat** Widespread in fairly long grass in open sandy habitats in the region and in Central Africa, rarely in gardens. Occurs on the Cape Peninsula (as does *A. turrita*). **Related species** There are 8 other species in the genus. Also easily confused with *Truxalis* and *Truxaloides*, but lacks stridulatory serrations on inner side of hind femora.

2 *Truxaloides*

Large (body length 50mm) but slender, tan, with 2 cream stripes running along the prothorax; hind wings magenta **(2A)**, with or without a mottled black pattern. Stridulatory serrations on inner side of hind femora and top of abdomen. Similar to *Acrida*. There are 3 regional species in the genus. **Biology** Make brief but strong flights, alighting on bare ground. **Habitat** Areas with some long grass.

3 *Kraussaria*

Large (body length 50mm), stocky, apple green, with characteristic brown markings on pronotum. Cream stripe along inner edge of tegmina. There are 2 similar regional species in the genus. **Biology** Egg capsules laid in shaded soil under shrubs, with spongy cap hardened with sand. Solitary, flying to agricultural land to feed, and resting in the shade of trees during the midday heat. Feeds on developing grass seeds. **Habitat** Open woodland, treed savanna.

4 *Cyrtacanthacris aeruginosa*   Green Tree Locust

Large (body length 55mm) and stocky, mottled green and yellow, pronotum with yellow margins. Tegmina have a thin white line along both front and back margins and also a thin red line. Hind wings pale yellow. **Biology** As with many other locusts, has breeding cycle in which sexually immature adults survive the dry season and lay eggs in response to the first rains. Eggs hatch simultaneously and nymphal development is completed during the rainy season. A minor pest of vegetables and tobacco, and a nuisance in gardens. **Habitat** Rank (often shrubby) vegetation with long grass. **Related species** *C. tatarica* **(4A)** and *Acanthacris ruficornis* (p.120).

5 *Abisares viridipennis*   Notched Shield Grasshopper

Large (body length 53mm), rufous brown and grey, with very long antennae, weakly banded legs, and large notched prothoracic crest. Eyes have characteristic vertical bands. Nymphs **(5A)** like adults, but grey or green. The only species in the genus. **Biology** Slow-moving. Sits on low herbs. **Habitat** Shaded subtropical forest and bushveld.

6 *Nomadacris septemfasciata*   Red Locust

Large (body length 51–72mm). Tegmina very long, brown or reddish-brown, often with pale dorsal stripe and a few parallel, oblique brownish bands where dark brown spots have coalesced. Hind wings clear, with red or purple base. Hind tibiae red. The only species in the genus. **Biology** A major pest in tropical Africa, where swarms (as long as 30–60km and 3–4km wide) can develop. Maize, wheat, sorghum, sugar cane and various grasses are favoured foods. Nymphs **(6A)** develop in floodplains after the first summer rains. One generation per year. **Habitat** Subtropical grassland. Swarms occur very rarely, under extremely humid conditions in coastal KwaZulu-Natal and on Mozambique floodplains, spreading into Limpopo, the Northern Cape and the Free State.

1 *Locustana pardalina*   Brown Locust

Occurs in 2 distinct phases (forms): the larger (body length 41–51mm), gregarious (migratory) form changing from yellowish-grey to yellow when sexually mature, with characteristically constricted pronotum. Wings very long, covered in black spots from about halfway along their length to the transparent tip, hind wings pale blue. In the smaller (body length 26–36mm), solitary phase, adults are green, mottled with brown and black. Black-and-orange nymphs (hoppers or *rooibaadjies*) aggregate in bands. **Biology** A major pest in the Karoo, incurring costs of approximately R50 million in the sole elimination campaign (1986). Eggs rest in a diapause stage over the dry Karoo winter until heavy summer rains synchronize hatching. May complete 3 generations in a wet summer, but usually 2–4 generations per year. Large swarms **(1A)** occur sporadically (every 7–11 years). Insecticides sprayed to control swarms kill harmless insects. Primarily a grass-feeder, attacking cereal crops or, in their absence, vegetables. Adults are preyed on by a sphecid wasp (*Prionyx*, p.490), and eggs are parasitized by bombyliid flies (*Systoechus*, p.332), and 2 other fly species (*Stomorhina lunata* and *Wohlfarhtia euvittata*), which also live inside the 'hoppers' and can kill as much as a third of the population. **Habitat** Semi-arid regions of the central Karoo. In plague years, swarms may spread to most parts of the region except the Western Cape, KwaZulu-Natal and the coastal Eastern Cape.

2 *Acanthacris ruficornis*   Garden Locust

Large (body length 65mm), familiar, with cream midline stripe down the dark brown thorax and wings. No brown-and-white marking at base of the prothorax (cf. *Cyrtacanthacris tatarica*, p.118) Hind tibiae have well-developed red-and-white spines. Hind wings pale lemon. Nymphs **(2A)**, by contrast, are soft and bright green (some later becoming brown or pink). The only species in the genus. **Biology** Feeds on the leaves, buds and flowers of various trees, but also on grasses. Life cycle similar to that of *Cyrtacanthacris*. Adults common in autumn, when approximately 120 very large eggs are laid loosely in a weak pod, often in freshly turned soil. Winter is probably spent in the egg stage, and nymphs appear in October. The wings of adults produce a crackling sound in flight. When these strong insects kick, the spines on their hind legs **(2)** can break human skin. Nymphs jump weakly. **Habitat** Found on bushes or trees in open ground, especially in stands of Port Jackson Willow (*Acacia saligna*) but also in gardens. Absent from very arid areas. **Related species** *Cyrtacanthacris tatarica* (p.118).

3 *Anacridium moestum*   Tree Locust

Very large (body length 50–90mm) but slender, with mottled reddish-brown or grey body and finely speckled grey tegmina. Hind wings pale blue with broad black band near base. Hind tibiae purple, whereas in similar Desert Locust (*Schistocerca gregaria*) below, they are yellowish. Nymphs (hoppers) are yellowish-green. **Biology** Swarms on occasion. Preferred foodplants are *Senegalia*, *Zizyphus*, *Capparis decidua* and gum trees. Can damage crops, specifically fruit trees in the Free State. **Habitat** Always in trees or shrubs in veld with scattered *Senegalia* trees. **Related species** There are 2 other regional species in the genus.

4 *Schistocerca gregaria*   Desert Locust

Large (body length 72mm). Exists in 2 forms: a yellowish swarming phase (shown), which is pink when freshly moulted, and a greenish-white solitary phase. Prothorax contrastingly marked with 3 cream stripes and a number of short brown stripes. Eyes have vertical stripes (cf. plain dark eyes in *Locustana pardalina*, above, which it resembles). Tegmina very long, whitish owing to cream-coloured veins, but otherwise covered in black spots, like those of *L. pardalina*. Hind wings yellow. The subspecies *S. gregaria flaviventris* (Southern African Desert Locust) occurs in South Africa. **Biology** Swarms in the region on a small scale (last in 2005), although present in both solitary and swarming phases. A major pest in North Africa, where it forms huge swarms. Areas of outbreak are unpredictable, but swarms move downwind with rain. Damages wheat crops in the Free State. **Habitat** Arid parts of the region (Gordonia district). Absent from humid, forested parts. Occasionally swarms along the Orange River, feeding on a wide range of crops.

1 *Gastrimargus*

Large (body length 52mm), stocky, with both green and brown forms. Head, thorax, wing bases and hind femora boldly patterned in bright green and dark brown. Hind wings banded in yellow and black. There are 9 regional species in the genus. **Biology** Strong fliers. Eggs over-wintered, 1 generation completed per year. Feed on both tough and soft grasses (mostly of the subfamily Andropogonae). **Habitat** A variety of habitats, including well-grassed areas and shrubby thicket.

2 *Heteracris*

Medium-sized (body length 30mm), with bold, elegant markings in olive green and canary yellow. Hind femora banded in yellow and black. Antennae very long. *Heteracris* includes 10 regional species. **Biology** Mostly alert, agile fliers, alighting on exposed branches and rocks. **Habitat** Grasses and herbs in the shade of riverine trees.

3 *Paracinema tricolor*   Vlei Grasshopper

Large (body length 47mm), bright green with a cinnamon band running along entire length of body, wings shading to brown towards the tip. Hind tibiae bright red. The only species in the genus. **Biology** Typically flushed from reeds or tall grasses in vleis, where it flies accurately and strongly, resettling on a grass stem with head upright. **Habitat** Very widespread. Apparently prefers wet habitats with tall grasses.

4 *Orthochtha dasycnemis*

Medium-sized (body length 22mm), slender, green with a cream stripe along top of wings. Tegmina dark brown with a thin green margin. Hind legs green, shading to red. Long antennae. **Biology** Eggs over-wintered, 1 generation completed per year. Feeds on tough and soft grasses, especially *Setaria sphacelata*. **Habitat** Common in areas of thick, tall grass. **Related species** There are 6 other similar regional species in the genus.

5 *Acanthoxia natalensis*   Grass-culm Grasshopper

Large (body length 60mm) and elongate, body cigar-shaped and straw-coloured, with a continuous pale yellow stripe running along the cheeks, thorax and femora. Wings and hind legs short. Antennae flattened and short. **Biology** Excellent grass mimic, generally resting with legs drawn tightly against the body. **Habitat** Grassland in open bushveld. **Related species** There are 2 other regional species in the genus.

6 *Cannula gracilis*   Grass-mimicking Grasshopper

Large (body length 60mm), long and slender, yellow-brown, with very elongate head and flattened antennae. Wings plain brown. **Biology** Superb grass mimic, always associated with grasses, on which it feeds, and found clinging tightly to grass stems. **Habitat** Common in open grassland.

7 *Scintharista saucia*   Rock Grasshopper

Medium-sized to large (body length 38mm), dark brown with fine black peppering, long antennae, bright red hind tibiae. Hind wings bright magenta with black border **(7A)**. **Biology** Brilliant hind wings displayed in short agile flights from large rocks. **Habitat** Only on rocky mountain slopes or mountain paths. **Related species** There are 2 other regional species in the genus.

1 *Acrotylus*
Burrowing grasshoppers

Medium-sized (body length 20–30mm), familiar, with large bulging eyes and middle legs elongated for digging. Hind legs banded with black. Hind wings red, blue or pale yellow at the base with a black band **(1A)**, or clear with a black border. There are 22 regional species in the genus. **Biology** Poor fliers, displaying banded hind wings in flight. Bury themselves when disturbed or in windy conditions, using the middle legs. Stridulate loudly. Egg pod with hard wall, lacking a cap. Rising spring temperatures break the resting period (diapause) of the eggs. There are 2 generations per year. *A. patruelis* is the commonest Karoo grasshopper and feeds predominantly on soft grass (in Mpumalanga it feeds mainly on the grass *Digitaria eriantha*); among the first invaders to recolonize an area after fire. **Habitat** Loose shifting sand or hard gravelly soil in open disturbed areas, e.g. road verges.

2 *Tmetonota*

Fairly small (body length 19–24mm), dark brownish-grey, with a rough surface. Large white patches on the head, base of hind femora and base of tegmina. Red-and-black band on hind wings. There are 3 regional species in the genus. **Biology** Poor fliers. Never occur in large numbers. Probably feed on short grass. **Habitat** Overgrazed and trampled open spaces.

3 *Sphingonotus scabriculus*
Blue-wing

Medium-sized (body length 24–32mm), colour varies to match soil or stone substrate. Lighter-coloured specimens occur on limestone. Head, thorax and first third of wings dark brown, with 4 brown bands over remainder of tegmina. Hind wings **(3A)** blue with very broad, completely black band; abdomen light blue. **Biology** Makes a characteristic click or snap of the wings during short, buoyant flights, and settles briefly after displaying the colourful hind wings. Never occurs in large numbers. **Habitat** Bare stony ground with sparse low vegetation. Widespread. **Related species** There are 4 other similar-looking regional species in the genus.

4 *Rhachitopis*

Medium-sized (body length 22mm), stocky, brown, with white markings on the head and metathorax. Tegmina brown with a tan stripe, continuous with a stripe along each side of the head and thorax. Hind wings pale yellow **(4A)**. There are 8 regional species in the genus. **Biology** Clumsy fliers, spending much time on the ground. **Habitat** Well-drained open areas with stubble (karroid) vegetation.

5 *Vitticatantops humeralis*

Medium-sized (body length 20–25mm), stocky, with characteristic black, white and grey hind femora, and large black patches outlined in white on the head and thorax. Nymphs uniformly coloured in shades of green or brown. **Biology** Adults common from late summer to July. Usually found in association with *Eyprepocnemis calceata*. Attracted to thick vegetation (e.g. *Euphorbia* thickets). Resettles on plants, never on open ground, when disturbed. Feeds mainly on leaves of shrubs such as *Burkea africana* and *Grewia flavescens*. Re-invades burned areas when herbs begin to sprout. One generation per year, with over-wintering adults. **Habitat** Long grass or, less often, short scrub. **Related species** There are 2 other regional species in the genus.

1 Eyprepocnemis plorans

Medium-sized (body length 30mm), stocky, predominantly grey with a brown stripe along top of body. Hind tibiae armed with well-developed white spines. **Biology** Egg pod straight, with rudimentary wall. Adults occur in late summer and early winter and are particularly common in the Western Cape. **Habitat** Long grass or short scrub. **Related species** The only other regional species is *E. calceata*.

2 Oedaleus   Yellow wings

Medium-sized (body length 28–42mm), stocky, occurring in a number of colour variations, but most commonly grey with crisp black-and-white markings. Tegmina largely black, banded with brown, hind wings banded in yellow and black. Abdomen yellow. There are 6 regional species in the genus. **Biology** Egg pods curved and constricted in the middle, and laid in firm or very hard ground. There are 2 generations per year. Have been observed to flash the yellow-and-black hind wings at the approach of a wasp. Feed on short grasses. Males call in flight using the hind legs. **Habitat** Open pebbly or sandy spaces with sparse grass cover, often heavily grazed. Avoid tall grasses. *O. nigrofasciatus* occurs on the Cape Peninsula.

3 Tylotropidius didymus

Fairly large (body length 47mm), dull-looking, uniformly grey with brown tegmina banded in black, occasionally with a green stripe. Hind wings pink basally. Pronotum green in some forms. Hind femora long and distinctive. **Biology** Active and mobile, moving half a kilometre a day. One generation per year, with over-wintering adults. Feed on a variety of soft grasses and are unusual in that they feed on dry grass in winter. Re-invade burned areas very slowly. **Habitat** Long grasses in bushveld and grassland. Locally abundant in dense stands of *Panicum maximum*.

4 Oraistes luridus

Medium-sized (body length 30mm), body uniform grey apart from a striking orange stripe running along length of body. Tegmina with faint blotching; hind wings clear. **Biology** Nymphs occupy low vegetation and then move into trees as they mature. **Habitat** Subtropical forest edges.

ORDER PHASMIDA • STICK INSECTS

Large, slow-moving insects with very elongate, slender bodies and legs; well known for their ability to mimic twigs or leaves.

Stick insects are unlikely to be confused with other insects, except perhaps Grass Mantids (which have much larger eyes and grasping fore legs). The head is rectangular to oval, with small eyes and strong biting mouthparts. The first thoracic segment is short, but the second and third segments are greatly elongated and cylindrical. The abdomen has 11 long segments and ends in a pair of 'tails' (cerci), which may be modified into claspers in the male. The very long, thin walking legs are sometimes ornamented with lobes or spines, but are otherwise unmodified and incapable of jumping. Wings may be present or absent, and are usually strongly developed only in males. Where present, fore wings are short and tough, and only partially cover the hind wings, which are longer, opaque and hardened anteriorly, but broad and membranous posteriorly.

Stick insects are nocturnal herbivores and are difficult to spot as they creep slowly around among their foodplants. If disturbed, they may perform rhythmic rocking and swaying movements, or fall to the ground and feign death. When provoked, some raise their brightly coloured hind wings in a threat display. Males are usually smaller, thinner and less numerous than females. Parthenogenesis is common, and many species are known only from females. Eggs are large and hard, resembling seeds, and are usually simply dropped to the ground, although some species cement them to the substrate. Nymphs look like smaller adults. There are some 2,600 species globally and about 50 in the region.

Family Diapheromeridae
Common walking stick insects

The most diverse family of stick insects, including more than 1,200 species globally. They are convincing stick mimics, are usually slender, and have long, indistinctly segmented antennae and weakly spined legs. Wings may or may not be present. The family includes the longest and most spectacular stick mimics in the region.

1 *Bactrododema tiaratum*
Giant Stick Insect

Large (body length of females up to 185mm, of males 130mm), body uniformly dark brown, with rough bark-like texture and pronounced but short double 'horns' on top of head (those of male curving backwards). Males have fully developed hind wings, the smoky brown membrane marked with darker veins and clear spots; females have shortened wings. Among the largest species in the region. **Biology** Sways rhythmically when disturbed. Males may flash their wings in a threat display and are capable of sustained weak flight. Often attracted to lights at night. **Habitat** Trees and shrubs, especially in warmer regions. **Related species** There are several other 'giant' stick insects in the region, including *Bactrododema krugeri* (up to 200mm, northeastern parts of country) and *B. hecticum* (up to 150mm, Cape provinces), which have larger, straight or forward-curved horns on the head. In the related *Bactricia bituberculata* (**1A**), both sexes are completely wingless.

2 *Maransis rufolineatus*
Grass Stick Insect

Medium-sized (body length 61mm), males extremely slender, delicate and wingless, with especially long fore legs; females stouter. Colour variable, usually pale brown with longitudinal red or darker brown stripes along the body and head. Legs usually green at base, becoming brown at the tips. Body surface sometimes granular, especially along the sides. **Biology** Feeds on grasses. Extremely well camouflaged and difficult to spot in its natural habitat, as body and limbs closely resemble grass stalks. **Habitat** Grassland and savanna regions.

3 *Zehntneria mystica*

Medium-sized (body length 60mm), slate grey with black-tipped spikes on body, especially along the posterior margins of the thoracic and abdominal segments. Black line down middle of the head. Tube-like genitalia of females project well beyond tip of abdomen. **Biology** Poorly known. **Habitat** Fynbos plants in the Western and Northern Cape.

Family Phasmatidae

Mainly winged, with toothed fore femora (mid and hind femora sometimes also toothed) and distinctly segmented antennae, which are usually shorter than the fore femora.

1 *Carausius morosus* — Indian Stick Insect, Laboratory Stick Insect

Medium-sized (body length 75mm), relatively fat and robust, uniformly yellow or green, with characteristic bright red flash on inner surface of front legs. Occasional knobs on body, sometimes forming a line along the sides. **Biology** Feeds mainly on ivy and privet, but accepts numerous foodplants. Reproduces parthenogenetically, only females reported. Eggs **(1A)** large and seed-like, hatching in 4–6 months, the fragile nymphs taking 4–6 months to reach maturity. **Habitat** Common in suburban gardens and natural vegetation, especially on the Cape Peninsula. Native to India, probably introduced to Cape Town for laboratory use and escaped into the wild.

Family Bacillidae

A cosmopolitan family, with males and females closely resembling one another. The antennae are generally short, and both sexes are usually slender and wingless.

2 *Phalces brevis* — Cape Stick Insect

Medium-sized (body length up to 80mm). Colour variable, ranging from green to brown, grey or speckled. Legs often a different colour from body. Best identified by 3 distinct blue or green marks on the back, 1 above each pair of legs. Abdomen ends in a pair of short white cerci in male (shown); in female, a boat-shaped, chute-like appendage. Previously known as *P. longiscaphus*. **Biology** Feeds on Australian Myrtle, and also on *Erica* and other plants. **Habitat** Probably the most common stick insect in the Western Cape.

3 *Phalces tuberculatus*

Female large (body length 110mm; male only half this length). Long and slender, uniformly brown, with numerous small tubercles on body surface. These are especially dense on the first thoracic segment and decrease in number posteriorly. **Biology** Strong sexual dimorphism, typical of stick insects. **Habitat** Scrubby vegetation in southern Namaqualand. Although named only in 2000, the species is apparently quite common within its restricted range.

4 *Macynia labiata* — Thunberg's Stick Insect

Medium-sized (body length up to 60mm), head and mouthparts pink or yellow in adults, end of abdomen with pink tip. Females usually vivid green speckled with white, or with a pink stripe along the side. Males smaller and thinner, brownish-green, with reddish-brown pincer-like cerci. **Biology** Entirely herbivorous. Eggs take about 6 months to hatch. Nymphs emerge April–May and moult 5 or 6 times before maturing, in early summer. Breeding season October–January, after which the spent adults die. **Habitat** On foodplants, especially Australian Myrtle and *Erica*.

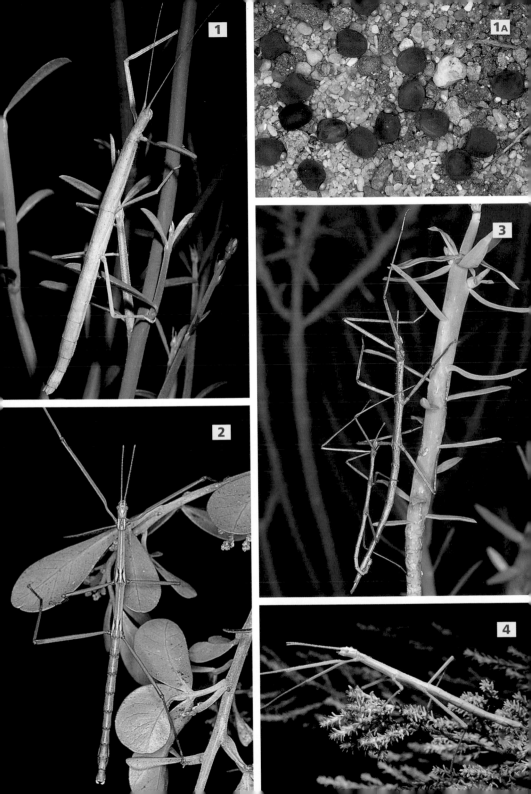

ORDER MANTOPHASMATODEA •
HEEL-WALKERS & GLADIATORS

Small to medium-sized wingless insects, with triangular mantid-like heads, large eyes, chewing mouthparts and long, thin antennae with hooked tip. An enlarged pad between the tarsal claws (arolium) always held upright. Spiny fore and mid legs used to subdue prey.

A small order of nocturnal, wingless, hemimetabolous insects currently known only from southern Africa. These predators resemble nymphal mantids and, owing in part to their secretive habits, were only discovered in 2002. The unique antennae have alternating long and short segments and end in a hook. The spined fore and mid legs are raptorial and used in capturing small insect and spider prey. The very large arolium between the tarsal claws assists in adhesion to vegetation. The tarsi are always held up off the ground, hence the common name, heel-walkers. In addition to the approximately 20 described species, many more species probably await discovery.

1 Family Mantophasmatidae
Northern heel-walkers

Medium-sized (body length 22mm), green with a thin yellow stripe running along the length of the body. Eyes small and yellow with a red stripe. In males cerci are narrow and pointed. *Sclerophasma* **(1)** includes 4 closely related green species. Apparently diurnal. Also known from the summer-rainfall parts of Namibia, from Windhoek northward.

Family Austrophasmatidae
Southern heel-walkers

Small to medium-sized (body length 10–20mm). Eyes generally mottled with pigment, except for *Viridiphasma*, which has a striped eye. Blunt claspers (modified cerci) at end of abdomen in males used for grasping females during mating. Largely restricted to the more arid parts of the winter-rainfall region of South Africa. Approximately 10 known species.

2 *Viridiphasma clanwilliamense*
Jade Heel-walker

Small to medium-sized (body length 11–16mm). Female jade green, male yellow-green, with a brown stripe running along length of body. Eyes yellow with a red stripe. **Biology** As with all heel-walkers, substratal tapping carried out by both male and female to locate mates; female often devours male after mating. **Habitat** Taller shrubs and small trees in mountain fynbos vegetation.

3 *Karoophasma biedouwense*
Namaqualand Heel-walker

Small to medium-sized (body length 14–20mm). Colour straw yellow to grey. Eye always grey and mottled. Male smaller and more slender than female. **Biology** Egg pods, each containing approximately 12 large eggs **(3A)**, laid in sand **(3B)**, hatching after the first autumn rains, but remaining dormant (in diapause) during drought years. **Habitat** Spiny bushes, grass clumps and succulent shrubs in arid fynbos and succulent karoo vegetation.

4 *Lobatophasma redelinghuysense*
Fynbos Heel-walker

Small to medium-sized (body length 11–18mm). Colour variable: light brown with a dark central stripe running along the body **(4)** or bright green with a pinkish central stripe **(4A)**. **Biology** Variable colour forms match the local vegetation. **Habitat** Hides by day in restio clumps in low strandveld and fynbos vegetation.

5 *Hemilobophasma montaguense*
Little Karoo Heel-walker

Small to medium-sized (body length 14mm), with mottled grey eyes. Body stout, heavily mottled in grey and brown, often with a dark brown central stripe. **Biology** Nocturnal, as with all southern heel-walkers. **Habitat** At the base of low shrubs and mesembs, and in grass clumps.

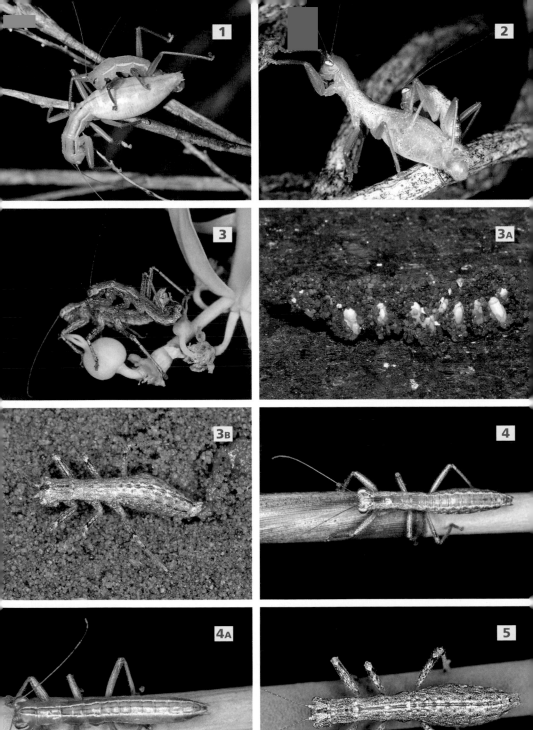

ORDER PSOCOPTERA • PSOCIDS (BOOKLICE & BARKLICE)

Small, soft-bodied insects with large, round, mobile heads, long slender antennae and (usually) membranous wings that are held tilted, roof-like, over the abdomen.

Psocoptera are also known as booklice or barklice, but the term 'psocids' is preferred. They are sometimes grouped with the true lice into the order Psocodea, but the traditional orders are retained here. Their compound eyes are usually big and bulging, but are small in wingless groups. Where present, the delicate transparent wings have reduced venation. Cerci are always absent, and the upper lip (labrum) is large and conspicuous. Although common, psocids are easily overlooked, as most are small and cryptic. The majority of species occur on bark, on karoo shrubs, in leaf litter, and under stones, where they feed on algae, fungi, lichens, and plant or animal debris. The most familiar species are wingless 'booklice' that are pests of stored products. Eggs hatch into nymphs that are initially wingless, but otherwise resemble adults. Of the more than 3,000 named species, 80 occur in South Africa.

1 Family Liposcelidae   Booklice

Small (body length 1–2mm) with flattened bodies, relatively short antennae, and greatly reduced eyes. Wings are almost always absent. Legs short, the femora of the hind pair enlarged. **Biology** The common domestic booklouse (*Liposcelis bostrychophila*) **(1)** has been dispersed by human activities worldwide and feeds mostly on stored grains, but also on dried meats, paste, glue or traces of mould. **Habitat** Buildings, especially among stored foods and old newspapers.

2 Family Psocidae   Common barklice

Large (body length 5mm), usually seem grey or black, but on close examination often have beautiful, intricately patterned wings. Most larger psocids belong to this family. **Biology** Nymphs generally live together in large groups (sometimes with adults), emerging at night to feed on lichens. Such groups scatter when disturbed, but quickly re-form. Common genera include *Pearmania* **(2)**, which contains a number of large and spectacular species, and *Psococerastis* **(2A)**. **Habitat** Tree trunks.

3 Family Amphipsocidae   Hairy-winged barklice

Medium-sized to large (body length 4–6mm). Attractive, delicate and hairy. Main wing veins with double rows of setae, margins with short setae. Fore wings broad, with branched, patterned veins posteriorly; held closer to the horizontal than in other psocids. Only 2 named species regionally (*Harpezoneura multifurcata* shown); undescribed species occur. **Biology** Unknown. **Habitat** Leaves of trees.

4 Family Ectopsocidae   Outer barklice

Usually yellow-brown with clear wings lacking setae, held close to the horizontal. Pterostigma (wing-spot) distinctly rectangular, outlined in pigment in the *Ectopsocus* species shown (body length 3mm). Two genera and 4 species occur in the region. **Biology** Not known. **Habitat** Dried leaves and leaf litter. Distribution poorly known.

5 Family Peripsocidae   Stout barklice

Medium-sized (body length 3mm), usually dark. Attractively patterned wings held steeply sloped over body and lack setae. Pterostigma (wing-spot) has distinctly curved margin. Only 1 species, *Peripsocus setosus*, is currently known from the region. **Biology** Not known. **Habitat** Bark or dried leaves.

6 Family Trogiidae   Granary booklice

Small (body length 1–2mm). Body smooth. Fore wings reduced or absent; hind wings absent. Antennae with more than 20 segments, all lacking secondary internal divisions. Tarsi with 3 segments; tarsal claws are untoothed. **Biology** Some species associated with humans and infect grain and flour. *Helenotropos abrupta* **(6)** was originally described from St Helena island, but was apparently transported there from its original habitat on the eastern forested slopes of Table Mountain. **Habitat** On rocks among forest litter.

ORDER PHTHIRAPTERA • LICE

Small, flattened, wingless insects that are permanent external parasites occurring on almost all bird and mammal species.

Sometimes grouped with psocids (previous page) into the order Psocodea, but the traditional orders are retained here. Lice are a diverse group of highly modified parasitic insects with reduced eyes, short antennae, and chewing or piercing-and-sucking mouthparts. Their legs are short and stout and usually end in strong claws. The entire life cycle takes place on the host. Eggs ('nits') are glued onto the host's feathers or hairs. They hatch into nymphs, which moult 3 times before reaching adulthood. Nearly 5,000 species are known worldwide, with more than 1,000 occurring in South Africa.

1 Family Pthiridae
Crab lice, pubic lice

Small (body length 1–2mm), broad and flat, with crab-like appearance. Mid and hind legs stout, with very large claws. Abdominal segments with distinct lateral lobes. Only 1 species, the Human Pubic Louse (*Pthirus pubis*), in the region. **Biology** Eggs or 'nits' attached to pubic or coarse body hair. Bites cause irritation and leave a characteristic mark on the skin. Spread by close (usually sexual) bodily contact. **Habitat** Usually confined to human pubic hair, rarely the eyelashes, beard or underarm hair; not normally on the scalp.

2 Family Pediculidae
Human lice

Small (body length 2–3mm), elongate, abdomen longer and wider than thorax and without lateral projections. All 3 pairs of legs equally strong and developed. The 2 species, *Pediculus capitus* and *P. humanus*, are sometimes considered forms of a single species. **Biology** Head lice (*P. capitus*) suck blood from the scalp and lay eggs or nits on the hair **(2A)**, particularly behind the ears. Common and easily spread by close bodily contact and by sharing of combs or brushes, but are not serious disease vectors. Body lice (*P. humanus*) are transmitted by direct contact or shared clothing, and can transmit typhus and other diseases, but are now uncommon and are easily controlled by insecticides. **Habitat** Human body only.

3 Family Haematomyzidae
Elephant lice

Only 2 species in region. The Elephant Louse (*Haematomyzus elephantis*) (body length up to 6mm) is easily recognized by its habits, rounded body, reduced legs and rigid, elongated, down-curved proboscis. *H. hopkinsi* occurs on warthogs. **Biology** The proboscis was originally thought to penetrate the host's thick skin to draw blood on which to feed, but the species is now thought to feed mostly on hair and skin debris. **Habitat** Elephants only.

4 Family Gyropidae
Guinea pig lice

Small (body length less than 2mm). There are 2 species – the more common and slender *Gliricola porcelli* and the oval *Gyropus ovalis* **(4)**. **Biology** Adults occur all over the body, but the large nits are particularly abundant behind the ears of infected guinea pigs. Spread by bodily contact. **Habitat** Fur of guinea pigs. Introduced from South America.

5 Family Menoponidae
Biting bird lice

Small to very large (body length 2–12mm). Antennae fold into grooves on the sides of the broadly triangular head. Tarsi with 2 claws. Over 300 species in the region. **Biology** Crawl in the plumage of birds, feeding on the feathers. Some species chew holes in the quills and live inside the shaft of feathers. *Piagatiella* live inside the throat pouches of pelicans and cormorants. **Habitat** On birds. Mostly host-specific.

6 Family Philopteridae
Bird lice

Small to large (body length 2–8mm), shape varying from short and rounded to long and slender. Antennae relatively elongate, not concealed in grooves on the head. The largest family of lice, with over 500 species in the region. **Biology** Feed on downy parts of feathers. **Habitat** On most bird species; generally host-specific. Sometimes different species occur on various body parts of the host.

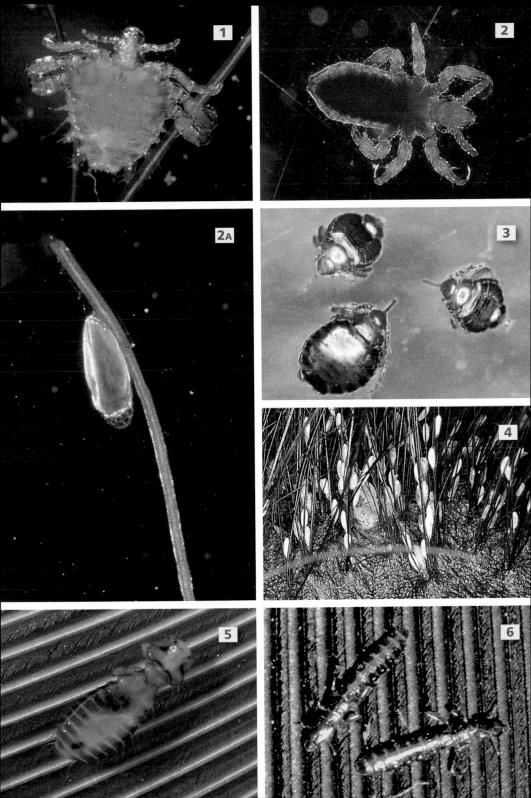

ORDER HEMIPTERA • BUGS

Bugs are characterized by having piercing-and-sucking mouthparts, 2 pairs of wings, and reduced hind wings. This large order is divided into 4 suborders, 1 of which (the moss bugs) does not occur in Africa. The suborder Heteroptera contains familiar bugs that are often carnivorous (stink bugs, etc), while the suborders Sternorrhyncha (aphids, scale insects) and Auchenorrhyncha (cicadas, leafhoppers) are herbivorous and are often agricultural pests.

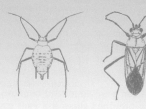

The Hemiptera, or true bugs, are the largest order of hemimetabolous insects (those in which the young develop into adults without going through a pupal stage – see introduction, p.12). All grades of wing reduction are apparent among adult bugs. Some, such as bed bugs, are totally wingless, others have winged males and wingless (apterous) females, while many have fully developed wings in both sexes. Nymphs are small wingless replicas of the adults. Many species have legs adapted for jumping or swimming. The long piercing-and-sucking mouthparts **1** are folded back beneath the head and between the legs. Bugs feed on a wide range of animal and plant foods. Because of their sucking habits, many are pests, especially those that act as vectors for viruses, bacteria and fungi that cause diseases in humans and livestock, or in crop and garden plants. There are about 3,000 described species in the region.

2 SUBORDER HETEROPTERA
Stink bugs, pond skaters, assassin bugs & kin

Heteropterans **(2)** are easily recognized by the presence of a hemelytron – a modification of the fore wing so that the basal part (yellow arrow) is hardened, while the apical part (white arrow) remains thin and membranous. They are often carnivorous and aquatic, and are well known for their ability to produce noxious secretions from repugnatorial glands. Many of these species can give a nasty bite and have bright warning coloration – typically a combination of red, yellow and black.

3 Family Enicocephalidae
Gnat bugs

Small (body length 2–7mm), fairly distinctive, plain light brown, with entirely membranous hemelytra. Head long, thin and cylindrical, with prominent compound eyes. Behind the eyes the head is constricted, then expands into a globular swelling bearing 2 simple eyes. Approximately 3 genera and 40 species in the region. **Biology** Feed on a range of arthropods. One species (found in an ant nest) presumed to feed on ants. Australian species form swarms of winged males at dusk, and persistently use humans as beacons to maintain the coherence of the swarm. Winged species may be attracted to lights. **Habitat** Leaves and logs in damp evergreen forests, moist coastal fynbos. Fairly scarce.

4 Family Cimicidae
Bed bugs

Small (body length 4–7mm), apricot-coloured, completely wingless, with circular body and flattened extensions of the prothorax behind the eyes. Most often seen is the cosmopolitan Common (Human) Bed Bug (*Cimex lectularius*). About 10 species known in region. **Biology** Adults and tiny nymphs emerge at night to feed on blood, but do not appear to be vectors of disease among humans. Bites can produce a strong reaction, with swelling and irritation lasting for days. Carry a wide variety of disease organisms in nature. Apart from 1 human- and 1 bird-associated species, all other regional species attack bats. Can survive starvation for months. Young are born live. **Habitat** Bat roosts, birds' nests and mammalian shelters, including human dwellings. Hide in crevices, walls, furniture and clothing.

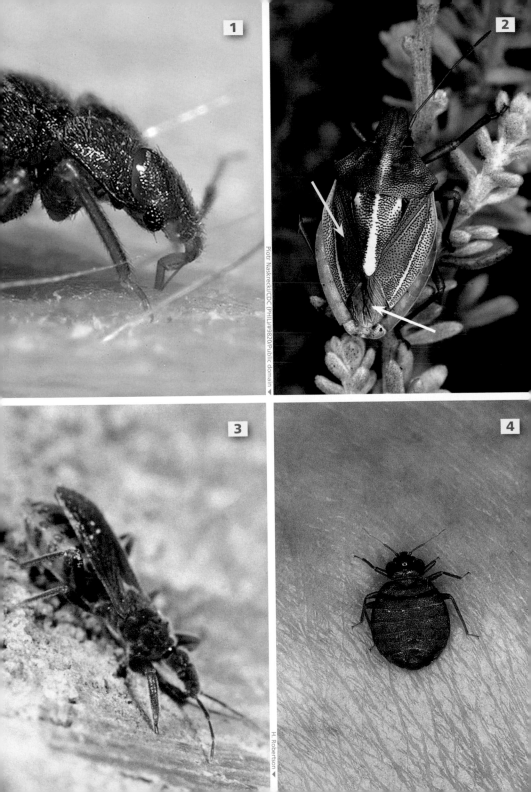

Family Miridae

Plant bugs, capsids

A large and economically important family of small and slender, medium-sized (body length 2–12mm) bugs. Present on virtually every plant. Microscopically identified by an obvious fracture in the leathery part of the fore wing, marking out a triangle (the cuneus). Females wingless in many species, some of them ant mimics. Most feed on plants; a few on insect eggs or blood. About 250 species known from the region.

1 Pameridea

Small (body length 4mm), shiny black with ivory-coloured markings and a pair of white spots on fore wing. **Biology** Walk with impunity on the sticky leaves of their host plant, the insectivorous Giant Sundew (*Roridula*), and feed on insects trapped by the plant to supply itself with nitrogen. **Habitat** Only on a few moist slopes in the Western Cape, where their host plant occurs.

2 Deraeocoris

Small (body length 8mm), dark brown, with a heart-shaped orange mark in the middle of the thorax. Membranous part of fore wing with 3 black circles. **Biology** Predacious (like all members of the subfamily to which they belong). Feed on aphids. **Habitat** Weedy urban and natural veld, and often in gardens.

3 Stenotus

Medium-sized (body length 12mm), tan with reddish legs, orange prothorax, and 2 dark brown stripes on each hemelytron. The genus contains many similar regional species. **Biology** Feed on developing grass seeds, secreting an enzyme that collapses the seed. Recorded from sorghum, in addition to native grasses. **Habitat** Grasslands, open woodland.

4 Aloea

Aloe bugs

Small (body length 3mm), body pale pink with darker head and thorax and a red triangular patch at ends of hemelytra. Femora red. Nymphs pink. **Biology** Pests of a range of aloe species, causing speckling of the leaves by sucking from them; damage persists even after eradication of the bugs. Bugs swarm rapidly over the leaf when disturbed. **Habitat** Occur in large groups on aloe leaves, both in gardens and in the veld, with certain aloe genera being favoured over others.

Family Tingidae

Lace bugs

Named for, and easily recognized by, the delicate lace-like sculpturing of their wings and thorax (better developed in some species than in others). Plant-feeders, living on the underside of leaves. However, their feeding causes white spots to appear on the upperside of leaves, while certain species cause galling of leaves. At high densities, they can cause leaves to curl and become discoloured. Often very host-specific. Eggs are laid in the tissue next to the leaf midrib, and sealed in a hard brown substance. Nymphs are spiny and gregarious, and develop lacy reticulation only after the final moult. One species that attacks olives is of economic significance. About 200 species are known from the region.

5 Ammianus wahlbergi

Small (body length 5mm), rectangular, with a constriction at the prothorax and 4 black patches on the lacy hemelytra. **Biology** Adults occur singly. **Habitat** Grassland and bushveld. **Related species** There are 2 other species in the genus, both smaller. Also, *Pogonostyla natalicola* (**5A**), from KwaZulu-Natal, which is more linear (body length 5mm), with a brown thorax and 2 brown marks at the base of the hemelytra.

6 Plerochila australis

Olive Lace Bug

Small (body length 4mm), dull grey-brown, with a faint network of wing veins where the wing tips meet. **Biology** Nymphs resemble aphids and feed gregariously on the underside of olive tree leaves or shoots, where their accumulated honeydew may promote sooty moulds. Damage takes the form of yellowed leaves or wilted shoots with bushy growth beneath the damage. **Habitat** Widespread on cultivated olive and Wild Olive (*Olea europaea* var. *africana*) trees in Africa and the Mascarene Islands. **Related species** *Compseuta ornatella* (**6A**), which is oval, very small (body length 3mm) and black, brown and white in colour. Can be found on the soft herbaceous leaves of shrubs; occasionally a pest on Indian mallows (*Abutilon*).

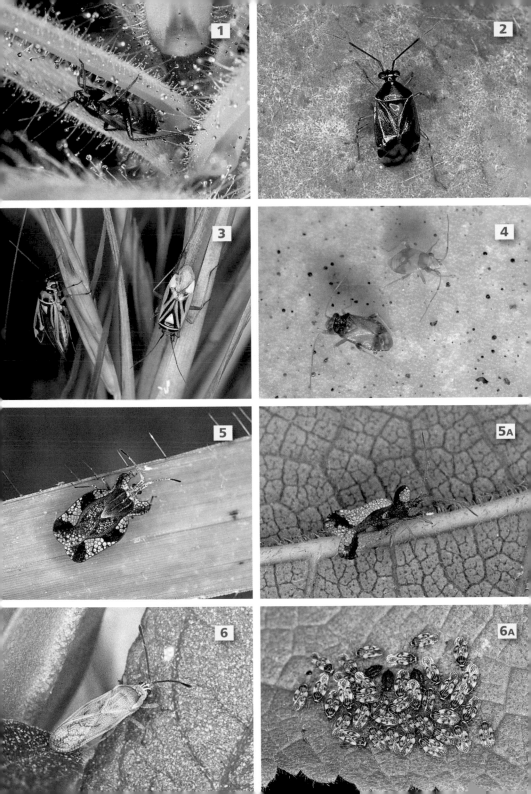

1 *Corythucha ciliata*
Sycamore Tree Lace Bug

Adults small (body length 3.5mm), white, with an expanded, lacy, dark pronotum. Elytra white and lacy with an extensive network of wing veins. Each hemelytron with a single dark spot. Black-and-white nymphs have small marginal spines. **Biology** An invasive alien that is restricted to London plane trees (*Platanus* hybrids) in the Western Cape. Adults over-winter under the flaking bark of large trees, emerging in spring. Nymphs feed on the undersurface of leaves, causing leaf bronzing in late summer. Probably spread by cars parked under infested trees. **Habitat** Plane trees in urban streets and gardens.

2 Family Nabidae
Damsel bugs

Medium-sized, dull-coloured, predacious bugs. Distinguished from assassin bugs by a 4-segmented beak (rather than the 3-segmented beak of assassin bugs) and the absence of the ridged groove present in Reduviidae. Widespread in the region, which hosts about 20 species. The Pale Damsel Bug (*Nabis capsiformis*) **(2)** is slender, medium-sized (body length 8–10mm), pale brown and long-winged. It flies well and thus occurs on many oceanic islands, and in most warm parts of the world. **Biology** Predaceous, feeding on insect eggs and caterpillars. **Habitat** All vegetation types, in gardens and in natural veld. Abundant in the tropical and subtropical parts of Africa and South America.

Family Reduviidae
Assassin bugs

Medium-sized to large, robustly built bugs that prey on other arthropods and have a powerful recurved beak (rostrum), the tip of which is rubbed against a ridged groove under the body to produce sound. They are ambush predators that move slowly towards prey before rushing out and grabbing it with the fore legs, then injecting a secretion from the rostrum that has a very quick paralytic action. The prey is then sucked dry. Their bite is very painful to humans, but no local species habitually bite mammals, unlike some South American species (which regularly take blood meals from humans and transmit Chagas' disease). A number target specific prey items. About 475 species are known from the region.

3 *Acanthaspis obscura*

Medium-sized (body length 16mm), dull black, with 2 white spots and a honey-coloured apical spot on each wing. **Biology** Nymphs cover themselves with debris and sucked prey, and often occupy deserted termitaria with the adults, feeding on ants. Adults can inflict a painful bite on humans. Attracted to houses by lights. Short-winged adults found together with full-winged individuals. **Habitat** Adults occur under bark or stones, or in termitaria. Subtropical, extending into tropical Africa. **Related species** There are 12 other similar regional species in the genus.

4 *Ectrichodia crux*
Millipede Assassin

Large (body length 22mm), stout, unmistakably patterned: shiny black, thorax dull yellow incised with a black cross, margins of abdomen yellow. Nymphs bright red, with black wing pads. **Biology** Rests during the day under stones or among debris, emerging at night to feed on an exclusive diet of millipedes. Nymphs and adults frequently feed gregariously on a single large millipede **(4A)**. **Habitat** Common in gardens and a variety of veld types.

5 *Cleptria rufipes*
Metallic Assassin Bug

Medium-sized (body length 12–16mm), robust, shiny, bluish-black or dark brown, with red, orange or cream markings. Females generally wingless, resembling nymphs. Adults have 2–3 spines arising from the last part of the thorax. Winged males (on right of image) have feathery antennae. **Biology** Emerges from hiding places to feed at night. Gregarious, possibly preying on millipedes. **Habitat** Under logs and stones in forests. **Related species** There are 6 other regional species in the genus.

6 *Holoptilus*
Ant wolves

Small (body length 7–9mm), light brown, with black markings on the wings and legs, the entire body (apart from wings) heavily covered in robust hairs. Wings almost entirely membranous. Seven similar-looking regional species occur in the genus. **Biology** Feed on ants that are attracted, intoxicated and, finally, paralysed by a substance secreted from a unique warty structure under the abdomen. **Habitat** In shade under bushes or litter, or on ant trails.

H. Robertson

1 *Lopodytes grassator* Grass Assassin

Medium-sized (body length 17mm), slender, grey, with elongate legs and antennae. Both sexes winged. **Biology** Excellent grass mimic. **Habitat** Grassland in open bushveld, highveld and Kalahari veld. **Related species** There are 5 other similar southern African species of *Lopodytes*. They are yellow or grey and may be confused with stick insects. Wingless forms are common. *Sastrapada baerensprungi* (18mm long) **(1A)** is very similar to *L. grassator* but has dark markings on the inner parts of the legs. *Thodelmus quinquespinosus* **(1B)** is medium-sized (23mm long) with a brown eyestripe and cream wing margin.

2 *Oncocephalus*

Medium-sized (body length 16–18mm), light brown, with cylindrical head elongated in front of the eyes, and wings with 2 central black marks. Thickened front legs with spined femora. There are 14 regional species in the genus. **Biology** Nocturnal. Attracted in numbers to lights. **Habitat** Drier vegetation types such as succulent karoo and valley bushveld.

3 *Rhynocoris segmentarius* Flower Assassin

Medium-sized (body length 18mm) and conspicuous. Black and shiny, with thorax and front part of wings red, and side margins of the thorax striped in yellow and black. **Biology** Frequently hunts honey bees and stink bugs (Pentatomidae) from exposed sunny positions on vegetation and flowers. As in many reduviids, the black, red and yellow coloration is a warning to potential vertebrate predators. Can deliver a very painful bite that produces numbness, nausea and general welts. A paste of bicarbonate of soda applied to the bite relieves the symptoms, which can last for months. **Habitat** Common in open bushveld, at forest edges and in gardens. **Related species** Approximately 24 other, similar regional species occur in this genus, including *R. neavii* **(3A)**. In some species, males exhibit parental care of the eggs and young.

4 *Platychiria umbrosa*

Medium-sized (body length 20mm), brownish, with black and tan spots. Abdomen and thorax with lateral spines, and fore tibiae with expansions. **Biology** Nocturnal. Attracted to lights. May occur on vegetation during the day. Does not bite when handled, but emits fluid from the thoracic glands while stridulating. **Habitat** Grass and low plants in subtropical bushveld. **Related species** Two other regional species occur in the genus, some of which feed on termites.

5 *Pantoleistes princeps*

Large (body length 27mm), shiny black, with yellow-and-black-banded legs and thin leaf-like expansions at the sides of the abdomen. Head very thin, with a robust beak. **Biology** Lives on tree trunks and branches at some height from the ground, and on termite mounds. Known to bite humans. **Habitat** Subtropical bushveld and forest. **Related species** *P. rex* (Limpopo Province) is the only other regional species in this genus.

6 *Phonoctonus* Cotton-stainer assassins

Large (body length 27mm), brightly coloured, with a red head, white neck, and white, red and black abdomen. Wings tan and black with black spots. There are 2 regional species in the genus. **Biology** Mimic, live with, and feed on various species of cotton strainers (*Dysdercus*). Eggs are laid in a ring around a plant stem. Can usually be found in the day on the host plants of their prey. **Habitat** Various vegetation types. Subtropical.

7 *Platymeris* Giant assassins

Very large (body length 36mm), hefty, dull black with bright orange bands on the legs, and an orange spot on the wings. There are 3 regional species in the genus. **Biology** The bite of these formidable bugs is extremely painful, and its effect may last for days. One tropical species is reported to feed on adult rhino beetles (*Oryctes*). **Habitat** Under bark. Subtropical, extending into equatorial Africa.

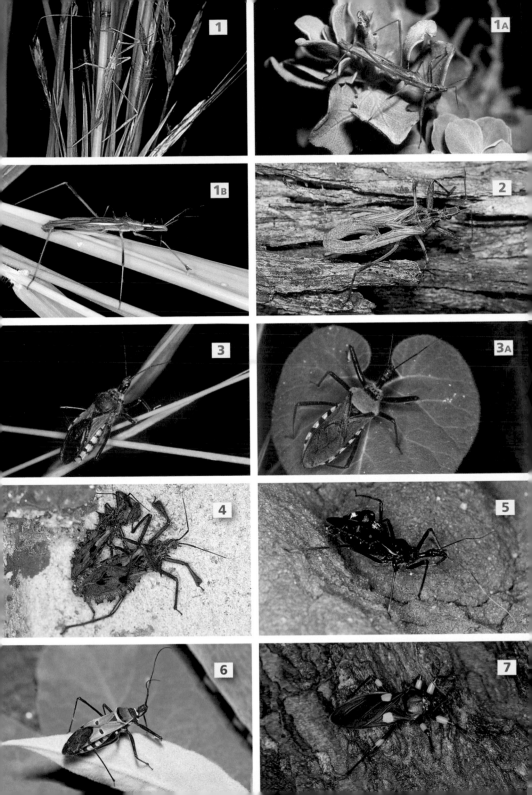

1 *Reduvius tarsatus*

Medium-sized (body length 15mm), with yellow-and-brown-banded legs, sparse white hairs on all appendages and a large dark spine on the last part of the thorax. **Biology** Nymphs attach sand and debris to their bodies using secretory hairs. Adults nocturnal and attracted to lights. **Habitat** Under stones and bark in most vegetation types, or in caves or buildings. **Related species** At least 4 other regional species in the genus. Also, the similar *Holotrichus* species (**1A**) are pale grey, with very setose legs, occur in the arid western parts of the region, and make a grating sound when touched.

2 *Parabotha*   Ambush bugs

Medium-sized (body length 10mm), body elongate and stout, green ventrally. Toothed fore legs modified for grasping live prey. Head long, with thickened antennae; last segment longest and reddish. Thorax with lateral expansions. **Biology** Wait on flowers by day for large prey such as bees, butterflies and wasps. **Habitat** Rarely seen, but frequent fynbos flowers, especially leucadendrons. **Related species** There are 4 other regional genera of ambush bugs (subfamily Phymatinae), with approximately 8 species.

3 Subfamily Emesinae   Thread assassins

Small (body length 9mm), very slender assassin bugs, with elongated appendages and raptorial fore legs. Hemelytra entirely membranous; wingless forms common. At least 25 species known from the region. **Biology** Apparently feed on small flies. Often attracted to lights at night. **Habitat** Common under stones, logs, bark, and in caves, often in very arid areas. Some apparently associated with spider webs. *Myiophanes* species (**3**) occur under stones and rocks in dry habitats.

4 Family Berytidae   Thread bugs

Small and very delicate, with elongate body, limbs and antennae. Ends of femora thickened, giving a knob-kneed appearance. Elbowed antennae also clubbed at tips. *Metacanthus* (**4**) are typical, being small (body length 5mm), with a green abdomen, tan thorax, and swollen black clubs at the ends of the antennae. Approximately 40 species occur in the Afrotropical region; southern African fauna is poorly known, with only 15 described species. **Biology** Species of *Gampsocoris* feed mainly on young shoots, but also on small insects, such as aphids. **Habitat** Mainly on grasses and tree trunks.

Family Aradidae   Flat bugs, bark bugs

Small to medium-sized (body length 2–12mm), easily recognized by their broad, flat appearance (an adaptation to living under loose bark). Mostly winged, but females may be wingless. Fine piercing mouthparts (stylets) coiled up inside the head, near the base of the beak. About 55 species are known from the region.

5 *Strigocoris*

Small (body length 10mm), very flattened, dull brown, with small, narrow wings surrounded by an expanded abdomen. **Biology** Feed on fungi – stylets well suited to following fungal threads. Some species guard their egg batches until they hatch. **Habitat** Commonly found under the bark of dead *Senegalia* branches. *Strigocoris* occur in tunnels constructed by other insects in rotting logs.

Family Coreidae   Twig wilters, squash bugs, leaf-footed bugs

Small to large (body length 5–40mm), light to dark brown, with a broad thorax (not as broad as that of stink bugs, Pentatomidae), and a head that appears quadrangular in shape. The numerous veins in the membranous part of the hemelytra run almost parallel to one another (see *Anoplocnemis*, p.148). Stink glands are visible between the middle and hind legs. In some of the larger species, the hind legs of males are enlarged and armed with spines. All species feed on young shoots, seeds or fruits, often those of squashes. About 150 species are known from the region.

6 *Acanthocoris*

Medium-sized (body length 13mm), stout, greyish-brown, with a rough body surface covered in small white bumps. A thin white line runs from the centre of the head to end of thorax. Hind legs enlarged. **Biology** Recorded on indigenous species of *Solanum* and exotic Bugweed (*S. mauritianum*). Often in groups on host plant. **Biology** Usually in semi-shaded forest.

1 Petalocnemis

Medium-sized (body length 13mm), stout, brown, with a rough body surface. Very similar to *Acanthocoris*, but darker, with enlarged hind legs, tan margins to the abdomen, and a cream spot on the sides of the wings. **Biology** Occur in aggregations, often on exotic Bugweed (*Solanum mauritianum*) and the grain legume, Pigeon Pea (*Cajanus cajan*). **Habitat** Subtropical forest margins.

2 Anoplocnemis   Twig wilters

Large (body length 24mm), dark brown, with a reddish wing membrane and orange tips to the antennae. Hind legs enlarged, with swollen femora, each with a stout spine. Genus contains 2 very similar regional species. **Biology** Pierce young shoots, macerating the tissue and injecting saliva with digestive enzymes that cause the shoot to shrivel beyond the puncture. *A. curvipes* attacks over 100 plant species, often shrubs, climbers and trees. Eggs are laid in rows in plant tissue. Females remain in the vicinity of the eggs and young nymphs. **Habitat** Lush vegetation, including gardens. Occur throughout tropical Africa.

3 Cletus

Small (body length 7mm), uniform light tan, with bulbous eyes, 2 white spots on the hemelytra and black tips to the antennae. **Biology** Host plants include indigenous potato (*Solanum*) and weed-like Amaranthaceae and Chenopodiaceae species. **Habitat** Stands of weeds in open, sunny situations. Widespread in Africa and Australia. **Related species** Species of *Latimbus* (**3A**) are similar, but uniformly orange, and occur in the tropics.

4 Prismatocerus auriculatus   Acacia Bug

Medium-sized (body length 19mm), stout, mottled in light brown, with dark brown on the membranous parts of the wings, and 2 rounded knobs arising from the thorax. Underside white. **Biology** While recorded on Australian Black Wattle (*Acacia mearnsii*), most *Prismatocerus* species apparently feed on indigenous *Senegalia*. **Habitat** Bushveld and forest. **Related species** *P. magnicornus* is similar, but all brown, while *P. discolor* (**4A**) is brown above and bright green below, with green legs.

5 Elasmopoda elata   Twig Wilter

Large (body length 23mm), with broad, flattened, forward-projecting, pointed extensions on the prothorax. Body dull brown with a whitish bloom, a white band on the sides of the thorax, and antennae banded in orange and black. Hind legs very enlarged and bent, but lack the spines seen in *Anoplocnemis* (above). **Biology** Lives on the young shoots of over 100 species of indigenous and other shrubs and trees, and often on creepers, causing typical wilting of the shoots. A pest of legume crops. **Habitat** A variety of veld types, gardens and cultivated fields. **Related species** A few other similar large species occur in the genus.

6 Leptoglossus membranaceus

Large (body length 16–22mm), black, with red-banded antennae and abdomen and a red stripe across the thorax. Readily distinguished from *Anoplocnemis* and *Elasmopoda* by flattened leaf-like projections on the hind legs. The only African species in the genus. **Biology** A pest of sorghum, granadilla, citrus, cucurbits like melons, squash and cucumbers, and of wild foodplants including the seeds of bitter melons and various wild calabashes (*Cucumis* and *Lagenaria*). Citrus fruits that have been fed on drop off the tree. **Habitat** Bushveld, subtropical forest, gardens and orchards. A cosmopolitan tropical genus, also occurring in Asia.

7 Petascelis remipes   Giant Twig Wilter

Very large (body length 36mm), bulky, velvety dark brown, with thin yellow lines along the sides and down the centre of the very dark brown thorax. Hind legs much enlarged, with leaf-like projections. Nymphs (**7A**) black with warning coloration comprising red spots on a whitish abdomen. **Biology** Gregarious and bold, walking towards intruders with the antennae vibrating when disturbed. Probably capable of squirting defensive secretions for some distance. Found on bush willows (*Combretum*). **Habitat** Bushveld and subtropical forest.

1 Carlisis wahlbergi

Medium-sized (body length 20–26mm), boldly marked, with tan-and-black fore wings, and black-and-white-banded antennae and abdominal margins. Hind legs enlarged. **Biology** As many as 9,000 individuals recorded on a single *Gardenia volkensii* shrub, which then failed to flower, but did not wilt. Large, oval brown eggs (visible in image) laid on the stems of the host plant. Can spray a defensive secretion up to 15cm. **Habitat** On various *Gardenia* species, in bushveld and in gardens.

2 Pephricus

Small (body length 11mm), unmistakable and bizarre, brown and grey with characteristic flange-like extensions of the thorax and abdomen, ending in long spines. The silvery wings lie flat in midline of the abdomen. Genus contains 4 very similar regional species, *P. paradoxus* being the largest and most common. **Biology** Associated with grasses. **Habitat** Grasses in bushveld.

Family Rhopalidae   Scentless plant bugs

Related to twig wilters (Coreidae) but lack repugnatorial glands, and are smaller and lighter in colour. They have numerous wing veins in the membranous portion of the hemelytron, and feed mostly on weedy plants.

3 Leptocoris hexopthalma   Litchi Bug

Small (body length 13mm), with a dark body and fore wings, and red eyes. Hemelytra have apricot-coloured markings along the wing veins, and an orange line separating the hardened from the membranous portion of the hemelytra. **Biology** Commonly found on introduced balloon vine (*Cardiospermum*); also on litchi, Jacket Plum (*Pappea capensis*), avocado, macadamia and impala lily (*Adenium*). Adults feed on seeds and plant sap. **Habitat** Fruit plantations, gardens and a range of veld types, in groups on host plants **(3A)**.

Family Alydidae   Broad-headed bugs

Medium-sized, easily recognized by the combination of a slender body, broad triangular head and enlarged hind legs. Microscopically, the curved last antennal segment can be seen to be the largest. Very alert and active bugs, taking flight almost as rapidly as house flies. Feed on grass seed and other seeds, and may be found in large numbers on the ground below trees or shrubs where seeds have fallen. Nymphs are often dark grey ant mimics. There are 25 species known from the region.

4 Hypselopus gigas   Giant Broad-headed Bug

Medium-sized (body length 18mm), brightly coloured, with an orange to dark red body and legs, banded antennae, and black fur on the lower part of the hind legs. **Biology** Feeds on dry seeds of various tree species, including Candle Thorn (*Vachellia hebaclada*), Umbrella Thorn (*V. tortilis*) and Sweet Thorn (*V. karroo*). **Habitat** Bushveld.

5 Stenocoris apicalis   Rice Seed Bug

Medium-sized (body length 14mm), slender, green, with highly reflective wings, and red markings at the leg joints. Characteristic white bar near tip of the antennae. **Biology** Defensive secretion very strong and offensive. Often occurs in very large numbers in shade on the surface of leaves of herbaceous plants (especially *Hypoestes*). A pest of rice in tropical Africa, where the developing seed is attacked. **Habitat** Subtropical forest and bushveld.

6 Mirperus jaculus

Medium-sized (body length 14mm), reddish-brown, bow-legged with enlarged hind legs. Similar to *Anoplocnemis*, but more slender, with large eyes and a white triangular marking at the wing bases. **Biology** Associated with grasses. Pierces the pods of cowpeas and soya beans, damaging the crop. Also recorded on cashew nut. **Habitat** Grassland and bushveld.

1 Nariscus cinctiventris

Medium-sized (body length 14mm), light tan, with dark brown hemelytra and a yellow stripe through the midline of head and thorax. Enlarged hind legs with a small spine on femur. **Biology** Non-specific, feeding on the seeds of most *Senegalia* and *Vachellia* species, as well as those of the introduced mesquite (*Prosopis*). **Habitat** Arid bushveld and fynbos. **Related species** *Hypselopus* species **(1A)**, which have orange markings on the pronotum, antennae and hind tibia, and serrations on the inner hind femora.

2 Tupalus fasciatus

Medium-sized (body length 19mm), brown, with dull orange band on swollen hind legs, faint orange bands on abdomen, and conspicuous cream band on antennae. Nymphs grey. **Biology** Feeds on seeds of *Bauhinia*, and recorded from cashew nut. Like many in the family, nymphs are ant mimics. **Habitat** Variety of vegetation types.

Family Pyrrhocoridae   Cotton stainers, red bugs, fire bugs

Small to medium-sized (body length 5–20mm), brightly coloured (often red or a combination of red, yellow and black). Under magnification the absence of simple eyes (ocelli) can be confirmed. Many species are short-winged or totally wingless. All are plant-feeders associated with species of wild hibiscus (Malvaceae) and related families, feeding chiefly on the seeds, but also on stems. Of economic significance for tropical crops such as cotton, on which they leave a yellowish stain. There are 35 species in the region.

3 Cenaeus carnifex   Red Bug

Small (body length 9mm), oval, uniformly red or orange. Legs, antennae and membranous part of the hemelytra black. Short-winged forms common. **Biology** Usually encountered in groups on the seed heads of herbaceous plants, often in mated pairs. **Habitat** Lush vegetation on forest margins and in gardens. Abundant under herbs, commonly invasive mallow (*Malva*) species, where it feeds on fallen seed.

4 Dysdercus nigrofasciatus   African Cotton Stainer

Large (body length 17mm) but variable in size, elongate, tan-coloured with a reddish head and legs, and a white band around the neck. Black stripe across wings; ends of hemelytra black. Nymphs more reddish. **Biology** Adults and nymphs often encountered together **(4A)** feeding on developing seeds of Malvaceae (hibiscus, Baobab, Kapok Tree) and related families, and on crops such as cotton, sorghum and okra. Adults remain mated for days, and the colourful nymphs are gregarious. Parasitized by tachinid flies (*Phasia*). All except 1 species in the genus are pests of cotton, damaging seeds and transmitting a fungus that stains the cotton ('lint') yellow or brown, hence their common name. **Habitat** Bushveld and forest, agricultural land and gardens. **Related species** There are 9 other regional species in the genus. The Welwitschia Bug (*Probergrothius angolensis*) **(4B)** is another large (body length 18–21mm) tan cotton stainer, boldly marked with 2 black bars and 2 black spots on the hemelytra. It feeds on seeds in the developing cones of the primitive Upside-down Plant (*Welwitschia mirabilis*) in the Namib, and does not appear to occur in the region.

5 Scantius forsteri

Small (body length 10mm), with crimson margins to the black body. **Biology** Flightless. Short-winged individuals commonly run around, feeding on fallen seeds. **Habitat** Very common under bushes. Largely ground-dwelling, often encountered under dry logs. Has a cosmopolitan distribution.

Family Lygaeidae   Seed bugs, ground bugs

A diverse family of small to large, usually quite slender or oval bugs capable of producing the classic bug stink. Under magnification can be seen to have only 3–4 (often faint) longitudinal veins on the soft membranous part of the fore wings, thus distinguishable from Coreidae, which have at least 6 veins. The wings may be reduced, especially in ground-dwelling species. Although most are herbivores, feeding chiefly on seeds, a few species feed on other insects. About 400 species are known from the region, with many cosmopolitan genera.

1 *Dieuches*

Large (body length 12mm), parallel-sided, dark brown, with a thin cream marginal stripe. Most of the 45 regional species have white spots on the wings. **Biology** Ground-dwelling. Feed on dassie (hyrax) pellets. Some foreign species damage strawberries, cotton and stored peanuts. **Habitat** Under logs or low vegetation.

2 *Pachygrontha lineata*

Small (body length 8mm), elongate and slender with very long antennae. Light brown, with a yellow stripe along the midline of head and thorax. Femora of fore legs enlarged. **Biology** Associated with grasses. **Habitat** Common on herbaceous vegetation. The genus occurs in all tropical parts of the world. **Related species** There are 2 other species of *Pachygrontha* in the region.

3 *Oncopeltus famelicus*   Milkweed Bug

Large (body length 14mm), unmistakable, boldly marked in black and orange. **Biology** Feeds on seeds of Urticaceae (*Pouzolzia*), *Triumfetta*, granadilla, and on cotton and sweet potatoes, which it damages. Most often found feeding in numbers on the pods of Common Milkweed (*Asclepias fruticosus*). **Habitat** Open grassland, road verges and forest margins; also found in gardens and on agricultural land. Range extends into tropical East Africa. The genus occurs in most tropical and subtropical parts of the world. **Related species** *O. jucundus*, which is known to feed on crushed millipedes in Zimbabwe.

4 *Spilostethus pandurus*   Milkweed Bug

Medium-sized (body length 13mm), easily recognized by the combination of red bars and dots on a black to grey background. **Biology** Feeds on seeds of various milkweeds (*Asclepias*), related *Calotropis* and on indigenous *Solanum*. A minor pest of sunflowers, feeding on developing seeds. Also recorded on cotton, sorghum, millet, citrus, mangoes, sweet potatoes, apricots and maize. **Habitat** Very common and widespread in various veld types (usually grassland, where it occurs on the flowerheads of yellow daisies), and in disturbed veld, gardens and agricultural lands. **Related species** There are 7 other regional species in the genus.

5 *Macchiademus diplopterus*   Grain Chinch Bug

Small (body length 4–5mm), black, flattened and parallel-sided, with pale orange tibia. Males slimmer than females, with black hemelytra and a glassy grey wing membrane. **Biology** An agricultural pest, feeding on the leaves and seeds of wheat, oats and other grasses, and then seeking sheltered spots (under eucalyptus bark or in fruit-packing sheds) in which to spend the dry summer months in the Western Cape. **Habitat** Very common in agricultural lands and disturbed areas; usually seen under tree bark.

6 *Oxycarenus*   Dusky cotton stainers

Small (body length 4–6 mm), with a shortened body; either dull brown with clear shiny wings, or various combinations of red and black. Enlarged, spined fore femora. There are approximately 55 African species. **Biology** Many, such as the Dusky Cotton Stainer (*O. hyalinipennis*), are pests of cotton and related malvacous plants. **Habitat** Disturbed and agricultural land where their host plants are common.

Family Thaumastocoridae

A small family, most common in Australasia. Feeds on *Eucalyptus*, wattles and palm trees. They can damage *Eucalyptus* trees and, when disturbed, scuttle about on the leaves without flying off.

7 *Thaumastocoris peregrinus*   Eucalyptus Thaumastocorid

Small (body length 3mm), light brown, with large bulging eyes. **Biology** The only representative of the family in Africa, and a recent introduction from eastern Australia. Fast-moving, but does not fly readily. Life cycle is completed in 35 days. Feeding causes foliage to turn yellow-brown and drop off, as well as reducing nectar production, thus posing a threat to both the timber industry and to beekeeping. **Habitat** On *Eucalyptus* trees.

Family Cydnidae
Burrowing bugs

Small to medium-sized, with a shiny, oval, dark brown or black body and spiny hind legs. They spend most of the time burrowing in soil or sand, where they feed on roots. There are 30 species in the region.

1 *Geocnethus plagiata*

Small (body length 11mm), black, with a smooth and shiny thorax. **Biology** Feeds on leaves, stems and roots. Attracted to lights in large numbers. Can release an unpleasant-smelling secretion from the thoracic glands, which is known to cause brown lesions when it comes into contact with skin. Minor pests of groundnuts and strawberries. **Habitat** Most common in summer-rainfall areas, where huge numbers may be attracted to lights; digs up to 2m underground.

Family Plataspidae
Pill bugs

Distinctive small to large, dull-coloured, oval and convex bugs. A shiny extension of the thorax (scutellum) is so enlarged it covers the entire body, including the abdomen and very long folded wings. Superficially, some might be confused with beetles, others with Scutelleridae (below), although they are more convex and rounded in profile. All are capable of flight. The 25 species of this mostly tropical family are restricted to the northern parts of the region.

2 *Libyaspis wahlbergi*
Leopard Pill Bug

Large (body length 9–13mm), rounded. Anterior often white, fading to a rich reddish colour towards posterior of the body. Entire body peppered in black pits. Nymphs **(2A)** are covered in a whitish bloom, and have conspicuous gland openings. **Biology** Sucks juices from various species of wild pea. Some *Libyaspis* species are thought to feed on fungal threads beneath bark. Adults live communally with nymphs on tree branches. Eggs laid in 2 rows and covered in a pellet-like secretion. **Habitat** Branches of trees; has been recorded on Australian Blackwood (*Acacia melanoxylon*) and Flat-crown Acacia (*Albizia adianthifolia*).

3 *Coptosoma affinis*
Legume Plataspid

Small (body length 5mm), oval and black. Scutellum enlarged, with a cream border and 2 cream spots. Abdomen with cream lateral stripe. **Biology** Gregarious, possesses stink glands and is capable of flight when the hemelytra beneath the scutellum are spread. A pest of legumes. **Habitat** Widespread in both agricultural land and undisturbed veld.

Family Scutelleridae
Shield-backed bugs

Easily recognized by the very enlarged extension of the thorax (scutellum), which covers the entire abdomen and wings. Medium-sized, oval bugs, more elongate and brightly coloured than plataspids (above), with which they may be confused. A few look superficially like beetles. Feed on a wide range of shrubs and trees. A subtropical family, most common in the northern parts of the region, where 30 species are known.

4 *Sphaerocoris*
Picasso bugs

Medium-sized (body length 10mm), oval and convex, intricately and clearly marked in various bold patterns (including concentric spots), in olive green, tan, red and black. Nymphs are yellow, oval and convex, and marked with black stripes on the abdomen. The Brown-spotted Shield Bug (*S. testudogrisea*) **(4A)** has a cream body dotted with numerous black spots and regularly spaced brown patches. **Biology** Feed in groups on cotton (*Gossypium*), coffee, citrus and *Vernonia*. **Habitat** Stands of weedy plants at the edges of subtropical forest and in bushveld.

5 *Alphocoris indutus*

Medium-sized (body length 10mm), spindle-shaped, whitish, with a brown scutellum. The only species in the genus. **Biology** Feeds on tender grass shoots (*Panicum, Pennisetum, Stipagrostis*) and ripening grass seeds. A pest of pearl millet in tropical Africa. **Habitat** Open grassland.

1 *Calidea dregii* — Rainbow Shield Bug

Medium-sized (body length 14mm), spindle-shaped, unmistakable, with 2 rows of black spots on the iridescent green, blue, yellow and red back. Underside red or orange, with green plates. **Biology** Sometimes a pest of cotton; also breeds on sunflower, sorghum, tobacco, castor oil and various other domesticated and indigenous plants. Shown here on Melkbos (*Euphorbia mauritanica*). Feeds by piercing young seeds, causing seed shedding and distortion of the seed head. Eggs are laid in a spiral on a stalk. **Habitat** Weed patches, gardens, agricultural and natural vegetation of various kinds. **Related species** *Calidea* includes a few other similar regional species.

2 *Solenosthedium liligerum*

Fairly large (body length 14mm), stocky, weakly metallic olive green, with 3 yellow blotches at the end of the scutellum. In some forms the blotches fuse into a wavy band. Abdomen shows red in flight. Nymphs dark reddish-brown with a thin, metallic green marginal band. **Biology** Usually encountered singly on shrubs. Aggregates to feed on civet dung. **Habitat** Bushveld.

3 *Graptocoris aulicus*

Medium-sized (body length 11–13mm), brightly orange and black with metallic blackish-green head and body markings. **Biology** Nymphs (**3A**) are gregarious, and feed on Malvaceae. **Habitat** Subtropical bush. *Graptocoris* also occurs in the tropics.

4 *Steganocerus multipunctatus* — Ladybird Bug

Small (body length 9mm), ladybird-like, usually black with evenly spaced bright orange spots, but variable in colour and occasionally brown without spots. Beak and fused wings forming shield-like scutellum distinguish it from a ladybeetle (p.270). **Biology** Like others in the family, appears to feed on a variety of flowering shrubs and trees, including cashew nut, cotton (*Gossypium*) and coffee. **Habitat** Bushveld.

Family Tessaratomidae — Inflated stink bugs

Large (body length 15–30mm), green or yellow, with distinctively flattened and expanded abdomen, and a short extension of the thorax (scutellum) that does not cover the wings. Very short beak, which does not quite reach the base of the front legs. They suck plant juices, usually from shrubs and trees, some species occurring in large numbers. A tropical family with 10 species known from the region.

5 *Encosternum delegorguei* — Edible Stink Bug

Large (body length 25mm), green-yellow or brown, with lateral margins of abdomen not exposed as in similar *Natalicola* species. Nymphs (**5A**) circular and green. **Biology** Flies in droning swarms on hot days, settling in the evening. Feeds on *Senegalia ataxacantha* and other shrubs and trees. Can eject an acrid fluid from the stink glands with some force. In Zimbabwe aggregate on mahobohobo trees (*Uapaca*), apparently arriving at favoured spots every winter in enormous swarms to feed on Wild Loquat trees (*Oxyanthus speciosus*), where they remain from April to September. The bug (*harugwa*) is a local delicacy, killed in hot water, squeezed to remove the almost nauseating secretion, and then roasted and dried. Also eaten in South Africa, raw or cooked. **Habitat** Subtropical open woodland and bushveld.

Family Dinodoridae

A small family of large, robust bugs resembling Pentatomidae, but most closely related to Tessaratomidae. They have small keels along the sides of the head and pronotum. Antennae with 4 or 5 segments, some of the lower ones flattened. Twelve species are known from the region.

6 *Coridius nubilis*

Large (body length 18mm), oval, mahogany brown, with the last 2 antennal segments bright orange. Nymphs lighter brown. **Biology** A pest of watermelons and muskmelons. Severe infestations may kill seedlings. **Habitat** In shrubs in bushveld and in agricultural land in summer-rainfall regions.

Family Pentatomidae

Stink bugs, shield bugs

A diverse family of small to medium-sized, often brightly coloured bugs varying considerably in shape, although all have an enlarged triangular scutellum ('shield') extending at least halfway along the abdomen and partially covering the wings. All are well armed with stink glands, which open on the top of the abdomen in nymphs. Generally plant-feeders, but some in the subfamily Asopinae (predatory stink bugs) feed on soft-bodied insects, such as beetle larvae. A number are pests of economic significance. About 300 species known from the region.

1 *Coenomorpha*

Bark stink bugs

Medium-sized (body length 20mm), flattened, brown or streaked with grey, sometimes with light orange marbling and stripes on edge of abdomen. Antennae with orange bands. Nymphs (1A) oval, convex, and covered in a loose bluish-white waxy layer. **Biology** Herbivorous in the region; known to feed on litchi, but apparently polyphagous. British species exhibit maternal care of the eggs and newly hatched young. **Habitat** Common in gardens, orchards and veld. Occur frequently on Brazilian Pepper tree (*Schizinus molli*). **Related species** *Pseudatelus* species (body length 15–22mm) (1B), which are similar, but have unbanded antennae, orange bands along the sides of the abdomen and 2 small orange spots at the edge of the widest part of the scutellum. In winter they develop a whitish powdery covering. Also, the Mottled Stink Bug (*P. raptoria*), which is a pest of pistachio nuts and also feeds on maturing pecan and macadamia nuts and avocado fruits.

2 *Antestia lymphata*

Medium-sized (body length 12mm), with white spots on the thorax, including the scutellum, and orange dots along margin of the abdomen. **Biology** Numbers effectively regulated by Strepsiptera (p.308) in the tropical parts of Africa, where it is sometimes a pest of coffee. **Habitat** Disturbed weedy fields and gardens.

3 *Antestiopsis thunbergii*

Antestia Bug

Small (body length 8mm), flattened, with highly variable coloration. Easily confused with similar species. Intricately marked in orange, black and grey or yellowish-green, with 1 orange spot on prothorax, 2 on scutellum, and 3 white stripes on the black head. **Biology** Feeds on young shoots, buds and fruit of most deciduous fruit trees and proteas, and is very destructive of coffee beans, citrus fruits and apples. Pines, willows, Port Jackson and other wattles are alternative hosts. Eggs are laid in neat clusters resembling tiny circular jars with lids. **Habitat** Urban and agricultural areas.

4 *Bagrada hilaris*

Bagrada Bug

Small (body length 6mm), body black with bright orange or red markings (often with rows of 3 spots on pronotum, scutellum and hemelytra). Young nymphs are shiny red, later becoming brownish-red. **Biology** In large numbers damages the shoots of young plants, inhibiting growth. Attacks cabbage and its relatives, potatoes, peanuts, garden flowers, papayas, sorghum and even rooibos tea bushes, leaving leaves shrivelled and dry. Adults often seen in copulation. **Habitat** Subtropical agricultural areas, gardens and forest margins. A well-known pest species in Africa, southern Europe and southeast Asia.

5 *Caura rufiventris*

Medium-sized (body length 14mm), chunky, brownish-black. Eyes, wing bases and stripes on exposed margins of the abdomen are bright red. The only species in the genus. **Biology** Recorded from indigenous potatoes (*Solanum*), coffee, citrus, sorghum, and sweet potato. **Habitat** Bushveld and subtropical forest.

6 *Nezara viridula*

Southern Green Stink Bug

Medium-sized (body length 12mm), familiar, uniform apple green or brownish-green, with a row of 3 small white spots between the prothorax and scutellum. Eggs laid in a neat series of rows on the host plant. Young nymphs red and black and gregarious (6A), later moulting into solitary green nymphs with cream and red spots (6B). **Biology** An introduced species recorded on more than 200 plant species, including most vegetables. Feeds on young leaves, growth points and fruit, causing a decaying wound in fruit. Parasitized by tachinid flies. Adults hibernate in compost heaps. **Habitat** Very common in gardens and agricultural land. Virtually cosmopolitan in warm and temperate regions of the world.

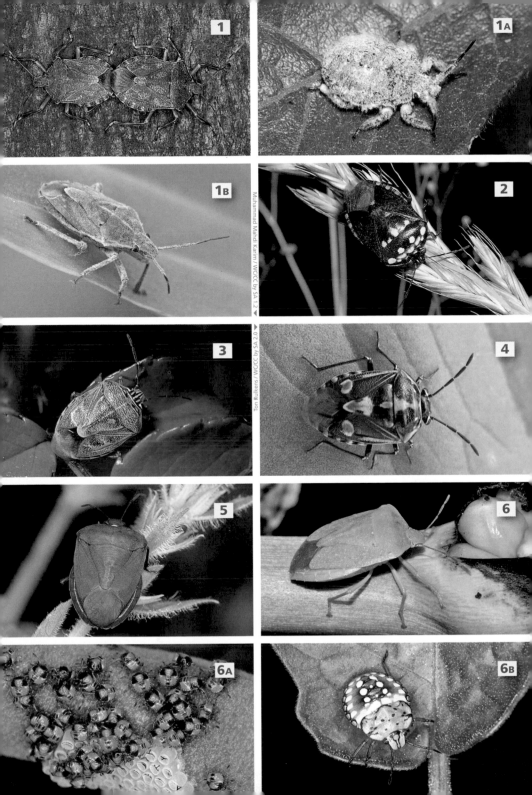

1 Basicryptus costalis

Medium-sized (body length 14mm), uniformly ochre, with 2 thin cream stripes on the margins of the hemelytra, and a black spot below each side projection of the thorax. **Biology** Recorded from cotton (*Gossypium*), but apparently also preys on caterpillars. **Habitat** At the base of grasses in open bushveld or grassland, or at the margins of subtropical forest.

2 Veterna
Grass stink bugs

Small (body length 9mm), shield-shaped, with tan body, maroon wings and thorax, and a large pale yellow scutellum. Characteristic sharp spines on margins of the thorax. **Biology** Recorded from indigenous *Solanum*, cotton (*Gossypium*) and sesame seed (*Sesamum*), and frequently associated with grasses. **Habitat** Common in grassland, bushveld and agricultural lands.

3 Acoloba lanceolata

Medium-sized (body length 14mm), with pointed rostrum, and green thorax and scutellum. **Biology** Recorded from sorghum, and probably feeds mainly on grass seeds. However, in tropical parts of its range also recorded from cashew, *Dracaena*, *Pandanus* and *Dioscorea*. **Habitat** Essentially a grassland species. **Related species** *Mecidea linearis* (**3A**) is a similar species (body length 10mm) from the same habitat and parts of the region, but is smaller and dull brown.

4 Pododus
Turtle stink bugs

Small (body length 6mm), circular and flattened. Dull brown or grey, with head expanded and indented in the region of the eyes. Body covered in fine punctures. Approximately 5 species in the region. **Biology** Scuttle rapidly on sand, and likely to feed on grass seeds. **Habitat** Arid parts with low vegetation, always found on the soil surface.

5 Agonoscelis versicoloratus
Sunflower Seed Bug

Small (body length 10–12mm), with a black-striped head, yellow-tipped scutellum, and yellow stripes on the exposed margins of the abdomen. Thorax and hemelytra heavily pocked with maroon spots. **Biology** Feeds on the developing seeds of sunflowers and sorghum, and may become a minor pest. **Habitat** Subtropical bushveld and forest margins.

6 Subfamily Asopinae
Predatory stink bugs

Medium-sized (body length 10–18mm), brownish-red with yellow markings. Fore legs with a short spine near end of femur. Nymphs bright blue and red. First segment of beak short and thick. Nine genera and approximately 20 species in the region. **Biology** All species are predaceous, feeding on the soft-bodied larvae of moths, butterflies and beetles, and are thus useful regulators of pest species. Nymphs feed on plants initially, later switching to an insect diet. Saliva has a paralysing effect on prey. **Habitat** Bushveld and subtropical forest margins.

SUBORDER HETEROPTERA (CONT.)
Aquatic, semi-aquatic & shore bugs

The following heteropteran families live either below or on the water surface, or in damp habitats alongside waterbodies. Many of the families are closely related to one another, and form distinctive evolutionary and ecological groups.

7 Family Saldidae
Shore bugs

Small (body length 2–7mm), oval, quite easily identified by dull dark brown coloration, often enlivened by small white or yellow markings. Head broad and large, eyes very large. *Capitonisalda* **(7)** (body length 5mm) are oval, shiny brown and somewhat convex, with long antennae, slender legs, and a number of small yellow markings on the wings. There are 8 species and 3 genera in the region. **Biology** Very agile and difficult to catch. Active predators of small insects. *Capitonisalda* feed on small organisms living on the surface of the moist ground. Mating takes place in a side-by-side position. **Habitat** At the sandy margins of fast-flowing streams or on moist paths. *Capitonisalda* occur among moss and stones bordering high-altitude mountain streams.

1 Family Ochteridae
Velvet shore bugs

Small (body length 3–6mm), very active, oval, occasionally marked with bluish or yellow spots. May be confused with saldids, but stouter, with spines on the hind tibiae. Only 2 species in the region. *Ochterus caffer* **(1)** is small (body length 5mm), light brown, covered in fine hairs that impart a velvety sheen; sides of the body marked with tan patches, the 3 largest on the thorax (especially in Western Cape forms). Antennae project from the side of the head. The widespread *O. marginatus* **(1A)** has sky blue spots and yellow shoulders. **Biology** Adults run and fly rapidly. Nymphs **(1B)** may stay underwater for brief periods. All species predaceous, feeding on aquatic fly larvae. **Habitat** In moist shady areas adjacent to flowing water, but also along the margins of clean standing water.

2 Family Gelastocoridae
Toad bugs

Medium-sized (body length 6–10mm), squat, robustly built, with a broad warty head and back, strongly protruding eyes, and raptorial fore legs. Only the widespread *Nerthra grandicollis* has been recorded from the region, but a number of undescribed species occur. **Biology** They leap upon insect prey, grasping with the enlarged fore legs. Nymphs cover themselves with sand granules, making them very hard to find in the clean wet sand in which they dig at stream margins. Winged forms are capable of flight. **Habitat** In wet mud, sand and vegetation next to slow-flowing streams.

3 Family Leptopodidae
Spiny shore bugs

Small (body length 2–7mm), elongate, grey-brown, with spiny fore legs and simple eyes on a raised platform. *Valleriola moesta* **(3)** is small (body length 7mm), slender, dull dark brown, with elongated limbs, very long fine antennae, yellow-streaked wings and thorax, and partially stalked compound eyes. Just 1 species known from the region and 2 from Angola. **Biology** Active fliers and difficult to catch. Probably feed on small insects. **Habitat** On rocks in fast-flowing subtropical mountain streams. Rare.

4 Family Hebridae
Velvet water bugs

Very small (body length 2–3mm), stout, winged or showing various stages of wing loss. *Hebrus* **(4)** (body length 2mm) are dull brown with 2 white spots on the hemelytra, fairly long antennae and short legs. They retain the tarsal claws of their terrestrial ancestors. There are 4 species in the region. **Biology** Thought to be carnivorous. Do not skate on the surface of water as much as Veliidae and Mesoveliidae, clinging instead to aquatic plants above the surface. **Habitat** Floating vegetation and very wet moss.

5 Family Hydrometridae
Water measurers

Small (body length 9mm), unmistakable, thread-thin and elongate. Resemble some very thin reduviids, but are even narrower (p.142). Eyes in the middle of the very long head. Winged forms rare. There are 5 species in the region, all in genus *Hydrometra*. **Biology** Feed on dead or freshly killed insects. Groups often seen feeding together on insects trapped on the surface of water. They also spear live food (mostly mosquito larvae and water fleas) just beneath the surface, holding their prey pierced on the beak while feeding. **Habitat** Very moist freshwater habitats, or the surface of standing water. Some prefer stagnant or brackish water. *Hydrometra* is a widespread cosmopolitan genus, *H. ambulator* being found in emergent vegetation in standing, often stagnant or brackish water.

6 Family Mesoveliidae
Water treaders, pondweed bugs

Small (body length 3mm), slender, pear-shaped, dull olive green, with silver-grey hemelytra and long legs and antennae. Seen under magnification, the first part of the thorax (pronotum) does not quite extend to the next (the mesonotum). Female's ovipositor is long and toothed, adapted for laying eggs in plant tissue. Males generally winged, females wingless. Closely related to the Hebridae. **Biology** They feed on small aquatic insects and crustaceans, as well as dying insects trapped on the water surface. **Habitat** Abundant near floating vegetation in standing water. The widespread *Mesovelia vittigera* **(6)** also occurs in South Africa's neighbouring countries and in Madagascar, South America, India and the Mediterranean.

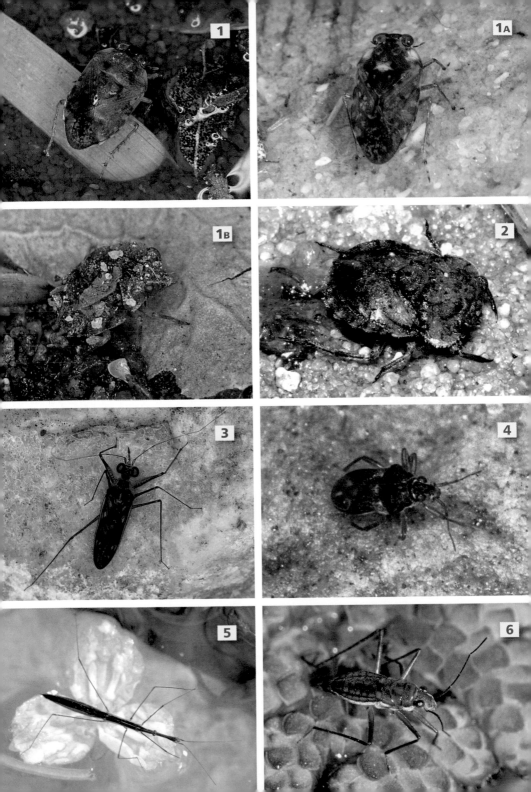

1 Family Paraphrynoveliidae — Gondwanan pygmy water bugs

Small (body length 2.5mm), dark-bodied, wingless and parallel-sided, with a furrow running down the middle of the head. Under magnification small gold reflective dots are visible all over the body. Only 2 species of *Paraphrynovelia* in the region. Family restricted to South Africa and Lesotho. **Biology** Adults slow-moving, feeding on small insects associated with mossy seeps. **Habitat** Wet rock surfaces covered in moss, such as at waterfalls and seeps.

Family Veliidae — Water crickets, small water striders

Small and stout. They look and behave like the larger pond skaters (Gerridae), but the legs are evenly spaced and much shorter (middle and hind legs very close together in Gerridae), and the thorax is far wider than the rest of the body. Generally wingless. Six genera and 15 species in the region.

2 *Rhagovelia* — Pygmy water crickets

Small (body length 5–6mm), uniformly black, with the body constricted near the end of the hemelytra. Head short, with large eyes. Winged and wingless forms occur. There are 3 regional species. **Biology** They move rapidly against currents by rowing the mid legs, which end in tufts of hair, and suck fluids from insects trapped on the water surface. **Habitat** In areas of reduced flow in streams and rivers.

3 *Tenagovelia sjoestedti* — Grey Water Cricket

Small (body length 4mm), fairly stout, usually wingless. Grey body marked with a pair of thin black and white abdominal stripes, leg bases orange. The only species in the genus. **Biology** Gregarious. They swim using only the middle pair of legs; hair tufts hidden in slits on the hind tarsi act as fans, making it possible to swim against currents. **Habitat** Subtropical, on wet sand and algal mats associated with rocky rivers and streams.

4 *Ocellovelia* — Cape water crickets

Small (body length 6–7mm), light brown with cream streaks at the base of the hemelytra. Head bent downwards, and thorax hairy beneath. There are 2 regional species. **Biology** Breeds in spring (October). **Habitat** In natural ponds, slow-flowing streams and rock seeps; occasionally occurs in swimming pools.

5 *Xiphoveloidea major* — Riffle Water Cricket

Small (body length 4mm), oval, with a pair of small white patches on the front part of the abdomen. Legs short. The only regional species in the genus. **Biology** Swims actively on water, often against the current. **Habitat** Open water such as large dams, lakes and big rivers.

Family Gerridae — Pond skaters, water striders

Familiar, medium-sized to large (body length 5–20mm), with compact bodies and very elongate legs that row them across water at high speed. Only the middle legs used in locomotion. Fore legs are held close to the body and used only for feeding and mating, vibrations from struggling prey being detected through the fore legs, as are vibrational signals from potential mates. Fore legs also used to subdue prey. Claws are bent behind the tarsi, and the tips of the legs carry a patch of water-repellent hairs. Often wingless, but winged forms fly well. Uniform texture of the fore wings is not common among heteropteran bugs. *Halobates micans* is unusual in that it is wholly marine, inhabiting the relatively calm waters off KwaZulu-Natal and Mozambique. There are 15 species in 9 genera in the region, many of them widespread.

6 *Gerris swakopensis* — Common Water Strider

Large (body length 8–11mm), plain grey-brown body with white sides. Short black stripes along mesothorax and metathorax, and a faint median stripe along thorax. As in most gerrids, females are larger than males and often wingless. The only regional species in the genus. **Biology** Feeds on large insects that have fallen onto the surface of water. **Habitat** Common in most bodies of standing water. Widespread.

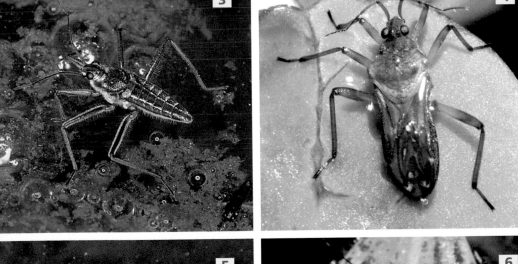

1 *Eurymetra natalensis* — Flat Water Strider

Small (body length 6mm), easily recognized by oval body with very abbreviated abdomen. Black, marked with paired dull yellow bars. Apparently always wingless. The only species in the genus. **Biology** Less aggressive than other gerrids, occurring in groups near marginal vegetation. **Habitat** On standing water, usually in shaded forest pools.

2 *Limnogonus hypoleuca* — Striped Pond Skater

Large (body length 10mm), easily recognized by narrow, elongated, dark grey body with thin orange midline stripe and white lateral stripes on the thorax and abdomen. Females larger than males, both sexes often wingless. **Biology** Bold, attacking any struggling insect it detects through surface vibrations in the water. **Habitat** Common in the marginal vegetation of suitable bodies of standing water. Subtropical. **Related species** There are 4 other regional species in the genus.

3 *Neogerris severini* — Yellow-ringed Pond Skater

Small (body length 6mm), males (3) smaller and slimmer than females (3A). Thin yellow band at edge of scutellum. Large protruding eyes. **Biology** Prefers to form small groups on water at margins of streams. **Habitat** Under marginal vegetation of shaded subtropical rivers and streams.

4 *Tenagogonus* — Yellow-striped pond skaters

Large (body length 15mm), with a central orange stripe that is bordered by 2 thin white stripes on thorax; often winged. Three regional species in genus. **Biology** Gregarious, feeding on insects trapped on water surface as well as on emerging aquatic insects. Territorial, communicate by transmitting vibrations on water surface using front legs. **Habitat** Slow-moving, shaded forest streams, including swamp forest.

5 *Aquarius distanti* — Orange Water Strider

Large (body length 12mm), with broad orange border to thorax and at bases of legs, body thin and long. Single regional species. **Biology** Forms groups in mid-water, eggs probably laid under floating vegetation and lily pads. **Habitat** Large permanent bodies of water, typically coastal lakes. **Related species** The Pencil Water Strider (*Rhagodotarsus hutchinsoni*) (5A) (body length 5mm) occurs in the northeastern parts of the region and also has an orange thorax and legs, but the last 2 abdominal segments extend backwards into a cylindrical cone.

Family Corixidae — Water boatmen

Back flatter than in notonectids (with which they can be confused), with brownish mottled or striped wings (clear light bluish-grey in notonectids). Do not swim upside down (unlike notonectids), and spend most of the time near the bottom of waterbodies. Tips of fore legs flattened to form scoop-like structures (palae), used for filtering algae and other debris. Males produce a very loud courtship song by rubbing fore legs against the face. Spend most of the time near bottom of a waterbody, rising occasionally for air, and are able to leap out of the water and fly. Five genera and 14 species in region.

6 *Sigara* — Common water boatmen

Small (body length 6mm), mottled; prothorax with a number of dark transverse stripes, and fore wings finely mottled in dark grey and tan. Five regional species. **Biology** Colour changes according to that of substrate. Both sexes make sounds by rubbing legs against mouthparts. Able to tolerate polluted water, and used as an indicator of poor water quality. Fly readily by night and often attracted to lights in large numbers on warm evenings. **Habitat** Almost any stagnant or slow-moving waterbody.

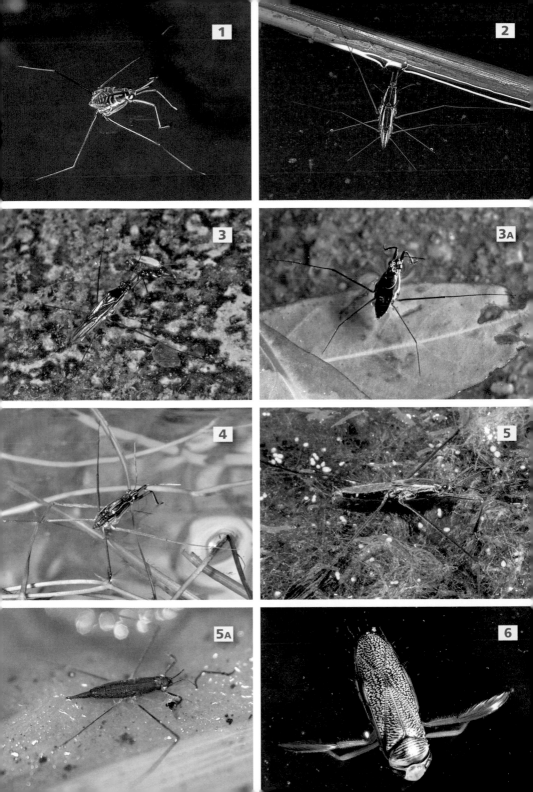

Family Micronectidae
Pygmy water boatmen

Very small, flattened, oval water bugs, resembling miniature water boatmen. Head and eyes large. Often winged. Large hind legs fringed with hairs used for swimming. Family represented regionally by 14 species of *Micronecta*.

1 *Micronecta*
Pygmy water boatmen

Very small (body length 1–3mm), light brown, occasionally with diffuse darker patches. **Biology** Males vibrate their reproductive structures to produce an audible underwater buzz, which attracts females and repels males. **Habitat** Oxygen-rich waters along the margins of rivers and streams. Prefer shallow warm water with algal scum or a crust on the substrate, and avoid ponds with submerged vegetation. *M. scutellaris* is among the region's most common water bugs.

Family Notonectidae
Backswimmers

Small to medium-sized (body length 4–12mm), voracious predators. Unlike corixids, they orientate with the convex, keel-shaped underside facing upwards. They swim using the fringed hind legs. Underside of body has a groove, replenished at frequent intervals with air, thus functioning as a lung. Often found in groups, in shaded positions in slow-flowing or stagnant water, facing at a gentle downwards angle. They dive if disturbed, but can also leap out of the water and fly away. Tackle a wide range of prey with their raptorial fore legs, including other aquatic insects, tadpoles and even small fish. Can pinch with the fore legs if handled.

2 *Anisops pellucens*
Common Backswimmer

Medium-sized (body length 11mm), parallel-sided and slender, with triangular orange scutellum, and milky blue opalescent wings. Underside **(2A)** dark, with 2 rows of white spots. **Biology** Aggregate in shaded areas of waterbodies, and often colonize swimming pools. Males able to stridulate underwater to attract females. Fly at night, and are attracted to lights. Feed on a variety of small prey, including mosquito larvae. **Habitat** Pans, pools, vleis and dams. **Related species** There are 21 other regional species in this genus.

3 *Enithares*
Stout backswimmers

Medium-sized (body length 12mm), robust, with dark red eyes, a white thorax, and grey wings. Body broadest at the head, tapering towards the end. Underside **(3A)** uniformly dark brown. There are 4 species in the region. **Biology** Formidable predators of large aquatic insects and tadpoles, often in very shallow water. **Habitat** Slow-flowing water and pools in streams and dry riverbeds. Also occur in other parts of Africa, Asia, Australia and South America.

4 *Notonecta lactitans*
Milky Backswimmer

Medium-sized (body length 11mm), fairly robust, with black wings marked by a pair of cream stripes. **Biology** Probably migrates to vleis on the Cape Flats to breed in winter, returning to mountain streams in summer. The only species in the genus in sub-Saharan Africa. **Habitat** Vleis and pools. This essentially temperate genus occurs mainly in Europe and North America.

Family Pleidae
Pygmy backswimmers

Minute aquatic bugs, which, like notonectids, swim on their keel-shaped backs. The legs show little modification for swimming. Under magnification, pit-like depressions are visible all over the body. Only 1 genus in the region.

5 *Plea*
Pygmy backswimmers

Very small (body length 2mm), broad and short. Only 2 species (*P. pullula* and *P. piccanina*) found in the region. **Biology** Predaceous, feeding on minute organisms and mosquito larvae, and small crustaceans such as water fleas. Females lay eggs by piercing plant tissue. Can fly to new waterbodies. **Habitat** Widespread, clinging to vegetation in still waters.

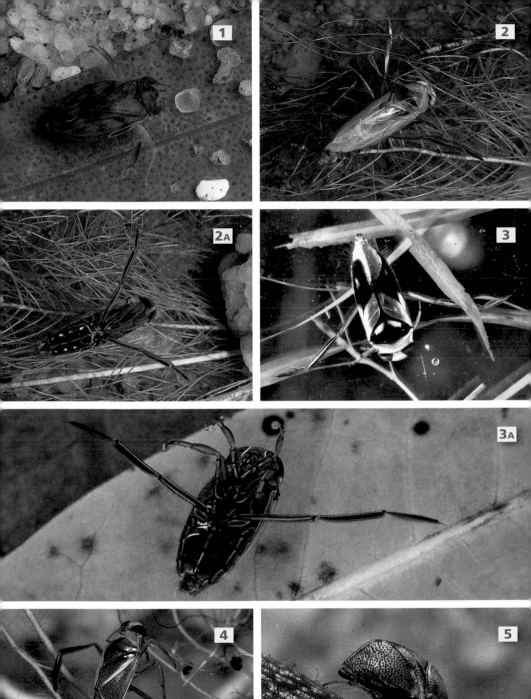

Family Naucoridae

Saucer bugs

Medium-sized (body length 9–14mm), oval, fairly distinctive. Not as flattened as belostomatids (with which they can be confused), but also with very well-developed raptorial fore legs, and a short, stout beak. The other legs are not flattened and fringed with hairs, being better adapted for crawling through vegetation than for active swimming. Winged and wingless forms occur. Five genera and 13 species in the region.

1 *Laccoris*

Flat saucer bugs

Medium-sized (body length 11mm), roundly oval, brownish bugs with yellow margins to the body, and raptorial fore limbs. Body green when viewed from the underside. There are 5 regional species. **Biology** They forage under rocks in mountain streams for the aquatic larvae on which they feed. Can inflict a burning bite if handled. **Habitat** In standing water among water weeds and (unlike belostomatids) under rocks in fast-flowing mountain streams. Fairly widespread.

2 *Macrocoris*

Creeping water bugs

Small (body length 10mm) convex body; green with brown fore wings. Head small, and femora of fore legs enlarged and powerful. There are 3 regional species. **Biology** Feed on a range of aquatic insects, including mosquito larvae. **Habitat** On vegetation in a range of slow-moving or stagnant subtropical waterbodies.

Family Belostomatidae

Giant water bugs

Includes giants of the insect world, some species reaching 80mm in length, although most are medium-sized and some are as short as 10mm. Fierce predators, with raptorial fore legs strengthened to capture and subdue prey, including insects and their larvae, tadpoles, fish, molluscs, nymphs of their own kind and anything else that can be overpowered. The other legs are flattened and fringed with hairs for swimming. All species are light brown or greenish, flattened, with a conical head projecting in front of the large (often red) eyes. There are 2 retractable breathing tubes at the end of the abdomen, which lead to an air-storage chamber beneath the wings. Often seen resting, rear end up, at the water's surface to replenish this 'lung'. Adults fly actively at night, and are attracted to lights. There are 4 genera and 13 species in the region.

3 *Limnogeton fieberi*

Fieber's Giant Water Bug

Very large (body length 52mm), broadly oval, dark brown, with narrow, tapering head and large raptorial fore legs. Eyes triangular. One of 2 huge belostomatids in the region and the sole member of the genus. **Biology** An obligate predator of aquatic snails and therefore useful in reducing populations of bilharzia snails. Flies strongly, and is attracted to lights on warm, humid evenings. Approximately 200 eggs are laid on the backs of males and take 2 months to hatch. **Habitat** Subtropical vleis and pans with dense plant growth.

4 *Lethocerus niloticus*

Wrestling Water Bug

Huge (body length 50–70mm), parallel-sided, with round protruding eyes and stouter raptorial fore legs. Both mid and hind legs bordered with swimming hairs. **Biology** Egg masses are laid at night on plant stems above water level; males guard and wet the eggs until hatching. Adults feed on frogs, tadpoles and fish. May fly in swarms on hot nights after the first rains. **Habitat** Heavily vegetated standing pools and dams; occasionally found in swimming pools.

5 *Appasus*

Brooding water bugs

Medium-sized (body length 12mm), oval, light brown, with faint tan stripes on the head, and raptorial fore legs. There are 5 regional species. **Biology** A predator of other water insects and tadpoles. Females glue eggs to the hemelytra of males (5), which carry the eggs until they hatch. **Habitat** Well-vegetated vleis and ponds.

6 *Hydrocyrius columbiae*

Great Water Bug

Huge (body length 50–70mm), with triangular eyes on a projecting head and flattened fore legs that can fold into grooves on the fore femora. Head with 2 distinct pale stripes; similarly coloured stripes along wing margins. **Biology** The adults feed on frogs, tadpoles and fish. Eggs laid in a mass on reed stems out of water. **Habitat** Well-vegetated standing waterbodies.

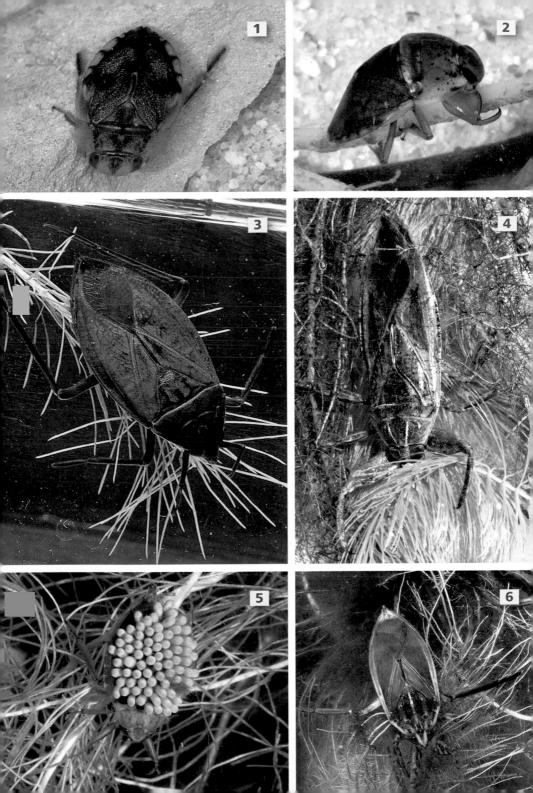

Family Nepidae
Water scorpions, water stick insects

Large to very large (body length 20–45mm, excluding breathing siphon), unmistakable, elongate and oval, with 2 very long thin breathing tubes. Robust and enlarged raptorial fore legs add to scorpion-like appearance and are used in prey capture and feeding. Ambush predators, hanging onto aquatic vegetation just below the surface of the water, breathing through the siphon and waiting for prey. Feed on a range of large aquatic insects and tadpoles. Weak swimmers, legs not adapted for swimming. Eggs are laid above the waterline in mud or on plants, and have long breathing processes. Widespread and common in dams, vleis and pans. There are 3 genera and 13 species in the region.

1 *Ranatra*
Water stick insects

Medium-sized to large (body length 27–42mm, excluding 50mm-long breathing siphon). Body and limbs very elongate, including weakly raptorial fore legs. Winged. Beak fairly small. Capable of producing a squeaky sound. There are 7 similar-looking regional species in the genus. **Biology** Sluggish, spending most of their time clinging to aquatic vegetation near the water surface. **Habitat** Common in standing waterbodies with submerged aquatic plants.

2 *Laccotrephes*
Common water scorpions

Large (body length 41mm, excluding 60mm-long breathing siphon), dark brown and parallel-sided, with barred, raptorial fore legs. Top of abdomen red, eyes round and protruding. There are 5 similar-looking species in the genus. **Biology** Cling to underwater vegetation, usually just below the surface, with siphon above water, awaiting large prey such as tadpoles. Although capable of flight they rarely fly. **Habitat** Most bodies of standing water, especially those with submerged aquatic plants.

3 *Borborophilus afzelii*
Dwarf Water Scorpion

Small (body length 15mm, excluding 7mm-long breathing tubes), dark brown, with short breathing tubes and rectangular body. Raptorial fore limbs with few or no teeth. **Biology** A rare, sluggish predator of aquatic insects. **Habitat** Subtropical standing waterbodies, including muddy dams.

SUBORDER AUCHENORRHYNCHA

The families within Auchenorrhyncha comprise highly varied terrestrial, small to large, plant-feeding bugs that differ from Heteroptera chiefly in their wing design and (generally) much smaller prothorax. Their fore wings are uniform in texture and held roof-like over the body (unlike the flat hemelytra of Heteroptera). They have short, bristle-like antennae, 3 tarsal segments and well-defined ovipostors. They jump and fly well. Many are agricultural and horticultural pests. Apart from damaging plants by direct feeding, some are important vectors of viruses that cause various plant diseases. Many also produce honeydew (surplus sugars from their liquid diet) from the anus.

Family Tettigometridae
Tettigometrid planthoppers

Small and stout, magnification of antenna revealing a ringed, whip-like antennal tip arising from a large, swollen basal segment. There are only 3 species in the genera *Tettigometra* and *Hilda* in the region.

4 *Hilda patruelis*
Rooibos Bug

Small (wingspan 10mm), squat, with a very broad head, brown or olive green prothorax, and olive or greenish-brown wings with 3 dark patches highlighted by a short silvery stripe (unpatterned individuals also occur). Nymphs have a thin spine projecting from the head. **Biology** Feeds on a range of indigenous trees and shrubs, including wild figs, Rooibos Tea Bush and proteas; also hibiscus, peanuts, legumes, sunflowers and maize. Recorded on 41 different hosts in Zimbabwe. Nymphs and adults occur together, invariably attended by ants, whose subterranean tunnels enable the bugs to live on plant roots. **Habitat** Protected parts of plant buds and shoots, often wild figs.

Family Cixiidae
Primitive snout bugs

Very primitive bugs, with 3 simple eyes and a V-shaped pronotal collar. Females have a fairly long ovipositor, superficially resembling that of Orthoptera. The nymphs are subterranean, probably feeding on plant roots. Adults feed on a range of woody plants. Approximately 24 regional species.

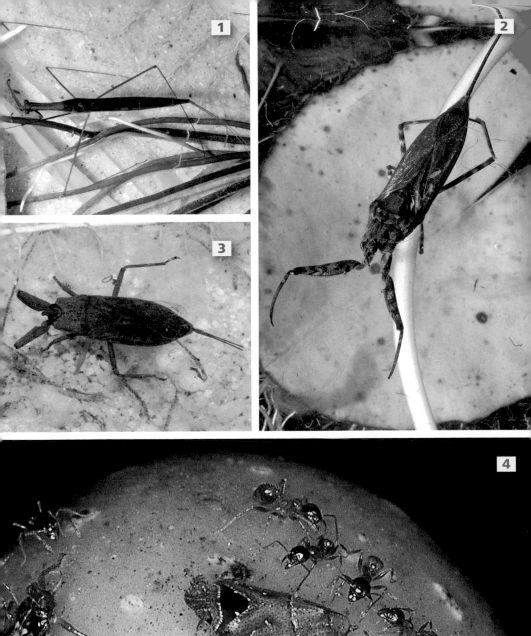

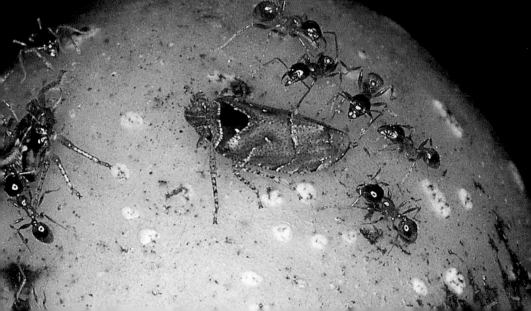

1 *Taomma rasnitsyni*   Four-spotted Planthopper

Small (wingspan 8mm). Wings large with translucent brown stripes; wing tip black with 4 white dots. **Biology** Feeds in groups on herbaceous Asteraceae. **Habitat** Shrubby vegetation at forest edges.

Family Meenoplidae

A small family whose members superficially resemble the Cixiidae and are restricted to the tropics and subtropics. Many species live underground in caves, feeding on plant roots. Veins on the fore wings appear granular in many species, due to the presence of sensory pits. Females have wax-producing plates on some of the abdominal segments.

2 *Anigrus sordidus*

Small (wingspan 14mm), uniformly whitish-grey, with elongated wings covered in a fine dusty powder. The only species in its genus. **Biology** Nymphs may feed on grass roots, adults on a range of woody vegetation. Apparently feeds on vines and herbaceous vegetation. **Habitat** Bushveld and subtropical forest margins.

Family Derbidae   Long-winged snout bugs

Easily identified by the very elongate fore wings and rudimentary hind wings. Appear to be closely associated with palms, where they occur in groups on the underside of the leaves. As in many snout bugs, the wings and body are covered in a fine layer of wax. Capable of jumping. Fourteen species of this largely tropical family are known from the region.

3 *Diostrombus abdominalis*   Lala Palm Snout Bug

Small (wingspan 16mm) and short. Wings very elongate and clear, with dark veins and whitish tips. Thorax brown, abdomen white with black dorsal patch. The only species in the genus. **Biology** Adults found only in groups, all facing in the same direction, on the underside of palm leaves, especially the Lala Palm (*Hyphaene natalensis*). Nymphs are fungal-feeders and live under bark and in rotting logs. **Habitat** Fairly common on palms in open grassland or sand forest in subtropical areas.

4 *Helcita wahlbergi*

Small (wingspan 24mm), covered in white wax. Wings whitish, very elongate and iridescent, with regular brown mottling. **Biology** Like others in the family, strictly associated with palms, resting with wings held upright and apart. Some host-specificity apparent and recorded from the African Oil Palm (*Elaeis guineensis*). **Habitat** Palms in subtropical forest and riverine vegetation.

Family Delphacidae   Delphacid planthoppers

Very small and inconspicuous. Large mobile spur at the end of the hind tibiae needs a little magnification to be seen clearly. The largest family (55 species) of planthoppers in the region.

5 *Perkinsiella saccharicida*   Sugarcane Planthopper

Small (wingspan 14mm), with a tan stripe on the head and along the back, smoky black wings with darker tips, and stout, elbowed antennae. **Biology** Feeds on grasses. Transmits viral diseases to various cereal crops, including maize. *P. saccharicida* is a minor pest of sugar cane in the region, but a serious pest in other parts of the world, as it transmits Fiji disease to cane. **Habitat** Subtropical grassland.

Family Fulgoridae   Lantern bugs, snout bugs

Among the most spectacular of all bugs. Small to large, with bright coloration and an often bizarre extension of the head into a snout. Can be distinguished from related families by the combination of a very swollen first antennal segment, bearing a small bristle, and a network of cross-veins in the fan-like basal (anal) area and apex of the hind wings. The nymphs, like those of other planthoppers, are covered in white wax. Adults often have a white waxy layer on the abdomen. About 22 species of this chiefly tropical family are known from the region.

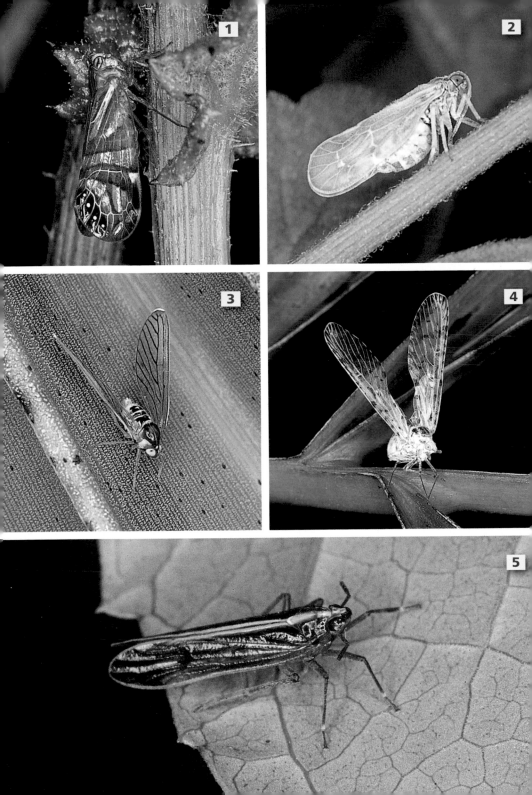

1 *Zanna*
Twig snout bugs

Medium-sized (body length of species shown here 31mm, but some are twice as large). Body and wings pinkish-grey peppered with black. A row of orange spots on the side of the snout. Genus contains about 8 similar-looking species. **Biology** Recorded on *Terminalia sericea*, and are attracted to lights. Jump when alarmed. Molecular studies suggest that *Zanna* may in fact be better placed in the family Dictyopharidae, but this change has not yet been formalized. **Habitat** Bushveld and the margins of subtropical and dune forest.

2 *Eddara euchroma*
Painted Snout Bug

Medium-sized (wingspan 36mm), stout, brilliantly coloured and unmistakable, with a green head, red abdomen, and black wings with green-ringed yellow spots. **Biology** Probably feeds on grasses or herbs. Can be locally fairly abundant. **Habitat** Grasses and herbaceous plants in the shade in subtropical bush.

3 *Druentia variegata*
Red-winged Lantern Bug

Large (wingspan 44mm) and squat, with a short upturned snout. At rest, mottled brown coloration provides excellent camouflage on tree bark. Faint mottling in brown, red, yellow and orange. Head very broad when viewed from the front. Hind wings boldly coloured in scarlet and black, with black and white dots. **Biology** Gregarious, occurring in small groups on tree bark. **Habitat** Usually seen at lights, but spends most of its time on tree bark.

Family Dictyopharidae
Dictyopharid planthoppers

A large family of small to medium-sized planthoppers, with an elongated head that forms a snout. Wings narrow, often clear, but without the network of tiny cross-veins in the anal area seen in lantern bugs (Fulgoridae). Wingless species are common. There are 41 species known from the region.

4 *Capenopsis*

Small (body length 8mm), rounded, with a large, curved, rhino-horn-like snout. Fore wings reduced, with a few raised parallel veins, ending in a straight margin (not rounded as in *Capena*). Abdomen pinkish-brown. Pronotum with 2 parallel ridges on either side of the midline. **Biology** Usually encountered on the ground. **Habitat** Fairly common in low vegetation (often weedy annuals) in arid regions. **Related species** *Engela minuta*, which is a small species with an inflated green snout; occurs in the arid parts of the Western Cape.

5 *Dictyophara*
Lantern flies

Medium-sized (wingspan 30mm), bright green, with a short, pointed snout and clear wings with green veins. Long and slender legs. Upright and alert posture. **Biology** Foodplants not known, but probably feed on herbs or grasses. Jump well. **Habitat** Fairly common in grasses under trees and herbaceous vegetation in subtropical bushveld. European species of *Dictyophara* have a broad range of foodplants. **Related species** *Aselgeia ramulifera* (**5A**) (wingspan 24mm), which is a duller green with broader green wings folded roof-like over the back. Wings with many finer wing veins. Occurs in more tropical areas.

6 *Putala rochetii*
Twig-snouted Bug

Small (wingspan 16mm), light brown, with a slender upturned snout and clear wings with a dark brown pattern towards the tips. Limbs long and slender. Tibia spined on the hind legs. **Biology** Appears to feed on herbaceous plants. **Habitat** Common in low vegetation at the edges of subtropical forest, and in vegetation under trees in bushveld.

7 *Raphiophora vitrea*

Medium-sized (wingspan 22mm), with an unusual appearance. Fore legs greatly elongated and flattened, thorax with pale green markings, and wings clear with brown-and-black marbling at the tips. **Biology** Occurs in numbers on *Strychnos* (small trees containing the poison strychnine), on which it appears to feed. **Habitat** Bushveld and subtropical forest.

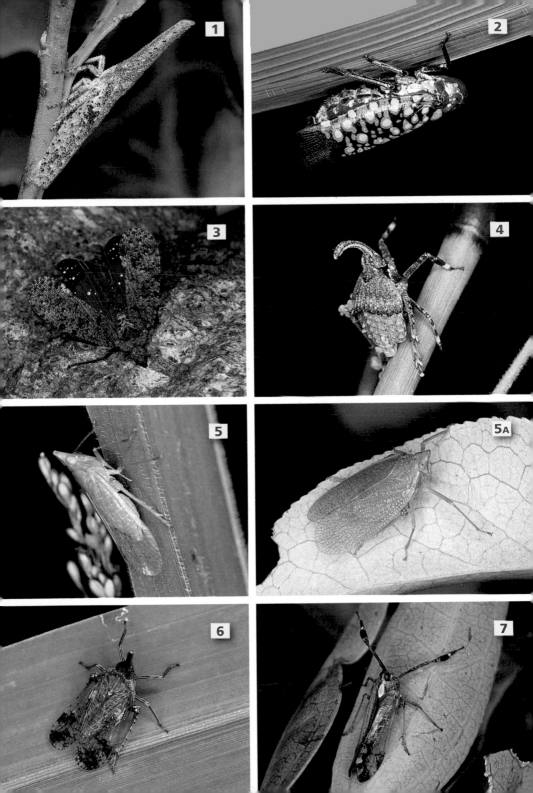

Family Tropiduchidae
Tropiduchid planthoppers

Variable in appearance, with the corners of the middle part of the thorax (the mesonotum) cut off by a fine line. At rest the wings, when present, are held roof-like over body. Some are small, dumpy planthoppers, with short hardened fore wings and no hind wings. Others have scale- or strap-like fore wings, with a roughened texture, or rounded and convex fore wings. Usually brown, although bright green species occur. There are 36 known species in the region.

1 Numicia insignis

Small (wingspan 12mm), with bright red eyes and broad greenish wings marked with 3 dark brown chevrons. Nymphs pale green, with a plume of wax filaments at the end of the abdomen. **Biology** Grass-feeders, recorded from 34 different grass species, including Kikuyu Grass. A pest of sugar cane where the cane is heavily irrigated and natural grasses have dried out. Causes yellowing of the leaves and flaccid cane stalks. Also occurs on sedges. **Habitat** Tropical parts, generally in grasses under trees.

2 Gamergomorphus

Small (body length 4mm), squat, brown, with elongate, strap-like fore wings and no hind wings. Several wingless species occur. A pale stripe runs along the head and thorax. **Biology** Often flightless. Jump well. **Habitat** Low, scrubby fynbos vegetation. Several wingless species occur in the Western Cape. **Related species** *Johannesburgia rossi* (**2A**) (body length 2–5mm), which is similar, but bright green, and occurs in Gauteng, Mpumalanga, the Western Cape and Limpopo. Also, members of the small tropical genus *Monteira* (**2B**) (7mm long), which have heavily sculptured fore wings and occur on grass under trees in Limpopo province; and *Afroelfus* species (**2C**) from Namaqualand, which have very long, narrow fore wings, and sensory pits on the underside of the abdomen.

Family Eurybrachidae
Broad-frons planthoppers

Squat leafhoppers with a broad, blunt head (3 times broader than its length) and no spines at the end of the second tarsal segment of the hind legs. Fore wings usually in mottled greens and browns, providing excellent camouflage against tree bark. At rest the wings conceal the brightly coloured abdomen. Adults feed on the sap of trees or shrubs, and lay their eggs on a circular waxy disc, which is then covered by another protective waxy layer.

3 Paropioxys jucundus
Mottled Avocado Bug

Medium-sized (wingspan 26mm), probable lichen mimic, with a very broad head. Wings emerald green with concentrations of white spots and blotches, overlaid with black spots. Underside of abdomen bright red. Nymphs (**3A**) bizarre in appearance, with remarkable extensions at the end of the abdomen, which apparently mimic antennae; they also have eyespots at the end of the abdomen and walk backwards, completing the false head mimicry. Presumably predators attack the false head, allowing the bug to escape by leaping away. **Biology** Feeds on tree sap, including that of avocados and macadamias. **Habitat** Adults occur on tree trunks in bushveld and subtropical forest, as far south as Pretoria. **Related species** There are 4 other similar-looking regional species in the genus.

4 Family Flatidae
Moth bugs

Easily identified by the very broad, generally green wings and large, fan-like, triangular hind wings, which are often covered in a dusty layer of white wax. The numerous cross-veins tend to line up in rows. At rest the hind wings are folded roof-like against the body. Nymphs (**4**) are equally diagnostic, the body covered not only in a fine layer of white wax, but with long curly tendrils of wax extending from it. Adults and nymphs are frequently gregarious. Nymphs are commonly observed in groups on the runners of milkweed vines (Asclepiadaceae) in forests. At least 21 species are known from the region.

5 Cryptoflata unipunctata
One-spot Moth Bug

Medium-sized (wingspan 26mm), pale green, with red eyes, and a distinct black dot on inner margin of fore wings. Hind wings end in a square that has sharply pointed edges. **Biology** Adults jump well, but are weak fliers, occurring singly on shrubs. Also attracted to lights. **Habitat** Bushveld and subtropical forest.

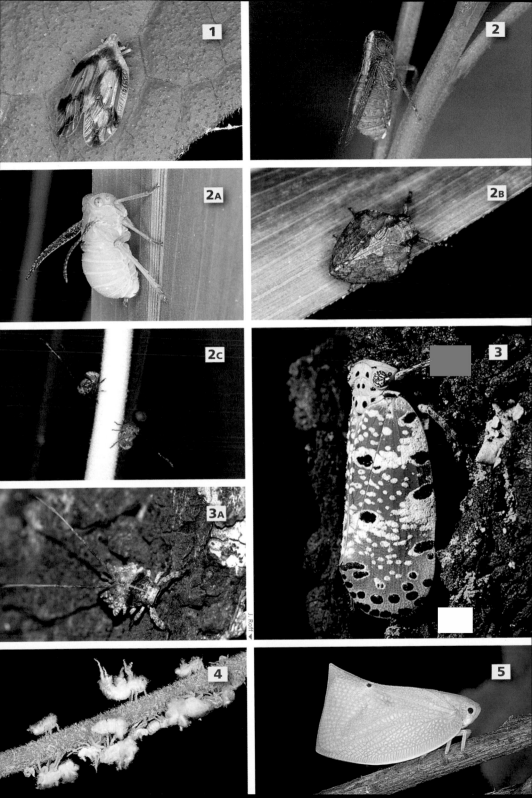

Family Acanaloniidae

Resemble Flatidae, in that they have broad green wings that are held roof-like over the body. However, there are fewer cross-veins in the fore wings, and they do not line up in regular rows (as in Flatidae). Feed largely on woody plants and shrubs. Only 3 species known from the region.

1 *Parathiscia*

Medium-sized to small (wingspan 20mm), yellow or lime green, with a sharp snout, and broad red-bordered wings that turn upwards at the tips when at rest. Two species in the genus. **Biology** Adults usually occur singly on shrubs and are attracted to lights. **Habitat** Bushveld; also often in gardens. Recorded from avocado and macadamia trees.

Family Ricaniidae
Ricaniid planthoppers

Small planthoppers, often with attractively patterned wings. Many species have striking black bands on the wings. They resemble small flatids, but are a little squatter, with shorter, triangular wings. Like flatids, they are capable of jumping. The second tarsal segment lacks spines. Ten species are known from the region.

2 *Lugardia mimica*

Small (wingspan 13mm), with very broad black, grey and bronze wings with a clear central window. Wings held horizontally. **Biology** Occurs on Wild Elder (*Nuxia*). **Habitat** Subtropical bushveld and forest margins.

3 *Mulvia albizona*

Small (wingspan 12mm), striking, with very broad, rounded black wings, each with 2 white transverse bands. Rests with wings spread horizontally. **Biology** Feeds on grasses, including *Sporobolus*; jumps well, but flies weakly. Always associated, in numbers, with grasses growing in the shade of trees. **Habitat** Grassland, bushveld and subtropical forest.

4 Family Cercopidae
Spittlebugs, froghoppers

Named for the nymphs' unusual habit of living in a ball of spittle **(4)**, produced by mixing air with copious anal secretions derived from plant sap. The spittle probably protects nymphs from predators, parasitoids and drying out. Some nymphs live underground on the roots of plants, others on grasses, and those of larger species on shrubs and trees. Adults medium-sized and stoutly built, with a cone-shaped head, almost round eyes, and often brightly coloured fore wings. Capable of jumping, but unlike planthoppers do not secrete wax. The eyes are slightly longer than their width, and posterior margin of the pronotum is straight. Thirty-three species are known from the region, most belonging to the genus *Locris*.

5 *Rhinaulax analis*
Cape Spittlebug

Small (wingspan 18mm), present in 2 colour forms: 1 yellow and red, the other plain yellow; both with a black head and black markings at the wing tips. **Biology** Found on small shrubs. **Habitat** Common in short fynbos vegetation.

6 *Locris arithmetica*
Red-spotted Spittlebug

Medium-sized (wingspan 23mm), familiar, bright scarlet, with characteristic black spots. **Biology** May be present in numbers in managed Kikuyu Grass pasture, which it can damage. Also numerous on sugar cane, along with the similar *L. areata*, but (unlike spittlebugs in the Americas) does not damage the cane. Typically encountered in small groups, the nymphs in association with their 'cuckoo spit' froth. **Habitat** Stands of indigenous and introduced lush grasses. Common in urban areas. **Related species** *L. areata* **(6A)**, which has an orange head and thorax, with sparse black markings, and red wings. It occurs in the Eastern Cape and KwaZulu-Natal.

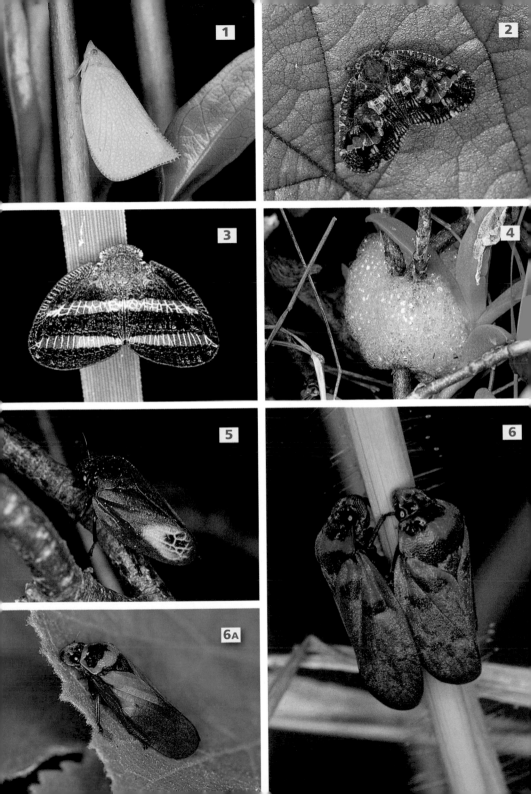

Family Aphrophoridae

Aphrophoridae nymphs, like those of the closely related Cercopidae, have mucus-producing cells that allow them to produce a mass of spittle from the anus. Nymphs live gregariously within this spittle. Eyes not longer than their width, and the posterior margin of the pronotum has a distinct indent.

1 *Ptyelus grossus*   Rain-tree Bug, Tipuana Spittlebug

Large (wingspan 30–35mm), with slate grey wings marked with 2 large cream-and-orange spots and smaller cream dots. Nymphs (1A) black with fine cream hieroglyphic-type markings and an orange central stripe. **Biology** Well known for the phenomenon of 'rain-trees', produced by the constant dripping of processed tree sap through the bodies of clustered nymphs and adults (1B). Rain trees (*Peltophorum*) and *Lonchocarpus capassa* are favoured hosts, but *Tipuana*, *Grevillea*, *Senegalia*, *Searsia*, *Strychnos* and even *Eucalyptus* are also attacked. Acrid-tasting exudate forms pools under the host tree. **Habitat** Subtropical bushveld, forest and gardens, up to equatorial Africa. **Related species** *Ptyelus* includes 2 other regional species.

2 *Poophilus*

Medium-sized (wingspan 22mm), oval, robustly built, body tapering at both ends. Cryptically coloured in brown or tan. Genus contains a number of similar-looking species. **Biology** They probably feed on larger grasses, such as Common Thatching Grass (*Hyparrhenia hirta*). **Habitat** Most species are widespread, common and well-camouflaged inhabitants of grassland. **Related species** *Cordia* species (2A), which are very similar, but smaller (wingspan 14mm) and less robust, and which are not broadest at mid body. They are light tan, with a cream stripe along the sides of the wings, and are common in grassland in the Cape, Mpumalanga, Free State and KwaZulu-Natal.

3 Family Clastopteridae

Small (body length 3mm), rotund spittlebugs, producing small isolated foam pools comprising finer bubbles. They lack the lobes at the end of the abdomen in Cercopidae that enable them to trap air effectively within the mucus. The large prothorax ends in a distinctive spine. Poorly represented in the region, with approximately 6 species in 3 genera. The Sunflower Spittlebug (*Clastoptera xanthocephala*) is a minor pest of sunflowers in North America. Clastopteridae occur in a range of vegetation types, from succulent karoo to subtropical grasslands.

Family Lophopidae   Lophopid planthoppers

Medium-sized, easily recognized by their flattened appearance, with patterned wings held flat over the body and not in roof-like fashion. Fore legs flattened and enlarged. Head longer than wide, bearing a ridge at each side. Nymphs have 2 very long antennae-like tails, which are held erect.

4 *Elasmoscelis*

Small (wingspan 16mm), comprising similar-looking, flattened species with wings marbled in brown, black and white, and enlarged flattened fore legs. Seven species are known from the region. **Biology** Probably grass-feeders, present in large numbers in herbage and grasses in the shade of shrubs and trees. Pests of rice in tropical Africa. **Habitat** Moist summer-rainfall regions, common in bushveld and subtropical forest.

5 Family Cicadidae   Cicadas, Christmas beetles

Members of this large well-known family produce a shrill buzzing that is a typical sound of summer. Males have a pair of circular sound-producing organs (tymbals) that appear as 2 round membranes on either side of the abdomen, each reinforced by a strong ring. A muscle attached to the centre contracts, and the recoil produces a click; rapid contraction of these muscles produces a continuous shrill noise. Both sexes have ears (larger in males) on the underside of the abdomen. Males have at least 1 specific call to which females are attracted. One calling male stimulates others to join in, forming a chorus. Cicadas have 2 pairs of transparent wings, often attractively marked in dull colours. Exotic species lay eggs in slits made in twigs, and the newly hatched nymphs dig into the ground using their very enlarged fore legs (5). Here they feed on roots, and it takes many years (sometimes over a decade) before they emerge, climb up a tree, and moult into adults, leaving the dry nymphal skin attached to the tree bark. About 140 species occur in the region, but the biology of most is poorly known.

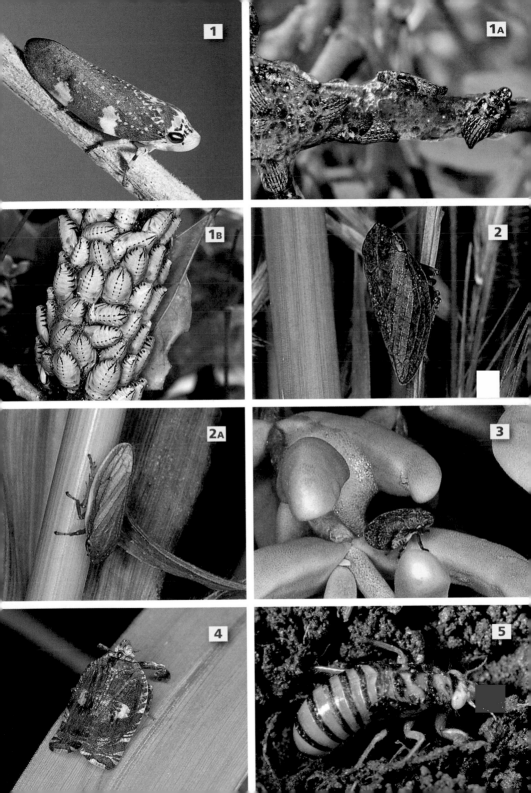

1 *Neomuda* — Cape emerald cicadas

Medium-sized (wingspan 40–50mm), with emerald green thorax and head, boldly marked in black, and long white hairs covering the entire body. Abdomen plain brown dorsally and green ventrally with yellow bands. Wings clear brown, with darker brown markings on the veins. *Neomuda* comprises 3 similar-looking species. **Biology** Adults emerge in spring and summer, when their intermittent clicking call (made in flight) is a familiar sound. Nymphal skins occur on Assegai (*Curtisia dentata*) in forests. **Habitat** Riverine forest and oak-planted suburbs in the Western Cape. Endemic to Afromontane forest and fynbos.

2 *Oxypleura lenihani* — Axe-head Cicada

Large (wingspan 90mm), olive green, with black markings surrounded by brown on the head and thorax. Body covered in short silvery hairs. Sides of prothorax expanded into triangular flanges. Wings clear, with green veins on the first third, otherwise black. **Biology** Associated with wild figs. Gives a sustained high-pitched call, alternating between a smooth note and a trill. **Habitat** Common in coastal forest and subtropical bushveld.

3 *Pycna semiclara* — Giant Forest Cicada

Large (wingspan 90mm), head and thorax green, with black markings surrounded by a yellowish-brown area. Wings green at base with short silvery hairs, otherwise brown, especially around the veins. **Biology** Adults emerge October–February, feeding on a range of indigenous and exotic forest trees. Produces a very loud continuous call, almost pure in tone, from a shaded position on a tree trunk. Also has a trill-like territorial call. **Habitat** Common in subtropical riverine forest and exotic plantations.

4 *Quintilia* — Karoo cicadas

Medium-sized (wingspan 50mm); body dark (often black) with brown markings. Wings clear with black veins, barred with yellow in some species. Genus contains over 20 similar species. **Biology** Some species emerge en masse in the Karoo in mid-summer, possibly in response to rainfall. Males produce croaking or bleating calls from the ground or low shrubs. Eggs are laid on a variety of shrubs, including Kraalbos (*Galenia*) and the mesem *Eberlanzia*, on which nymphs probably feed. Adults live for about 2 weeks. **Habitat** Fairly common in arid succulent karoo and bushveld. **Related species** *Trismarcha sirius* (**4A**), which is small (body length 26mm), uniformly brown, with a dark reddish-brown thorax and clear wings. Occurs in KwaZulu-Natal, Mpumalanga and Limpopo.

5 *Platypleura capensis* — Cape Orange-wing Cicada

Large (wingspan 76mm), mostly grey, with grey prothorax, brown-and-green mesothorax, and black abdomen. Grey fore wings with green veins, hind wings pale orange with a weak black border. **Biology** Emerges in November and again in January. Call is a single, continuous high-pitched note, almost smooth, but changing to a trill. Like others in the genus, adults appear to have a few host plants, including *Maytenus* and Coastal Milkwood (*Mimusops caffra*). **Habitat** Common in mountain and coastal fynbos in the southwestern Cape and, in the east, on daisy bushes.

6 *Platypleura haglundi* — Orange-wing Cicada

Medium-sized (wingspan 60mm), with a grey prothorax, brown-and-green mesothorax, and black abdomen with incomplete white terminal band. Fore wings cryptically mottled, with green veins. Hind wings rich ochre with a complete black border. **Biology** Associated with large shrubs and trees (*Senegalia*, *Dichrostachys cinerea* and *Delonix regia*). Able to raise its body temperature allowing it to call at cooler dusk temperatures. **Habitat** Common in a variety of habitats, including agricultural land and gardens. **Related species** There are at least 18 similar related regional species in the genus. Also, the similar *Systophlochius palochius* (**6A**) (wingspan 62mm), which has a black stripe down the middle of the thorax and head, and a complete white band at the end of the abdomen. Call is continuous. Occurs in Limpopo, where it feeds on *Cassia* and Knobthorn (*Senegalia nigrescens*).

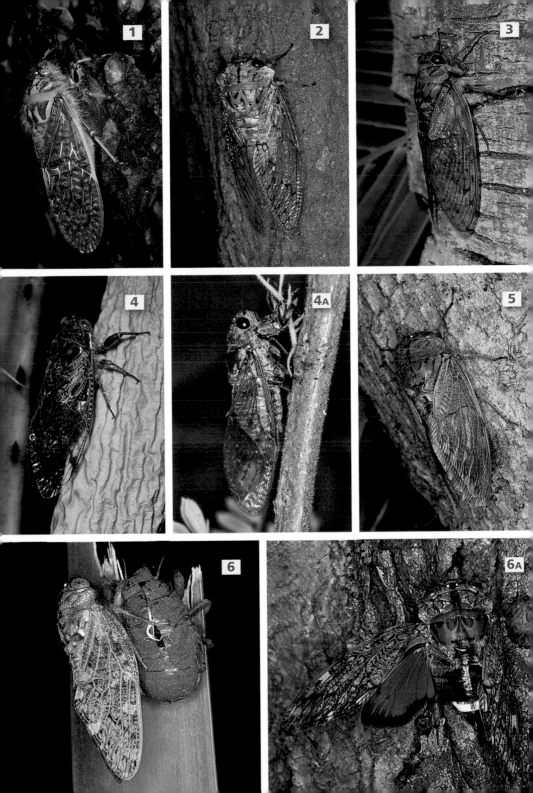

1 *Stagira* — Green-wing cicadas

Small (wingspan 32mm), uniformly green, with clear wings (some with red wing veins), and red or orange eyes. In some species females are wingless. *Stagira* contains 39 similar-looking species. **Biology** Adults emerge October–February. Most produce clicking calls as well as a low 'churr' from short vegetation. **Habitat** Common in summer in grassland, open forest and bushveld. All species are endemic to the region.

2 *Xosopsaltria punctata* — Bladder Cicada

Medium-sized (wingspan 52mm), unmistakable, with an inflated abdomen. Male **(2)** has a black-and-yellow-barred abdomen and clear wings with black shading over the veins. Female **(2A)** uniformly green. **Biology** Emergence time unpredictable. Produces a repetitive grating call, somewhat like that of a frog. May be common in summer. **Habitat** Valley bushveld. All 4 species in the genus are Eastern Cape endemics.

Family Membracidae — Treehoppers

Small, but easily recognized by enlargement of the pronotum into a pair of dark thorn-like structures (occasionally reduced) that extend right back over the abdomen. Bizarre blackish nymphs have an elongate abdomen tapering into a whip-like structure. Nymphs are gregarious, often found with adults, and usually attended by ants. Some species are of minor agricultural importance, but most are uncommon and found in low densities. They occur on a wide range of leguminous trees, shrubs and creepers. About 120 species are known from the region.

3 *Centrotus*

Small (wingspan 16mm), stocky, with 2 stout lateral horns and a curved backward-projecting process. Wings clear brown, thorax white. Sexes differ in appearance. Nymphs **(3A)** are tan, without lateral fringes, and have a single flat dorsal crest on the thorax. **Biology** Feed on *Terminalia sericea*. **Habitat** Open bushveld. **Related species** *Leptocentrus ugandensis* (8mm long) **(3B)**, which is distinctive, with black wings and body, white thorax, and a red dot on the leading edge of each fore wing. Occurs in the warmer parts of South Africa.

4 *Oxyrhachis*

Small (wingspan 18mm), with 2 forward-projecting horns and a long curved backward-projecting process. Wings clear, thorax brown. *Oxyrhachis* comprises 80 African species. **Biology** Associated with indigenous *Solanum*, *Senegalia*, *Indigophora*, *Psoralea*. *O. biformis* has been collected from Umzibeet (*Milletia caffra*). **Habitat** Bushveld to fynbos.

5 *Anchon*

Small (wingspan 12mm), dorsal horns are very elongate, with a thin white stripe at the base, and end in a blunt club and a backward-projecting process with an initial 90° bend. Wings rich brown, thorax with a white stripe. Nymphs pale green, covered in fine hairs, and with a large, thin vertical crest. **Biology** Recorded on Australian Black Wattle (*Acacia mearnsii*). **Habitat** Bushveld and subtropical scrub. *A. ximenes* is found on Wild Plum (*Ximenia caffra*) and another species of *Anchon* on *Rubus ludwigii*. **Related species** *Tricoceps* species are small (5mm long) and brown but heavily dusted with white powder (including the wing veins). Also, *Xiphopoeus horridulus* (6mm long) **(5A)**, which is brown, with fine white markings, and has a rough surface on the prothoracic process. Occurs in the warmer parts of the region.

Family Cicadellidae — Leafhoppers

A highly diverse and important family of small, often attractively coloured (many bright green) leafhoppers. May resemble small, slim froghoppers (Cercopidae), but have very enlarged leg bases and at least 1 row of small spines along the thin part of the hind tibiae. Excellent jumpers. Highly host-specific, feeding on trees, shrubs and grasses; particularly abundant on grasses. Western Cape fauna is rich and distinctive. A characteristic white patch appears where they destroy the leaf chlorophyll after feeding: in high numbers, they may cause the entire leaf to dry out. Many are pests of economic significance, since they transmit viral (e.g. maize streak virus), fungal and bacterial diseases. About 350 species are known from the region.

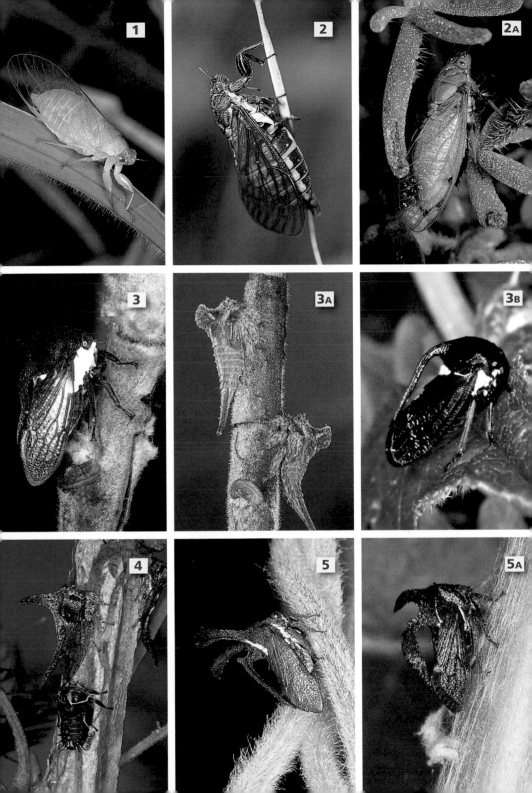

1 *Acia lineatifrons*
Grapevine Leafhopper

Small (wingspan 3mm), dull green, with a cluster of dark grey cells at base of wing; other wing cells faint brown. Nymphs yellow. **Biology** An indigenous African species with a wide host range, including a species of wild blackberry and, in the Western Cape, commercial grapevines. Feeding causes leaf margins to go brown, usually in late summer, which can result in premature leaf drop and, at the time of grape ripening, can lead to sunburn of grapes. **Habitat** Vineyards and forest margins. Common.

2 *Cicadella cosmopolita*
Blue-striped Leafhopper

Medium-sized (wingspan 20mm), with distinctive bright coloration. Wings striped in black and sky blue. Abdomen bright red. **Biology** Rests in an exposed position on broad-leaved grasses. **Habitat** Common under trees in subtropical forest. **Related species** *Poecilocarda minuscula* **(2A)**, which is small (wingspan 8mm), grey and bright orange, and feeds on soft-leaved forest plants, including Cape Ivy (*Delairea odorata*), in the Knysna area.

3 *Cofana spectra*
White Leafhopper

Medium-sized (wingspan 16mm), elongate, whitish to pale green. **Biology** Feeds on grasses. Often a pest of rice, barley, wheat, sugar cane and sorghum, as it transmits plant viruses. **Habitat** Tropical grassland. Cosmopolitan in the tropics.

4 *Cephalelus*
Restio leafhoppers

Small (wingspan 10mm), slender, with highly elongated snout. Genus comprises 15 similar species. **Biology** Each species is closely associated with a single host of the reed-like Restionaceae and may occur in large numbers. Rest flat against restio stems, being superb mimics of the leaf bracts. Jump weakly. Females generally wingless. **Habitat** Restio 'grassland'. **Related species** *Hecalus* **(4A)** have a large, spoon-shaped extension of the head, and short striped wings. Occur in summer-rainfall grasslands, and recorded on rice in tropical Africa.

SUBORDER STERNORRHYNCHA

The families within Sternorrhyncha comprise highly varied, terrestrial, small to large, plant-feeding bugs, differing from Heteroptera chiefly in their wing design and (generally) much smaller prothorax. The fore wings are uniform in texture and held roof-like over the body (unlike the flat hemelytra of Heteroptera). The antennae are fairly long; the tarsi are made up of 2 segments, and the ovipositor is generally rudimentary. Many are agricultural and horticultural pests. Apart from damaging plants by direct feeding, they are important vectors of viruses that cause various plant diseases. Many also produce honeydew (surplus sugars from their liquid diet) from the anus.

Family Psyllidae
Jumping plant lice

Small to minute bugs with clear wings (often patterned). Closely resemble small cicadas. Under magnification many wing veins appear to arise from a single stalk, with no cross-veins visible. Powerful jumpers, but weak in flight. Nymphs are flattened and feed on hard leaves, often causing pit-like galls or deforming leaves. Restricted to woody plants. There are 61 species known in the region.

5 *Retroacizzia mopanei*
Mopane Psyllid

Small (wingspan 8mm), pale apricot with transparent wings, a green abdomen and red eyes. **Biology** Occurs on Mopane (*Colophospermum mopane*); massive population outbreaks occur. Nymphs feed beneath a creamish-brown waxy scale, deforming the leaves. Attracted to lights in great numbers. **Habitat** Mopane veld. Abundant in Limpopo.

6 *Blastopsylla occidentalis*
Eucalyptus Shoot Psyllid

Very small (wingspan 3mm), yellowish, with dark bands across the abdomen and lines on the thorax. Resembles a miniature cicada. **Biology** An alien, originating from Australia and living on *Eucalyptus*. Nymphs are covered in a white ball of wax filaments, and feed on young leaves, causing leaf drop and abortion of flower buds. **Habitat** Various *Eucalyptus* species; in high densities can damage the trees.

P. Webb ▶

1 *Platycorypha nigrivirga* — Tipu Psyllid

Very small (wingspan 8mm), body yellow, boldly marked with black lines and dots, adults with 2 black stripes across the head; wings clear. **Biology** An alien species that attacks Tipuana trees (*Tipuana tipu*), causing leaf curling and drop. Produces large amounts of honeydew, promoting unsightly black mould growth on leaves. **Habitat** Streets and gardens where Tipuana trees occur.

Family Triozidae — Jumping plant lice

The wing pads of the nymphs are expanded forward into a pair of humps lying against the head. A number of wing veins converge at a single point, and there is no pterostigma (pigment spot) on the margin of the fore wing. A large family that includes many plant pests.

2 *Trioza erytreae* — African Citrus Psylla

Very small (wingspan 8mm), green when young, turning dark brown later, with large glass-like wings, and an upright and alert posture. Nymphs **(2A)** circular, with a marginal fringe of fine waxy filaments. **Biology** Infests all varieties of citrus, as well as indigenous plants of the lemon family Rutaceae (e.g. White Ironwood, Wild Lemon and Perdepis), transmitting a virus that causes 'greening' (die-back and bitter fruit). Older nymphs make and live in cavities in leaves, causing galling. **Habitat** On trees of family Rutaceae. Favours cooler, moist parts. Less common in Limpopo.

Family Aleyrodidae — Whiteflies

Very small (wingspan 2–4mm), body and wings covered in fine wax. Very few wing veins, some species with thin red-banded wings. Mature nymphs resemble scale insects and cannot move. Sixteen species known from the region. Unusual among bugs in having a resting ('pupal') stage. At this time they lose their legs but continue feeding. Produce large amounts of honeydew. They generally live in groups underneath leaves, often in very large numbers.

3 *Aleurothrixus floccosus* — Woolly Whitefly

Very small (wingspan 8mm), all white. Nymphs sedentary and covered in waxy filaments. **Biology** An alien whitefly. Appeared in South Africa in 2007, and now a serious pest of citrus in the Western Cape, Eastern Cape and Gauteng. The underside of an infested leaf is covered in a woolly layer derived from the wax filaments of the nymphs **(3A)**. Honeydew secretions attract unsightly black mould growth beneath the leaves. Appears to have a preference for lemons. **Habitat** Gardens and orchards with citrus trees.

4 *Trialeurodes vaporarium* — Greenhouse Whitefly

Minute (body length 1mm), body pale yellow with white wings. Resembles a very small moth when fluttering. **Biology** An agricultural pest, attacking a wide range of plants (especially tomatoes), causing them to dehydrate. Causes yellowed leaves and stunted growth, but is not a major vector of viruses. **Habitat** Typically on crop plants (e.g. tomato) or ornamentals (e.g. fuchsia) in gardens.

Family Aphididae — Aphids

Familiar, very small, pear-shaped bugs. Wings (when present) are clear, hind wings often reduced. Many species have 2 large tubes protruding from the end of the abdomen. Usually feed on young shoots and buds, and are generally detrimental, as they remove vast quantities of plant sap, produce honeydew (which forms a substrate for sooty mould) and act as vectors for viral diseases. Many, such as the Green Peach Aphid, Black Bean Aphid and Russian Wheat Aphid, are agricultural pests. Different generations utilize different host plants. Life cycles are complicated and involve winter resting stages, with males being produced in autumn and fertilizing over-wintering eggs. The emerging generation produces parthenogenetic females, which reproduce through summer. Preyed on by ladybeetles, hoverfly and lacewing larvae, and other insects, including numerous parasitic wasps. About 165 species occur in the region, many of them introduced.

5 *Aphis gossypii* — Cotton Aphid, Melon Aphid

Small (body length 1.5mm), pale yellow-green to olive green or dark green, depending on the host plant. Winged or wingless, with long thin legs. Head, legs and tubes arising from abdomen all very dark. **Biology** Feeds on buds, damaging the growing tip. Outbreaks may occur when *A. gossypii* is attended by ants. **Habitat** Common on a wide range of plants including potato, cucumber, citrus, cotton and diverse garden plants. In the veld, occurs on *Searsia*. Cosmopolitan.

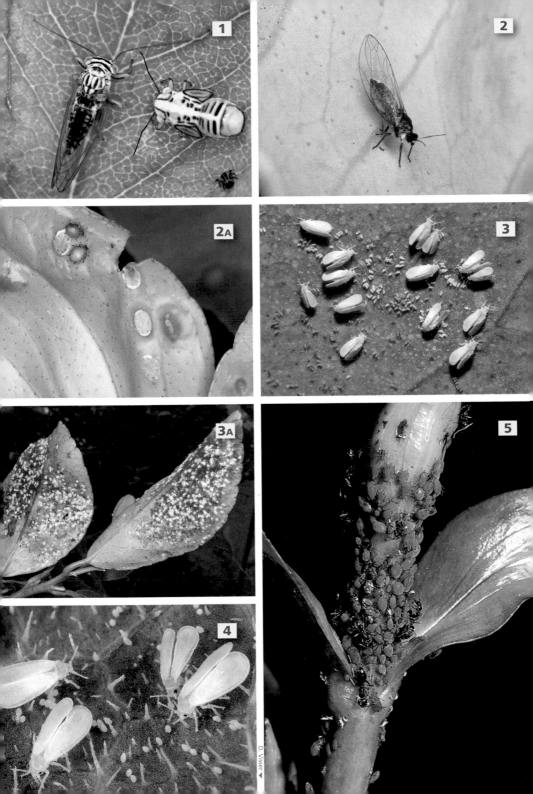

1 *Tuberolachnus salignus* Giant Willow Aphid

Very large (body length 4–5mm), well-known, dark grey aphid, with rows of black spots and a large black protuberance on the back between the short black abdominal horns. **Biology** Occurs on *Salix* (willows and pussy willows), and only reproduces asexually. Secretes vast amounts of honeydew, encouraging the formation of black sooty mould. **Habitat** Fairly common in gardens and on river verges where willows occur. Cosmopolitan.

2 *Aphis nerii* Oleander Aphid

Small (body length 2mm), unmistakable, bright yellow. Head, appendages and tubes from abdomen black. Winged forms have black wing veins. **Biology** Occurs on various milkweeds (Asclepiadaceae) in the veld, and on oleander and periwinkle (Apocynaceae) in gardens. Colonies usually occur on stems. **Habitat** Widespread in veld and gardens in the tropics and subtropics.

3 *Macrosiphum rosae* Common Rose Aphid

Fairly large (body length 3.5mm), pale green to pink or reddish-brown. Knees, antennae and very long horns on abdomen black. Legs yellow. Winged forms much darker. **Biology** A pest, especially in spring, when colonies develop on the shoots and buds of roses, deforming the flowers. Usually controlled by natural enemies such as ladybeetles, hoverflies and White-eyes (birds). Also occurs on apple and pear, and may alternate hosts. **Habitat** Gardens.

4 *Pemphigus* Poplar leaf-stem gall aphids

Small (body length 2mm). Wingless forms yellow, with fluffy white wax deposits on the end of the abdomen. Winged forms darker, with wax on the wings. There are 3 regional species. **Biology** They live in a purse-like gall **(4A)** with a slit-like opening, which they induce on the leaf stems of poplars, especially Cottonwood (*Populus deltoides*) and, at some times of year, on the roots of a secondary host (species of the mustard family Brassicaceae). **Habitat** Urban and agricultural areas.

5 *Uroleucon sonchi* Sow-thistle Aphid

Medium-sized to large (body length 3mm), very dark and shiny maroon, almost black. Knees, antennae and abdominal horns black. **Biology** Occurs in large colonies on the flower and bud stems of Sow-thistle (*Sonchus*), lettuce and chicory. Colonies invariably attended by various ladybeetle species. **Habitat** Roadside verges, agricultural and weed-infested areas. Cosmopolitan.

Family Diaspididae Armoured scale insects

Adult females are featureless, but under magnification the body can be seen to end in a thin projection. The body shell (scale) is made up of many layers, comprising shed skins from earlier stages. Diaspidids do not secrete honeydew, but their anal secretion helps to bind the various layers of the scale together. They live on the surface parts of plants and can be very destructive. Some species live in harmony with *Melissotarsus* ants, in chambers made by the ants. There are 270 species of this large family known from the region.

6 *Aonidiella aurantii* Red Scale

Small (1.5mm diameter). Adult females reside under round, translucent and flattened scales that take on the reddish-brown colour of the insect. Each scale has a nipple-like central protuberance. **Biology** A pest of citrus, banana, granadilla and 200 other host plants. Also occurs on at least 12 indigenous tree species. Reduces the vigour of the host, the apparently toxic saliva causing little yellow spots. Virtually every citrus tree is infested with this scale. **Habitat** The leaves, stems and fruit of many garden and agricultural plants. Originally introduced from China, now cosmopolitan.

7 *Duplachionaspis exalbida* White Aloe Scale

Small (body length 1mm), with elongate white scales narrowed at 1 end. Usually cover aloe leaves almost entirely. **Biology** Populations generally kept under control by parasitic wasps and ladybeetles, but severe infestations can cause leaf tips to wither. **Habitat** Aloe leaves in gardens and veld. **Related species** The Regius Scale (*Dynaspidiotus regius*) (up to 1.4mm long), which is much narrower, elongate and buff-coloured. Both the narrow males and broader females are shown here **(7A)**.

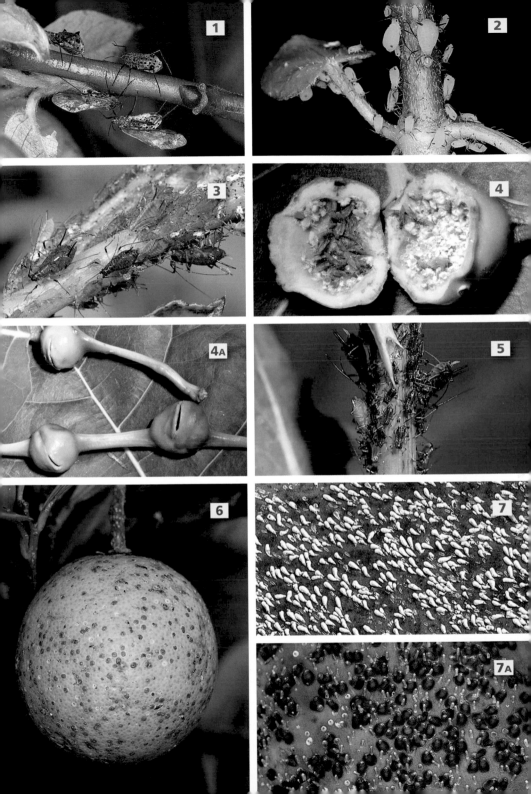

Family Coccidae
Soft scale insects

Vary considerably in appearance, some having only a thin veneer of wax, others with sheets of wax or a thick 'shell'. As in all scale insects, adult females are typically small (1–35mm diameter), wingless, with little distinction between head, thorax and abdomen, and lack functional legs, the body being just a bag for eggs. Females attach to a single site and extract plant sap with their very long, thin mouthparts (stylets). They release 'crawlers' that actively seek out new host plants, attach and begin feeding, utilizing all plant parts. Males look like typical winged insects, and are usually red with smoky grey fore wings and no hind wings. They often have waxy appendages arising from the end of the abdomen. Colonies produce honeydew, often in very large quantities, and are attended by ants. Pest species feed on citrus trees and Jacaranda. They are of great economic significance, some species being destructive and a few useful, providing red dyes, wax and shellac. Coccidae is diverse, with 130 species known from the region.

1 *Ceroplastes*
Wax scales

Medium-sized (10mm diameter), with an irregular roundish shape, resembling a blob of melted white wax, marked in some species with brown streaks. Nymphs **(1A)** flat and oval and surrounded by a fringe of wax scales. **Biology** Nymphs live on the veins of leaves for about a year, until they switch to twigs, where they remain fixed for the rest of their lives. Adult females lay reddish-yellow eggs under their waxy cap (the wax releasing water when squeezed). Damage to trees is minimal, but black sooty mould develops around infestations. **Habitat** A variety of host trees in the veld. Indigenous species have moved to cultivated hosts, such as citrus and Jacaranda. The common garden species *C. destructor* **(1B)** is a cosmopolitan African pest species.

2 *Coccus hesperidum*
Brown Soft Scale

Small (4.5mm diameter), oval, light brown and flattened when young, becoming darker brown and domed when older, with a pattern of radiating stripes. Parasitized individuals turn black. **Biology** They can infect a wide variety of hosts, such as citrus and other subtropical fruits, garden trees, shrubs, and even ferns. Inconspicuous in low numbers, but their presence is revealed by ant trails leading to their honeydew secretions. **Habitat** Indigenous and exotic plants. A ubiquitous cosmopolitan species. **Related species** The Soft Green Scale (*Pulvinaria aethiopica*), which is similar, but bright green and less numerous. Mature females have a thin loop of black spots on the back. It is an indigenous pest of citrus and coffee.

3 Family Dactylopiidae
Cochineal insects

Superficially resemble mealy bugs and feed exclusively on cactus species, congregating where the spines attach to the pad **(3)**. Males (see *Dactylopius opuntiae* **(3A)**, top left) are small (wingspan 2mm), delicate and winged; females (bottom right) are small (body length 2–3mm), soft-bodied, covered with wax filaments and filled with a red dye (carminic acid). There are 6 regional species, all of which are introduced. **Biology** *D. opuntiae* was introduced in 1937 to control the spread of the Prickly Pear (*Opuntia ficus-indica*). Large infestations cause pads to yellow, and then rot. Spineless varieties of the Prickly Pear may also be attacked by *Dactylopius* species. Females produce huge numbers of eggs, which hatch into very small 'crawlers' that disperse in the wind (females) or (if destined to form males) do not disperse. Females harvested elsewhere for red food dye (cochineal). **Habitat** Prickly Pear plants in arid regions.

Family Monophlebidae
Giant scales

Large to very large, with an elongated body. Adults have dark legs and fairly long antennae. A wax covering may be present in some species, and an egg sac may also be present. The family is unique in that pre-pupae are mobile. All species (including pests, such as *Icerya purchasi*) feed on woody trees or shrubs throughout southern Africa.

4 *Icerya purchasi*
Australian Bug, Fluted Scale, Cottony Cushion Scale

Large (body length 8mm), easily identified by conspicuous, large, fluted white egg sac at the end of the light brown body, filled with red eggs. **Biology** A cosmopolitan pest from Australian acacias that has successfully been controlled by the ladybeetle *Rodolia*. A self-fertilizing hermaphrodite that can produce winged males. Egg sac may contain as many as 1,000 eggs. Nymphs migrate from the midribs of leaves to stems as they mature. **Habitat** House plants, gardens and veld, where it attacks a variety of trees and shrubs (especially *Senegalia*).

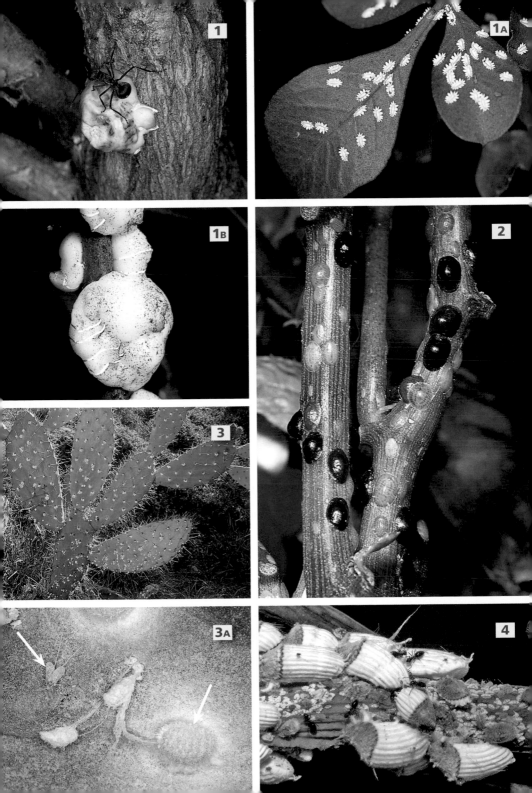

1 *Icerya seychellarum* — Seychelles Scale

Large (body length 8mm), body covered in white wax topped with yellow-tipped waxy knobs arranged radially and long, fine, glass-like hairs. Well-developed legs under wax-covered body. **Biology** Destructive, sucking fluids from a very wide range of woody plants, including tropical fruit trees. They excrete honeydew, causing black mould to develop on plants and blocking the stomata. **Habitat** Originated from the western Indian Ocean islands (the Seychelles, Madagascar and Mascarene Islands), but now cosmopolitan in tropical and subtropical parts of the eastern Pacific.

2 *Aspidoproctus maximus* — Mammoth Scale

Females huge (body length 35mm), oval and domed, dark brown, but covered in fine whitish-yellow wax; margins of body scalloped, with long wisps of waxy filaments. Males small (wingspan 5mm), red, with a fine wax covering. Nymphs oval, very flat, and wrinkled. **Biology** Adults secrete a thin stream of honeydew through the dorsal anal opening, especially when disturbed. This collects under trees, forming a sticky layer, overgrown with sooty mould. Females form a very large 'pre-pupa', covered in fine, fleecy, white wax filaments, at the base of a tree. Recorded on M'sasa Trees (*Brachystegia spiciformis*), Flamboyant, Silky Oak, casuarina and hibiscus. After they moult, nymphs migrate to thicker tree branches. **Habitat** Subtropical bushveld and gardens.

3 *Aspidoproctus mirabilis* — Horned Soft Scale

Large (body length 10mm), blackish-brown, with a dry, brittle shell bearing a central spike (7mm long) and 2 smaller, blunt lateral spikes. **Biology** As in all *Aspidoproctus* species, underside of body is attached to the host tree, a small patch of bark remaining on the underside if the insect is removed from the tree. Recorded from *Senegalia*. In all species in the genus, 2,000–3,000 larvae (adorned with long, glassy wax filaments arising from a central node) emerge from the brood pouch of the female. **Habitat** Bushveld.

Family Margarodidae — Ground pearls

Named after the subterranean *Margarodes*, which form cysts ('ground pearls') at depth in the soil near the roots of their host plants. The subterranean species feed on the roots of a wide variety of plants, and many are crop pests – attacking a range of plants, from sugar cane to grapevines. They form a legless cyst in the second nymphal instar, often with a metallic lustre. The fore legs of the nymphs are adapted for digging. They take a number of years to mature, with the flightless females and winged red-and-black males digging to the surface to mate. Neither feeds.

4 *Margarodes prieskaensis* — Northern Cape Ground Pearl

Cyst small (diameter up to 6mm), made up of small blocks of yellow wax (**4**); adult female (body length 3mm) circular and covered in hairs, the first pair of legs strongly hooked for digging. The chunky male has a circular red body and short greyish wings. **Biology** Eggs hatch into nymphs that pierce a plant root, and remain attached until development is completed. After the first moult they develop the cyst and continue to moult and grow within it. This stage is very resistant, surviving for years without feeding. Females either produce eggs that develop without fertilization, or can move to the surface in synchrony with the winged males for mating. Can destroy table grapevines on the Orange River. **Habitat** Below soil, feeding on roots; also in vineyards. **Related species** There are 12 other regional species in the genus. Also, orange or red *Monophlebus* species, which live above ground, feeding on grasses. The females (**4A**) (body length 5mm) are wingless, while the males are winged and have 2 red terminal filaments (**4B**).

Family Ortheziidae

Ensign scales

Easily recognized by the thick waxy ovisac, elegantly carried by females at the end of the abdomen. Have dark legs and antennae, and stalked eyes. Upperside of body carries a waxy secretion that forms fluted plates. Attack a very broad range of host plants, from fungi to mosses and shrubs. The species below is the only regional representative.

1 *Insignothezia insignis*
Lantana Bug

Small (body length 2mm), with spindly legs and black antennae. Nymphs have neat tufts of wax sticking up from the body **(1)**. Adult females carry the characteristic uptilted fluted ovisac **(1A)**, and have patches bare of wax on the top of the body. **Biology** Females reproduce parthenogenetically. In South Africa considered as a (weak) control agent for the invasive weed Lantana (*Lantana camara*). Feeds on commercial vegetable crops, as well as on tea and coffee shrubs. **Habitat** Mostly in agricultural habitats and gardens.

Family Pseudococcidae

Mealy bugs

The least specialized of all scale insects. Females of most species retain legs and antennae, and are capable of limited locomotion until egg development. Body segmentation visible through the waxy layer. Many are pests of a wide range of crops and fruit trees. There are 110 species in the region.

2 *Planococcus ficus*
Grapevine Mealy Bug

Small (body length 4mm), oval, segmented and pink, covered in a fine waxy layer, with a fringe of waxy hair-like extensions around the body, and a thin dark stripe devoid of wax along the back. **Biology** In the Western Cape found in spring and summer on new growth on grape shoots, its copious honeydew secretions causing black mould. Heavy infestations can cause grape bunches to wither. Up to 6 generations per season in warmer regions. Controlled to some extent by lacewing larvae and ladybeetles. Attended for their honeydew by ants, which often protect them from predators and parasitoids. **Habitat** Orchards and farmland.

3 *Pseudococcus longispinus*
Long-tailed Mealy Bug

Small (body length 3–5mm), white, with 4 very long thin wax filaments at the end of the body. **Biology** Attacks pears, apples, quinces, grapes and *Ficus natalensis*. Heavy infestations can cause leaves to yellow, and infested fruit cannot be marketed. The only *Pseudococcus* species in which females do not produce an egg sac; instead, the pale yellow eggs are brooded under the body. As in other mealy bugs, males are minute (body length 1mm) and winged. Normally natural enemies such as ladybeetles, brown and green lacewings and parasitic wasps hold populations in check, obviating the need for insecticidal control. **Habitat** All parts where fruit and ornamentals are grown, outdoors as well as in houses and greenhouses.

4 *Planococcus citri*
Citrus Mealy Bug

Small (body length 2–5mm), covered in fine white wax deposits, with a faint line running lengthwise along centre of the body. Winged males small, with 2 pale white filaments at end of abdomen. **Biology** A worldwide pest of citrus. Infested fruit can become deformed, develop red-brown blemishes and may eventually drop off the tree. Attacks all parts of citrus trees, including leaves, which grow into deformed shapes. Sooty mould develops on infested fruit as a result of the honeydew given off by the mealy bugs. White egg sac develops behind the body of the female. Natural enemies effectively curtail numbers. **Habitat** Citrus orchards and gardens where citrus is grown.

ORDER THYSANOPTERA • THRIPS

Minute insects with very slender bodies, piercing-and-sucking mouthparts and (usually) 2 pairs of narrow wings fringed with bristles. Usually found in flowers or on plant leaves.

Thrips are a very diverse and common insect group, but attract little attention because most species are less than 2mm long and are slender and inconspicuous. Both winged and wingless forms occur, sometimes within the same species. Where present, the 2 pairs of long, very narrow wings are fringed with long bristles and usually have few longitudinal veins. Antennae are short, with 6–10 segments, and the peculiar asymmetric mouthparts are arranged in a conical beak used for piercing and sucking individual plant cells. The legs are short and each ends in a bladder that can be extended to adhere to smooth surfaces and withdrawn when not in use.

Most thrips are weak fliers and feed by piercing plants and sucking up the sap, but some are predatory on mites, aphids and other small arthropods, while yet others feed on fungi or decaying plant matter, or form galls. Because of their feeding habits and transmission of viral diseases, some species are significant pests of cultivated plants, but they can also have beneficial effects as pollinators or biocontrol agents. They often target buds and new growth, and so can damage large trees in spite of their small size. Reproduction may be sexual or parthenogenetic. Eggs are generally laid on or in the tissues of plants. The wingless nymphs resemble adults in appearance and feeding habits. Of the approximately 6,000 species known globally, about 230 occur in South Africa.

1 Family Phlaeothripidae
Tube-tailed thrips

Named for the characteristic tubular and cylindrical last abdominal segment seen in both sexes. Adults **(1)** mostly dark and flattened; wings narrow, smooth, and lacking longitudinal veins. Nymphs **(1A)** wingless, sometimes brighly coloured, as in the unidentifed species shown here, feeding on fruit pods of the Bladdernut tree (*Diospyros whyteana*). Predatory species **(1B)** generally recognized by strong grasping fore legs. A diverse family, with about 112 species in the region, also containing the largest species. **Biology** Most often seen feeding on pollen, often in dense groups, but numerous species consume green leaves or fungal spores or are predatory, especially on mites. **Habitat** Widespread on flowers and plants, and also in more specialized habitats such as termite and birds' nests and leaf and bud galls.

2 Family Thripidae
Common thrips

The most diverse family of thrips, with almost 100 species in the region. Mostly minute (body length 1.5mm or less), dorsoventrally flattened. Hard to detect without a hand lens or microscope. Recognized by last abdominal segment in females, which is divided ventrally to reveal a prominent saw-like ovipositor that is curved downwards. Wings, when present, are very narrow and pointed, with 2 longitudinal veins. **Biology** Almost always flower- and leaf-feeding. Many are serious pests of cultivated plants. For example, the yellowish-orange Citrus Thrip (*Scirtothrips aurantii*) is a significant pest not only of young citrus fruit and leaves, but also of a wide range of other plants, including mangoes and grapes. **Habitat** Cosmopolitan, some species having been transported around the world with their host plants. The species shown here were found feeding on rose blooms **(2)** and onion leaves **(2A)**.

3 Family Aeolothripidae
Banded thrips

Small (body length about 2mm). Last abdominal segment in females conical and ventrally divided to reveal a saw-like ovipositor, which is curved upwards. Wings relatively broad and rounded at the tip, with 2 conspicuous longitudinal veins and several cross-veins. Fore wings usually conspicuously banded or striped, hence the common name. A relatively small family with 13 species in the region. By far the most common genus is *Aeolothrips*. **Biology** Feed mostly on plant tissue or pollen, but some also prey on small insects and mites. **Habitat** Inside flowers.

ORDER MEGALOPTERA •
DOBSONFLIES & ALDERFLIES

Fairly large, unspecialized insects with 2 equal pairs of opaque membranous wings, long beaded antennae, biting-and-chewing mouthparts, unmodified legs and a soft flexible abdomen.

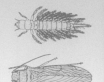

The dobsonflies and fishflies (Corydalidae) and the alderflies (Sialidae) are a small group of primitive insects with an unspecialized body form and a wingspan ranging from 20 to 170mm. Wing veins do not branch near the margins. The wings are usually held wrapped around the body when at rest. The short-lived adults emerge in summer, and are usually found clinging to the stems of plants near streams. Eggs are laid on rocks or vegetation hanging over water, and the larvae drop straight into the water after hatching. Larvae, sometimes referred to as toe-biters, are large, flattened, predatory and aquatic, and have prominent lateral abdominal gills and formidable jaws. They are found under submerged rocks and debris, and live for 1–5 years before undergoing a brief pupal stage. There are only about 300 known species in the world, and nearly all of the African species occur in South Africa, which has 10 species.

Family Corydalidae
Dobsonflies

Large dull brown insects, which perch on marginal vegetation around streams in summer. Although they have large wings, they are awkward fliers. They have 3 simple eyes (ocelli) on the top of the head. The larvae ('toe-biters') have 8 pairs of thin extensions along the sides of the body.

1 *Chloroniella peringueyi*
Peringuey's Dobsonfly

Large (wingspan 160mm), with sharp projections behind the eyes and a pair of black stripes on the sides of the head and prothorax. The upper lip (labrum) of all stages is cone-shaped (1) (see arrow). Larvae (1A) have 7 pairs of grey feathery gills at the base of the feathery abdominal extensions. **Biology** In mid-summer adults fly to lights at night, but probably do not feed. **Habitat** Larvae occur under stones in mountain streams, adults in the marginal vegetation of streams.

2 *Taeniochauloides ochraceopennis*
Cape Dobsonfly

Large (wingspan 130mm), body brown. Wings extend well beyond tip of abdomen and are tinged with grey or brown and dotted with darker brown along the veins. Simple eyes (ocelli) almost meet on top of the head (2A); cf. the related *Platychauliodes*, where the ocelli are widely spaced (2B). Larvae up to 40mm long, lacking gills at the base of the 8 pairs of smooth lateral abdominal appendages, and with a pair of stout hook-like structures at the end of the abdomen. **Biology** Flies weakly and erratically, on overcast days. Larvae live in fast-flowing streams and bite readily when handled; adults occur on vegetation or rocks at the water's edge in summer. **Habitat** Common in mountainous areas, in fast-flowing water. **Related species** There are 7 other regional species in the genus.

Family Sialidae
Alderflies

Smaller than Corydalidae and lack simple eyes (ocelli) on the top of the head.

3 *Leptosialis africana*
Alderfly

Small (wingspan 23mm), uniformly dark and smoky brown, with wings that extend beyond the tip of the abdomen and wrap around the body when at rest. Large-jawed larvae with 7 pairs of long lateral abdominal gills and a single long filament extending from end of abdomen (3A). The only member of the family in Africa (except for 1 species in Egypt). **Biology** Adults perch on vegetation along the banks of still streams and are active between November and February. Larvae are active predators of a variety of invertebrates, and occur in mud on the bottom or along the banks of slow-flowing reaches. **Habitat** Mountain streams and slow-flowing waters. Relatively common in KwaZulu-Natal, but rare in the Western Cape.

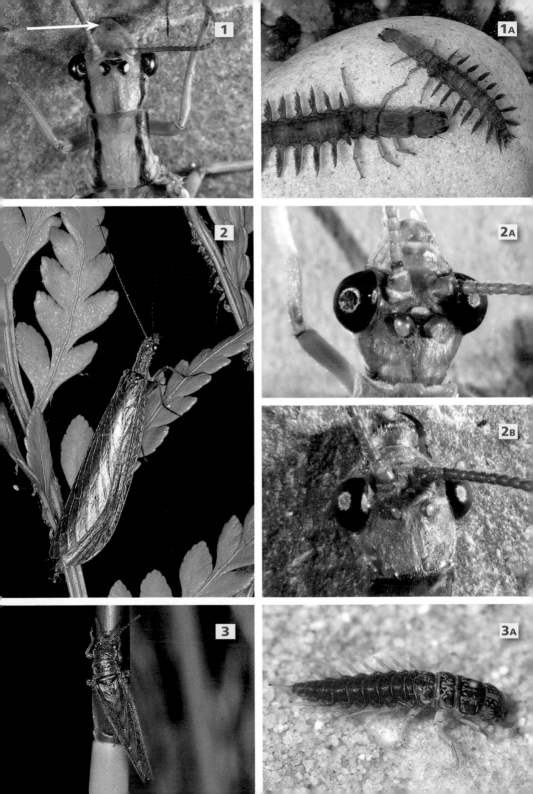

ORDER NEUROPTERA •
LACEWINGS, ANTLIONS & KIN

A diverse group of mostly medium-sized to large, soft-bodied insects with 2 pairs of large, membranous, often clear wings that are typically held roof-like over the body when at rest.

The wings of Neuroptera are supported by a network of veins, and the wing membrane is often attractively blotched and iridescent. The adults have chewing mouthparts and are mostly predators, although a few feed on nectar and pollen, and these have elongated mouthparts forming a rostrum. Many have a characteristic weak flight, with slow wingbeats. Larvae are all carnivorous (a few are parasitic) and differ widely in appearance and habits, but all have elongated mouthparts modified for sucking out the fluids of their prey. The anus is sealed, so the waste products of their liquid diet are stored until excreted as a white pellet (the meconium) after emergence from the pupa. Larvae are mostly plant- or ground-dwelling, but a few are aquatic (spongilla flies) or semi-aquatic (Osmylidae). All spin a spherical cocoon for pupation. The order is well represented in the region by 13 of the 16 known families, and by approximately 400 species.

1 Family Con_iopterygidae   Dusty-winged lacewings
Tiny (wingspan 4–8mm), body and wings entirely covered in a powdery white exudate. Antennae long and bead-like. Coniopterygidae includes 7 genera and 14 species regionally. **Biology** The tiny larvae live on leaves and devour mites, scale insects and other pests. **Habitat** Common in citrus orchards and often in gardens. Usually found in sweep-net samples from flowering shrubs and trees. Widespread, though often overlooked because of their small size.

Family Hemerobiidae   Brown lacewings, aphid wolves
Mostly small (wingspan 6–16mm), dull brown, with long bead-like antennae and, commonly, dark markings on the wings. Larvae feed on aphids and other small plant bugs, being thus of economic importance. The silken cocoon of the pupa is wedged into a crevice in bark. The larvae are more slender and less hairy than those of green lacewings. Adults are usually associated with flowering trees such as *Searsia*. The family includes 22 species in 7 genera regionally.

2 *Micromus*   Narrow-winged brown lacewings
Small (wingspan 14mm), with broad uniformly brown wings, held roof-like over the body. Fore wings with dark brown spots along the margins and a dark brown bar near the wing tip. There are 4 regional species in the genus. **Biology** Larvae feed on aphids and other small plant bugs. Adults common at lights. **Habitat** In a variety of vegetation types, and in gardens and orchards.

Family Osmylidae   Pleasing lacewings
A small, cosmopolitan family restricted to the warm, eastern parts of the region, where its members occupy moist forest. The semi-aquatic larvae live in wet moss and under stones bordering streams. Thin and elongate, the larvae lack gills, have long, straight jaws and feed on various insects, particularly larval flies. Eggs are generally laid in clumps, and larvae take almost a year to reach adulthood. The medium-sized to large adults have broad, patterned wings, and a fairly long prothorax. Family represented in the region by 3 species of *Spilosmylus*.

3 *Spilosmylus interlineatus*   Barred Pleasing Lacewing
Medium-sized (wingspan 40mm), with blotched markings on the broad, clear wings. Fore wing with a small dark brown spot near the trailing edge and barred markings on the veins that run near the leading edge. **Biology** Adults emerge in summer. They characteristically rest on the underside of leaves and are attracted to lights. **Habitat** Subtropical coastal forests with streams. **Related species** There are 2 other species of *Spilosmylus* in the region, including an unnamed species (possibly *S. laetus*) with less distinct barring near leading edge of the fore wing, which has been recorded from Sodwana Bay (**3A**).

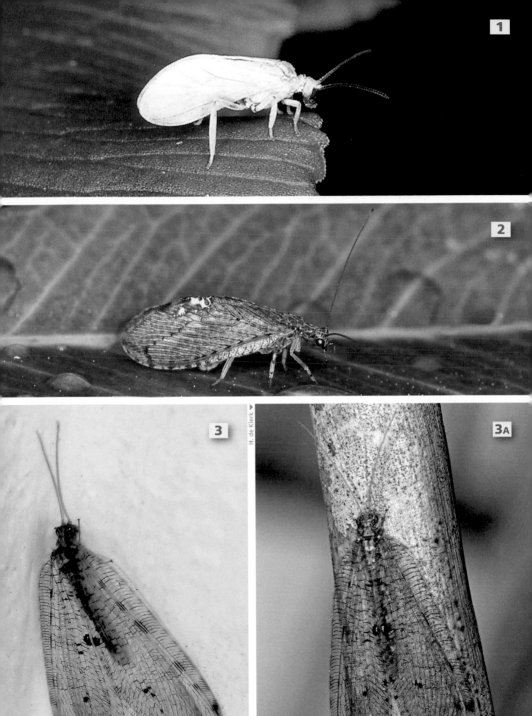

1 Family Chrysopidae Green lacewings, golden eyes

A very large family (79 species in the region), mostly light green, but sometimes yellow, or marked with red. Eyes metallic-looking, hence the common name. Adults medium-sized (wingspan 15–50mm) and feed on aphids, small homopteran bugs and various agricultural pests, but also on nectar and pollen. The voracious, slender larvae have a pair of short sickle-shaped jaws, and some carry sucked-out prey on their back **(1)**. Eggs are laid at the end of slender mucus threads. Chrysopidae have been used successfully in biological control programmes. They are often found on trees, shrubs or grasses, and many frequent gardens and agricultural areas.

2 *Chrysoperla* Green lacewings

Medium-sized (wingspan 24mm), plain yellowish-green, without spots or markings on the wings. **Biology** Adults feed on honeydew and pollen, not homopteran bugs. The larvae are elongate, move abruptly, and do not cover their body with debris. They feed on mealy bugs (*Planococcus*) and other homopterans. Adults attracted to lights. **Habitat** A variety of vegetation types. Common in gardens.

3 *Ceratochrysa antica* Yellow Lacewing

Relatively large (wingspan 37mm), with very long antennae (broken in the photograph), yellow overall, without wing markings. **Biology** Adults attracted to lights. **Habitat** Subtropical coastal vegetation and forest.

4 *Chrysemosa jeanneli* Grey Lacewing

Small (wingspan 20mm), with a grey body and wings. Diagnostic black spot in mid-hind margin of each wing, the spots meeting when wings are closed. **Biology** Adults probably pollen- and nectar-feeders; larvae very likely carry sucked-out prey. **Habitat** Common and widespread in agricultural areas, gardens and indigenous vegetation. Like others in the family, it is often collected in sweep nets from flowering trees.

5 *Pseudomallada baronissa* Broad-winged Green Lacewing

Medium-sized to large (wingspan 30mm), green, with very broad wings, each with a small black blotch near the end. Wings held rather flat when at rest. **Biology** Attracted to lights. **Habitat** Indigenous subtropical coastal or montane forest.

6 *Ankylopteryx* Speckled forest lacewings

Medium-sized (wingspan 38mm), body bright green, with very broad fore wings and narrow hind wings. Wings heavily speckled with black and spread flat at rest. Antennae short. Genus comprises 6 similar species. **Biology** Adults characteristically rest on the underside of large forest leaves, where they are easily detected. **Habitat** Subtropical forest. Occasionally on large-leaved garden trees in warmer parts of the region.

7 *Dysochrysa furcata* Red Lacewing

Medium-sized (wingspan 34mm), brightly coloured in red and yellow, with black shading over most of the wing veins. Eyes very large. **Biology** Not known. **Habitat** Swept from *Senegalia* grassland. **Related species** There is 1 other, similar-looking species in the genus.

Family Nemopteridae Thread-wing lacewings, spoon-wing lacewings

Unmistakable, the hind wings forming very long threads or streamers, often with terminal dilation. Adult mouthparts elongated into a beak to probe flowers for pollen and nectar. Fore wings generally clear and iridescent, but hind wings in the spoon-wing group are marked in white, brown and black. The larvae of the thread-wing group occur in rain shadows, and are easily recognized by their elongate prothorax and scimitar-like jaws. Larvae of the spoon-wing group are poorly known. They have short jaws without internal teeth and inhabit soil. Most (60) of the world's species occur in the most arid parts of Namaqualand and the Karoo.

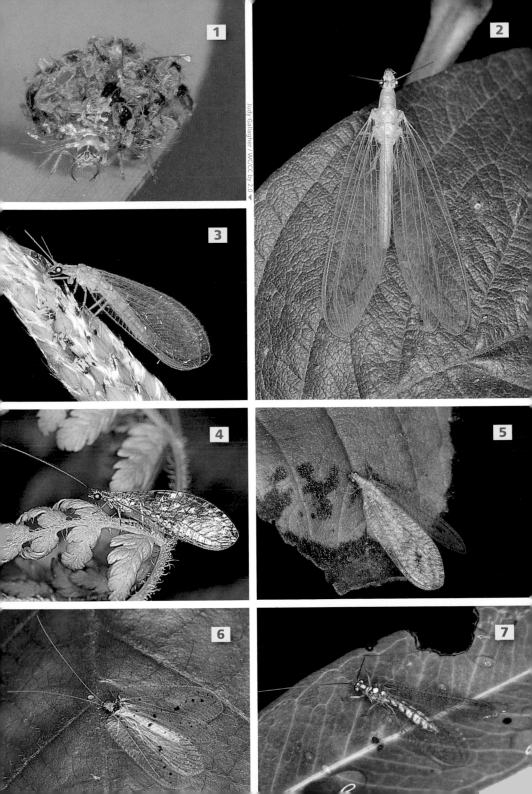

1 *Laurhervasia setacea*
Thread-wing Lacewing

Small (wingspan 25mm, length of hind wings 40mm), fragile, typical of subfamily Crocinae, members of which have thread-like hind wings without terminal dilations. Males of the genus have a knot-like scent gland on each hind wing. Prothorax of larva **(1A)** very long. **Biology** Adults appear in late summer, and rise and fall in flight in the same spot near rock walls. Attracted to lights. **Habitat** Rain shadows, rock ledges or cave mouths in arid and semi-arid regions. Larvae occur on the surface of very fine dry sand.

2 *Halterina*
Scarce spoon-wings

Medium-sized (wingspan 42mm, length of hind wing 36mm). Fore wing with a pale yellow stripe on leading edge. The brown expansions of the hind wing have simple venation running from centre to tip, without cross-veins. There are 3 species in the genus, but *H. purcelli* from Cape Town is probably extinct. Undescribed species shown. **Biology** Rest on sand or low vegetation, active by day and fly weakly when disturbed. **Habitat** Sandy areas with fynbos or restios. All species rare and found in only a few places.

3 *Knersvlaktia nigroptera*
Phantom Ribbon-wing

Medium-sized (wingspan 30mm, length of hind wing 30mm), with relatively short hind wings lacking terminal dilation. Thorax disproportionately large. Both male and female are unusual among Nemopteridae in having rounded black- (male) **(3)** or brown- (female) **(3A)** pigmented fore wings. **Biology** Adults emerge in early spring, and are capable of very strong rapid flight, always landing on bare ground. **Habitat** Restricted to the white quartz plains (Knersvlakte) of the Northern Cape.

4 *Sicyoptera dilatata*
Regal Spoon-wing

Medium-sized (wingspan 50mm, length of hind wing 40mm), fore wings clear with a brown tint, and hind wings with large, black, double apical dilations. Body yellow below, brown on top. **Biology** Active by day, flying weakly and settling on rocks or low vegetation. **Habitat** At high elevation in rocky mountain fynbos. Recently rediscovered, nearly 2 centuries after the first specimen was collected. **Related species** There are 2 other species in the genus.

5 *Barbibucca*
Stout spoon-wings

Medium-sized (wingspan 50mm, length of hind wings 50mm), robust with large, hairy thorax, a broad brown band near leading edge of fore wing, and no dilations at the thin banded tips of the hind wings. There are 2 known species, with more awaiting description. **Biology** Diurnal, powerful fliers that always come to rest on the ground. **Habitat** Associated with very sandy soils, in arid succulent karoo and restio veld.

6 *Nemia costalis*
Ribbon-wing Lacewing

Large (wingspan 100mm, length of hind wings 52mm). Fore wings clear with brown spots between the cross-veins, forming a brown margin to the fore wing. Hind wings thin and barely expanded terminally, with twisted white-and-brown-banded tips. Larvae unknown, but thought to resemble the unidentified nemopterine larva shown **(6A)**. **Biology** Adults are pollen-feeders, often attracted to lights in large numbers in late spring. **Habitat** Succulent karoo and arid grassland. More common on rocky slopes. **Related species** *Nemopterella* species, which are very similar, but with pigment markings forming drops over the cross-veins on the margin of the fore wing. They are most diverse in the Karoo and Namaqualand.

1 *Palmipenna aeoleoptera* — Rock Spoon-wing

Medium-sized (wingspan 32mm), spectacular, with bizarre hind wings expanding from a short stalk to form a very large (20mm long) iridescent brown paddle. Males have much larger and darker hind wings than females. Antennae short and thick. **Biology** Adults rest on rocks and the ground, often 'batting' the hind wings. They are pollen-feeders with very rapid, short flight. **Habitat** Restricted to localized koppies and slopes littered with quartzitic debris. **Related species** The rare Black Spoon-wing (*P. palmulata*) **(1A)** (wingspan 32mm) has thin hind wings (17–21mm long) ending in dramatic metallic black paddles. Found on flat-topped koppies in Namaqualand.

2 *Semirhynchia brincki* — Ribbon-wing Lacewing

Large (wingspan 80mm, length of hind wing 44mm) with a long abdomen. Hind wings elongate with somewhat expanded and twisted white-and-brown-banded tip. Fore wings clear with brown leading edge. **Biology** Larvae similar to those of *Nemia costalis*, p.210, living in sand and organic debris at cave entrances. **Habitat** Hot, arid sandy or rocky regions.

3 *Nemeura gracilis* — Common Spoon-wing

Large (wingspan 50–57mm, length of hind wing 50–62mm), with long hind wings banded in white and black, with a dilated and twisted tip. Fore wings clear. Top of body dark brown, underside cream. Small thorax, long abdomen. **Biology** Adults fly low and weakly between scrubby vegetation and usually settle on the ground. Often attracted to lights in late summer. **Habitat** Common in rocky fynbos and renosterbos.

Family Berothidae — Beaded lacewings

Small (wingspan 12–20mm), superficially resembling brown lacewings. In *Rhachiberotha*, the fore legs have developed into raptorial limbs. There are 11 species known from the region. Stalked eggs give rise to active larvae that feed on termites and ants, which they immobilize with a toxin. Attracted to lights. Bushveld or open woodland. Uncommon, except for *Podallea* (below), which are often seen at light traps.

4 *Podallea* — Hook-winged berothids

Small (wingspan 16mm), pale brown, with characteristic hooked wing tips and fairly hairy body and wings. There are 2 species in the genus in southern Africa. **Biology** Lays stalked white eggs in groups, which hatch into small, slender larvae with long, straight, piercing jaws. Adults rest with the head on the substrate and body and wings tilted upwards. **Habitat** Open grassy woodland.

Family Mantispidae — Mantidflies, Mantispids

Unusual, small to large (wingspan 12–60mm), resembling small mantids, with an elongate prothorax and raptorial (enlarged and spiny) fore limbs for capturing insect prey. Wings generally clear, sometimes with brown markings. Antennae short and bead-like. Eggs, laid in very large batches on short stalks (*Afromantispa*), hatch into active elongate larvae that search out spiders' egg sacs and feed on the eggs, or develop on young spiders, changing into fat grubs. Pupation occurs inside the host's egg sac. Adults frequent flowering shrubs and trees and are attracted to lights. The rich South African fauna is poorly known, but at least 35 mantispid species occur in the region; most found in the moister eastern parts. This number will change as the group is better studied.

5 *Pseudoclimaciella* — Wasp-mimicking mantispids

Large (wingspan 38mm) and convincing paper wasp mimics (see *Polistes*, p.474). Body brown with yellow markings, wings brown-tinted with a brown patch at the tips, neck elongated, with a yellow V-shaped marking at the base. There are 4 species in the genus. **Biology** May be seen during the day on the underside of tree leaves and probably parasitize large rain spiders. **Habitat** Restricted to subtropical forest and woodlands in the eastern parts of the region.

1 *Sagittalata*
Striped-leg mantispids

Small (wingspan 10–20mm). Conspicuous red or black margin to fore wings (the pterostigma); wings usually clear. Inner margin of first segment of the fore limbs marked with an unbroken brown line. Eleven species in the region. **Biology** Attracted to lights, but also congregate on a range of flowering trees such as Karee (*Searsia*) and Buffalo Thorn (*Ziziphus mucronata*). **Habitat** Open woodlands, forest.

2 *Afromantispa*
Banded-antennae mantispids

Small (wingspan 10–25mm), with rough, elongated prothorax. Antennae with a pale yellow band. Brown line on inner margin of first segment of the fore limbs is short and not continuous. Body colour variable; wings always clear. At least 18 species in the region. **Biology** Adults variable in size, depending on the size of the larva's spider prey. **Habitat** Common across a range of habitats.

3 Family Myrmeleontidae
Antlions

The largest lacewing family, comprising medium-sized to very large (wingspan 26–160mm) insects that superficially resemble dragonflies, but have short antennae usually ending in a club. Wings iridescent, often intricately marked in brown and black. Characteristic lazy, flapping flight. Most are attracted to lights in summer and autumn. Funnel-shaped pits of the larvae (3) are well known and occur in rain shadows. Most, however, have free-living larvae that roam under the sand. There are 180 species in 50 genera in the region, almost half occurring in arid western areas.

4 *Centroclisis*
Bark antlions

Large (wingspan 84–110mm), stout-bodied, grey and hairy, with short, broad wings that have rounded tips. Larvae have short, powerful jaws with few internal teeth. There are 8 species in the region. **Biology** Adults sluggish, resting with flattened wings on bark during the day. The larvae of some species live in sand at the base of trees and resemble those of *Syngenes* (below). **Habitat** A variety of vegetation types with trees. Extend into equatorial Africa.

5 *Syngenes longicornis*
Bark Antlion

Large (wingspan 102mm), hairy and less stout than *Centroclisis*, with broad, black-veined wings held flat against tree bark when at rest. Legs and abdomen banded in yellow and dark brown. Larvae pale. **Biology** Larvae travel freely in deep sand (often beach sand), leaving welt-like trails, and rush forward to grab prey on the surface. **Habitat** A variety of vegetation types, where there is loose, deep sand associated with trees. Extends into equatorial Africa.

6 *Cymothales*
Tree-hole antlions

Small (wingspan 56mm), delicately built, with very long, thin legs and highly iridescent wings, intricately patterned in shades of brown and ending in an elegant hooked tip. There are 3 species in the region. **Biology** Larvae live in detritus in tree holes or in fine sand on rock ledges below overhangs. **Habitat** Restricted to the region; more common in arid areas.

7 *Hagenomyia tristis*
Gregarious Antlion

Medium-sized (wingspan 70mm), with broad, iridescent blue wings, yellow thorax marked with black, black abdomen, and yellow legs. Wings clear, with very distinct white wing-spot. **Biology** Highly gregarious. Flight short and slow. **Habitat** On soft vegetation under trees. Genus extends outside Africa. **Related species** *H. lethifer* (**7A**), which is very similar, medium-sized (wingspan 60mm), but with the entire top of the thorax and abdomen black, cream sides and undersides, red legs, and a less distinct wing-spot. It is fairly common in grassland.

8 *Macronemurus tinctus*
White-tip Grassland Antlion

Small (wingspan 62mm), superficially resembling *Hagenomyia* (above), but with very long orange abdomen, narrower wings, and hind wings that project beyond the fore wings at rest. Wings with veins heavily marked in black and a distinctive white wing-spot. **Biology** Resettles rapidly and cryptically on grass stalks when flushed in grassland. **Habitat** Moist grassland. **Related species** There are 6 other similar-looking regional species in the genus.

1 *Cueta*
Pink pit-building antlions

Small (wingspan 58mm), often flesh-coloured, with a striped thorax. Wings peppered with brown dots, with a pink or cream wing-spot. Six similar-looking regional species in the genus. **Biology** Larvae usually construct conical pits in sand in rain shadows, with a vertical shaft beneath the pit to dissipate heat. They generally move backwards, but are capable of moving forwards as well. **Habitat** Most vegetation types. Extend into equatorial Africa.

2 *Creoleon diana*
Large Grassland Antlion

Medium-sized to large (wingspan 80mm), body cream and brown, with pink markings on the thorax, and pink legs. Wings flesh-coloured, long and pointed, with long yellow veins along fore margins and hooked tips. Hind wings project beyond fore wings at rest. **Biology** Rests on grass stalks with wings furled around the body. **Habitat** Grassland with scattered *Senegalia*. Extends into equatorial Africa. **Related species** *C. africanus* **(2A)**, which is medium-sized to large (wingspan 58mm) and similar, but a uniform chalky brown. It is usually flushed from low shrubs and fynbos and occurs in the western parts of the region.

3 *Banyutus pulverulentus*
Grassland Antlion

Small (wingspan 58mm), with cream head and thorax and a black abdomen. Antennae with reddish tips. Wings with white veins and evenly peppered with black dots. **Biology** Larvae live in fine shallow sand. Adults fly weakly when flushed, settling on grass. **Habitat** Usually in moist grassland.

4 *Neuroleon*

Small (wingspan 36mm), superficially resembling *Myrmeleon* (below). Body mottled grey, abdomen short. Wings with quite heavy black markings over the veins and a black V-shaped marking at each tip. *Neuroleon* includes 7 regional species. **Biology** Larvae live in fine sand under rock overhangs or in sand at the base of a tree. **Habitat** Scantily vegetated arid areas. They extend into equatorial Africa.

5 *Myrmeleon*
Grey pit-building antlions

Small (wingspan 44mm), dull, with body brown or grey, sometimes cream underneath. Wings generally unmarked. The most primitive antlions. There are 12 species in the genus in the region. **Biology** Larvae move backwards, and construct characteristic conical pits (usually in rain shadows in areas of high rainfall). **Habitat** More common in drier areas. Extend into equatorial Africa.

6 *Golafrus oneili*
Phantom Kalahari Antlion

Very large (wingspan 120mm), robust, with a tan stripe on the thorax. Wings narrow with pinkish margin, posterior margin incised, forming a raised ridge when at rest. Fore wings streaked in black, hind wings have black blotches. Larvae white. **Biology** Adults cryptic on dune vegetation. Larvae live in deep sand and are active in the cooler hours of the day, burrowing deeper at midday. **Habitat** Arid grassland and bushveld.

7 *Crambomorphus sinuatus*

Very large (wingspan 122mm), with body and legs grey and abdomen orange-red. Hind margin of wing very scalloped, ending in a hooked tip. Fore wings largely grey, with complex black-and-white markings, hind wings heavily marked in black. Larvae largely black, with a reddish head. **Biology** Larvae live in deep sand and come to the surface at night to feed. Adults flushed from low vegetation fly off with a strong swooping flight. **Habitat** Very arid regions with scant vegetation.

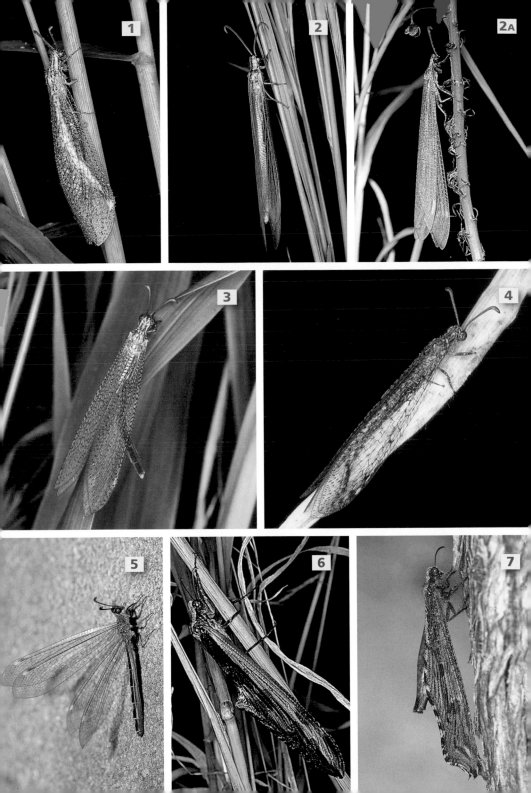

1 *Palpares sobrinus* — Dotted Veld Antlion

Large (wingspan 108mm). An even peppering of black dots on the wings and an inconspicuous black patch near each wing tip distinguish it from other *Palpares*. **Biology** Larvae in this genus travel freely in loose sand and lie just below the surface to ambush and drag their prey under. Adults of this and other *Palpares* species are also predacious. **Habitat** Open grassland in mapped area. **Related species** There are 15 other similar-looking regional species in the genus.

2 *Palpares lentus* — Black-banded Veld Antlion

Large (wingspan 150mm), wings each with 3 complete broad black bands, broader on the hind wings. **Biology** Larvae are large, free-living predators, found just below the surface, often in sand deposited along riverbanks. **Habitat** Open grasslands with sandy soils, in the eastern parts of the region.

3 *Palpares caffer* — Mottled Veld Antlion

Large (wingspan 112mm), with 3 yellow stripes on the furry thorax and yellow abdomen. Yellow wing veins heavily marked in black; 3 evenly spaced black patches along each wing. **Biology** *Palpares* larvae do not form pits. **Habitat** The most widespread species in the genus, extending into tropical Africa, and often found on mountains.

4 *Palpares speciosus* — Spotted Veld Antlion

Large (wingspan 112mm), similar to *P. caffer*, but with 4 large, irregularly spaced lead-grey spots on its wings. **Biology** Adults are frequently flushed from open fynbos, flying some distance before settling. **Habitat** Western Cape, in fynbos and succulent karoo vegetation.

5 *Palparidius concinnus* — Hook-tailed Antlion

Medium-sized (wingspan 80mm), body yellow and chalky grey. Wings clear, with sparse irregular black blotches, yellowish wing tips and conspicuous white wing-spots. Males (**5A**) easily identified by the very long, paired copulatory structures at end of abdomen. **Biology** Larvae live and feed in deep sand. **Habitat** A range of vegetation types in arid areas. **Related species** There are 2 other *Palparidius* species; the genus is endemic to the arid western and eastern parts of the region.

6 *Palparellus festivus* — Ocellate Antlion

Large (wingspan 80–100mm), wings boldly marked in cream and metallic blue-back, body smooth and small. Head bears large bulging eyes. **Biology** Sometimes flies by day, settling on short grass clumps among rocks. **Habitat** Rocky grasslands in savanna and bushveld. **Related species** There are 7 other species in this genus in the region, including *P. nyassanus*, which is similar, but has 2 (not 1) white bands on the fore wing and which occurs in the northwest of the region.

7 *Pamexis karoo* — Butterfly Antlion

Medium-sized (wingspan 60mm), body black with yellow markings, broad cream-and-yellow wings marked with black spots. **Biology** Adults in the Nama karoo occasionally flushed from sparse fynbos or karroid scrub. Flies rapidly by day, settling on low vegetation; unlike other antlions not attracted to lights. **Habitat** Sparsely vegetated arid areas. **Related species** There are 4 other species in the genus, but they are less common than *P. karoo*.

W Tarboton ▲

Bernard Dupont/WCCC by SA 2.0 ▲

1 *Tomatares citrinus*   Painted Grassland Antlion

Medium-sized (wingspan 100mm), brightly coloured, with yellow, cream and black body and opaque milky yellow wings marked boldly with black lines, some comprising adjacent small black circles. **Biology** Apparently diurnal, often flushed in grasslands. **Habitat** Restricted to the eastern grasslands. **Related species** The rare diurnal *Exaetoleon obtabilis* (wingspan 60mm) **(1A)**, which occurs in coastal fynbos in the Western Cape.

Family Ascalaphidae   Owlflies

Medium-sized (wingspan 44–80mm) and unmistakable because of their very long, clubbed antennae. Adults typically fly at dusk (but 1 group flies by day), hawking flying insect prey in the manner of dragonflies. Rarely numerous, although widespread. Larvae resemble antlion larvae, but are flattened from top to bottom and may have fringed edges to the body. They have huge sickle-shaped jaws that can open a full 180°. Larvae occur under stones or on tree bark. About 50 species are known from the region.

2 *Tmesibasis lacerata*   Blotched Owlfly

Medium-sized to large (wingspan 70mm), unmistakable, with very long thin antennae without obvious clubbed tips. Stalked wings boldly marked in metallic brown, with hooked tips. Larvae **(2A)** oval, mottled in white and brown, with prominent fringed edges to the body. **Biology** Adults rest near the base of grass stumps, and are attracted to lights. Larvae found under stones. **Habitat** Fairly common in open *Senegalia* woodland. Extends into tropical Africa.

3 *Proctarrelabis capensis*   Cape Owlfly

Medium-sized (wingspan 56mm), grey body mottled in brown and cream, thorax covered in long, dense, white hairs. Antennae clubbed and banded in black and grey. Wing veins marked in black. Larvae **(3A)** oval, with massive jaws; abdomen bordered with branched appendages. **Biology** Adults use the same resting place over extended periods, often flying low across beaches at dusk in late summer, hawking insects. Larvae live on tree trunks and branches. **Habitat** Rain shadows (e.g. rock shelters) in fynbos or urban areas. **Related species** There are 4 other less common species in the genus.

4 *Nephoneura capensis*   Red-dotted Cape Owlfly

Medium-sized (wingspan 60mm), with brown-shaded wing veins and red shading over the veins in leading edge of wings. Antennae and legs banded red and cream. Body white and black with yellow spots; thorax with tufts of black hairs. **Biology** Rests in typical owlfly fashion, with abdomen held at 90° to body. **Habitat** Fynbos, rocky scree, coastal milkwood forest and gardens. **Related species** Another *Nephoneura* species occurs in the warmer eastern parts.

5 *Eremoides bicristatus*   Horned Owlfly

Medium-sized (wingspan 60mm). As the name indicates, males easily recognized by the unique paired, flanged processes on the thorax, which end in a long hook; flanges absent in female (illustrated here). Fore wings clear, with a thin brown line along leading edge. Body largely yellow, with a few brown stripes running lengthwise. **Biology** Attracted to lights, but also flushed from grass. **Habitat** A common inhabitant of grasslands.

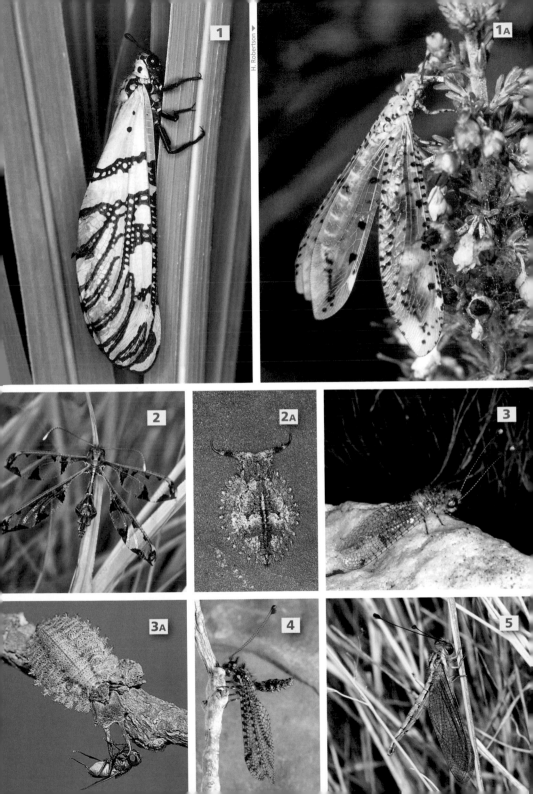

1 *Allocormodes junodi* Giant Owlfly

Large (wingspan 80mm), with a brown body and scattered black markings on long narrow wings, which have golden veins at the base and a white patch at the tip. **Biology** Hawks insects in powerful flight at dusk, much like a dragonfly. **Habitat** Open woodland and forest.

2 *Proctolyra* Tailed owlflies

Medium-sized (wingspan 50–60mm). Males have conspicuous, curved external genitalia projecting from the end of the abdomen. Thin orange or yellow bands on abdomen. Wings clear, occasionally with a thin brown stripe along the leading edge; membrane at wing base orange or yellow. Head and thorax densely covered in white hairs. There are 6 species in the genus. **Biology** Fly to lights. **Habitat** Arid succulent karoo or grasslands.

3 *Neomelambrotus* Hairy owlflies

Medium-sized (wingspan 45mm), very hairy, with tufts of stiff hairs on the top, bottom and sides of the abdomen. Seven regional species in the genus. **Biology** Like many others in the family, adults hunt small flying insects at dusk. Day-time resting posture mimics appearance of a short twig. **Habitat** Succulent karoo in rocky areas in arid western parts, dry bushveld elsewhere.

4 *Melambrotus* Blackline owlflies

Medium-sized (wingspan 60mm), antennae short, thorax with a pair of round, velvety black spots. Prominent blackish stripe along fore margin of wings. There are 4 similar species in the region. **Biology** Attracted to lights. **Habitat** Fairly common in arid succulent karoo and semi-desert, and in subtropical coastal forest and bushveld.

Family Psychopsidae Silky lacewings, moth lacewings

Medium-sized (wingspan 20–40mm) and unusual. Wings very broad and velvety, covered in hairs, mottled or intricately marked, with rounded tips. Small dark dot near apex of each wing marks the convergence of 3 wing veins. Antennae very short and bead-like. Eggs covered in a fine plant or mineral powder before oviposition. Larvae elongate, free-living, with slightly curved elongate jaws and fairly long antennae.

5 *Silveira jordani* Mottled Silky Lacewing

Medium-sized (wingspan 25mm), with narrow brownish wings. Larvae probably resemble Australian species, which have huge square heads and sickle-shaped jaws. **Biology** Larvae thought to live under the bark of trees. *S. jordani* adults are attracted to lights. **Habitat** Widespread in arid forest and savanna with scattered *Vachellia* trees, but generally rare. **Related species** The genus includes 4 other similar regional species.

6 *Zygophlebius zebra* Zebra Silky Lacewing

Medium-sized (wingspan 30–40mm), uniformly cream with faint brown lines running across the wings, forming bars. A small dark spot towards the apex of each wing. **Biology** Attracted to lights and rests with wings spread and held flat against substrate. Squat larvae live in bark crevices. **Habitat** Moist open woodlands and forest. **Related species** There are 2 other regional species in the genus. Another large, light-coloured silky lacewing in the region is *Cabralis gloriosus*, with white wings intricately marked with small black and brown spots. It occurs in the Northern Cape province.

ORDER COLEOPTERA • BEETLES

Minute to very large insects, usually with strongly hardened bodies. Fore wings characteristically modified into rigid cases (elytra) that fold over and protect the membranous hind wings when at rest. Mouthparts adapted for chewing.

Beetles vary tremendously in size, shape and colour, but most have a hard exoskeleton. The rigid fore wings or elytra are the most characteristic feature of the group. These are not used for flight, but instead cover and protect the hind wings. In flightless species, the elytra may be permanently fused together. The hind wings are longer and fold away underneath the elytra when not in use. All beetles have chewing mouthparts. Many feed on plant material; others are predatory or parasitic. Larvae vary widely in form, but all have a hardened head and biting mouthparts, and most are active and roam around freely. They may have 3 pairs of thoracic legs, but never additional fleshy ventral legs (prolegs) on the abdomen. Coleoptera is the largest and most diverse order in the entire animal kingdom, with about 390,000 species worldwide and some 18,000 known from the region. They comprise around 20% of all species on the planet.

1 Family Cupedidae
Reticulated or net-winged beetles

Medium-sized (body length 10–20mm), greyish-brown, elongate, parallel-sided, with thick, dark brown antennae almost as long as the body. Eyes prominent. Elytra with conspicuous, transversely interconnected, longitudinal veins framing large window-like punctures, giving them a reticulated appearance. *Cupes capensis* (body length 18mm) **(1)** is the only species in the region. **Biology** Adults occasionally attracted to lights. Larvae bore into rotten logs and fungi. **Habitat** Southern Cape and east coast forests.

2 Family Carabidae
Ground beetles

Small to very large (body length 3–60mm), mostly dull-coloured, with a somewhat flattened body. Head, thorax and abdomen clearly differentiated, thorax usually narrower than elytra. Antennae long and thin, eyes prominent, mandibles sharp and strongly developed. Elytra often grooved longitudinally. Many are flightless, with the elytra permanently fused together. Most species are ground-dwelling predators, others are arboreal. Some species, often with warning colours, squirt acids or quinones from the end of the abdomen in self-defence. Carabid larvae **(2)** are active predators, with legs adapted for running or burrowing; a few are parasitic on other insects. Mature larvae pupate in the ground or below logs or stones. Over 1,400 species occur in the region.

3 *Graphipterus limbatus*
Velvet Ground Beetle

Small (body length 10mm). Head black, with 2 longitudinal bars of flattened white hairs between the compound eyes. Pronotum and elytra densely covered with flattened yellowish-brown hairs, outlined with a margin of whitish hairs. **Biology** A diurnal predator of small insects. **Habitat** Arid areas with sparse vegetation.

4 *Graphipterus lineolatus*
Lined Velvet Ground Beetle

Small (body length 14mm), body surface finely punctured. Head black, with 2 stripes of white hair between the compound eyes. Remainder of dorsal surface of body is covered with flattened hairs. Pronotum black in the centre and white at the sides. Elytra black, with fine longitudinal lines and a margin of white hairs. Abdomen oval. **Biology** Diurnal. Very active and agile. Often digs. **Habitat** Arid subtropical areas with open spaces. **Related species** *G. cordiger* (12mm long) **(4A)**, which has a keyhole-shaped black marking on the abdomen and a reddish-brown thorax.

1 *Graphipterus trilineatus*   Three-lined Velvet Ground Beetle

Small (body length 11–14mm). Head black, with 2 stripes of pale hairs between the compound eyes, often merging in the middle. Pronotum with black hairs in centre, brownish hairs on sides. Elytra densely covered with brownish hairs, and with 3 longitudinal stripes of black hairs joined at the rear to form a 'W'. Sparse covering of white hairs on underside. Abdomen oval. Legs long, black and slender. **Biology** A diurnal predator, active at the hottest time of day. **Habitat** Clearings and paths in forest and woodland.

2 *Passalidius fortipes*   Burrowing Ground Beetle

Large (body length 32mm), black, parallel-sided. Antennae bead-like. Mandibles very strongly developed. Elytra with longitudinal grooves. Fore tibiae broad, flattened and strongly toothed, adapted for digging. **Biology** Adults burrow in soil, where they hunt insect larvae and worms. **Habitat** Forest to arid savanna.

3 *Pachyodontus languidus*   Table Mountain Ground Beetle

Large (body length 28mm). Black, with disproportionately large head and pronotum, very large mandibles and bead-like antennae. Fore tibiae toothed. Elytra have indistinct striae. **Biology** Adults burrow in loose soil, hunting for worms and insect larvae using their sharp mandibles. **Habitat** Occur in rotten logs and under stones; common in forests of the Western Cape.

4 *Acanthoscelis ruficornis*   Beach Ground Beetle

Medium-sized (body length 18mm), black. Head and thorax finely 'wrinkled', antennae elbowed and reddish in colour, mandibles prominent. Elytra with longitudinal grooves. Legs black, tarsi reddish. Fore tibiae broad, flattened and strongly toothed, adapted for digging. **Biology** Voracious nocturnal predator, feeding on isopod and amphipod crustaceans and various larval and adult insects. **Habitat** Sandy beaches, mostly under cast-up seaweed. **Related species** *Scarites rugosus*, which is similar but smaller, occurring in sand below shore plants. Several other related species occur within the genera *Scarites*, *Haplotrachelus* and *Macromorphus*.

5 *Anthia homoplatum*   Two-spotted Ground Beetle

Large (body length 35mm), black with smooth surface. Elytra have yellow patches at anterior corners and a yellow marginal line. Previously known as *Thermophilum homoplatum*. **Biology** Diurnal. Preys on other ground-dwelling insects. As with other *Anthia* species, known to transport small larvae of the blister beetle *Cyaneolytta*. Juvenile *Heliobolus* lizards in the Kalahari mimic *Anthia* species, both in coloration and posture, likely gaining some protection from vertebrate predators. **Habitat** A variety of subtropical vegetation types.

6 *Anthia decemguttata*   Ten-spotted Ground Beetle

Large (body length 28mm). Head black, pronotum reddish. Elytra black with 10 white spots, and with brownish hairs between longitudinal ridges. Previously known as *Thermophilum decemguttatum*. **Biology** Preys on other ground-dwelling insects, including Pugnaceous Ants (p.482). **Habitat** Common in sandy areas of karroid and fynbos vegetation and arid savanna.

7 *Anthia fornasinii*

Very large (body length 48mm), black. Head and pronotum lightly punctured. Elytra with longitudinal grooves and scattered bristles, and pale-coloured hairs around margins. Legs relatively stout. **Biology** Nocturnal, sometimes attracted to areas illuminated by lights. Ground-dwelling. **Habitat** Subtropical forest and savanna.

8 *Anthia maxillosa*   Tyrant Ground Beetle

Very large (body length up to 45mm), black with white hairs on 3 basal antennal segments. Mandibles large. Pronotum with characteristic raised concave 'dish'. **Biology** Sprays acetic acid from abdomen when molested. **Habitat** Most common in spring in drier western areas. **Related species** *A. thoracica* (Cape Town eastwards into Mozambique), which is even larger (up to 52mm), with distinctive yellow patches on the lateral flanges protruding from the pronotum.

1 *Atractonotus mulsanti* — Ant-mimicking Ground Beetle

Small (body length 10mm), black, with ant-like appearance resulting from combination of elongate head (separated by narrow neck from pronotum), relatively narrow, elongate pronotum and bulbous abdomen. Antennae longer than head and pronotum together. Elytra with longitudinal ridges and 4 spots of white hairs. Sparse covering of white hairs on legs and underside. **Biology** Runs about on the ground apparently mimicking large ants. Ground-dwelling. **Habitat** Subtropical woodland, sand forest and karroid vegetation.

2 *Calosoma* — Starred ground beetles

Medium-sized (body length 22mm), black overall, with a bronze sheen. Head and pronotum finely sculptured, last 7 segments of antennae brown. Pronotum and elytra with upturned margins, the latter with serrate longitudinal ribs and widely spaced rows of bronze dots. **Biology** Adults are nocturnal predators, feeding on caterpillars and other insects. Ground-dwelling, often attracted to lights. Produce a foul-smelling spray from glands at the tip of the abdomen. **Habitat** A wide range of vegetation types.

3 *Craspedophorus bonvouloirii* — Yellow-spotted Ground Beetle

Medium-sized (body length 17mm), distinctive, uniformly black, with 4 large yellow patches on elytra. Elytra with longitudinal grooves. **Biology** Preys on ground-dwelling insects. **Habitat** Subtropical woodland and savanna. **Related species** Several other species with similar warning colours occur in this genus and in *Epigraphus*.

4 *Tefflus* — Peaceful giant ground beetles

Very large (body length 46mm), black. Pronotum irregularly sculptured, somewhat flattened on top, hexagonal in outline. Minute head and jaws. Elytra with conspicuous longitudinal tuberculate grooves. Larvae have legs adapted for running or burrowing. Two species occur in the region. **Biology** Fast-moving, mostly nocturnal predators of land snails, including giant land snails (*Achatina*). Able to inflict painful bites. Activity pattern during rainy periods matches that of snail prey. **Habitat** Subtropical forest and savanna.

5 *Boeomimetes ephippium*

Small (body length 12mm), smooth and shiny brown. Elytra finely grooved longitudinally, with a broad, diffuse, longitudinal black band running down the centre. Legs brown, with bristles. **Biology** As for family (Carabidae). **Habitat** Subtropical forest and savanna.

6 *Chlaenius tenuicollis*

Small (body length 10mm), with grooved black elytra, iridescent reddish-green head and thorax, and orange legs. One of many regional species in this diverse genus. **Biology** Prefers damp localities. Eggs are laid singly on the undersides of leaves, and then covered with a thin layer of mud. Larvae of European species of *Chlaenius* known to attach to much larger frogs, feeding on them and eventually killing them; adults also paralyse and feed on frogs. In South Africa a ground beetle larva (possibly *Chlaenius*) was recorded feeding on an adult Painted Reed Frog (*Hyperolius*) **(6A)**. **Habitat** Subtropical forest and savanna.

7 *Bradybaenus opulentus* — Marsh Ground Beetle

Small (body length 12mm), elongate beetles, with large jaws and strong running legs. Fawn-coloured, with characteristic metallic green markings on pronotum and elytra. Elytra with fine longitudinal grooves. Legs fawn-coloured, with bristles. **Biology** Nocturnal. Often in shallow burrows. **Habitat** Treed savanna, woodland; often on wet sand.

Subfamily Cicindelinae
Tiger beetles

Medium-sized to large, slightly flattened, fast-running beetles with long, slender legs. Many have bright metallic colours; others are brown or black, with yellow or white patterns. The body is clothed in flattened hairs. The 4 basal antennal segments are usually metallic and shiny, the remainder dull. Eyes are prominent, and the head, which is broader than the thorax, is equipped with large, curved, sharp-toothed mandibles. Most species are winged and fly readily for a short distance if disturbed. All species are voracious predators of other insects, and the majority are found on bare ground and beaches. Larvae are also predatory, ambushing their prey from within vertical burrows in the ground. Over 150 species in the region.

1 *Lophyra*
Leopard tiger beetles

Small (body length 11mm). Head and pronotum copper-coloured, with white hairs on pronotum and behind the compound eyes. Lower part of face and base of mandibles white. Elytra yellowish, with characteristic black markings. Underside with white hairs. Legs copper-coloured, with some metallic green, especially on tarsi. More than 20 similar species in the region. **Biology** Active, fast-moving predators of other insects. **Habitat** Diverse. Open sandy areas, especially near water. **Related species** *Habrodera* species, which are all similar in appearance, with expanded lobes on the base of the pronotum that give a trapezoid shape when viewed from above; dark areas on the pronotum are distinctly impressed. They occur on beaches. Also, the larvae of *H. capensis* **(1A)** show the large head (used for blocking the burrow entrance) and well-developed mandibles typical of the family.

2 *Lophyra barbifrons*
Pale Tiger Beetle

Small (body length 11mm). Head and pronotum metallic copper, with a conspicuous covering of white hairs, especially on the pronotum. Elytra yellowish-white, with black border on inner margins, each with more or less conspicuous longitudinal black central line and a posterior spot. Legs metallic copper, but densely covered with white hairs. Face, underside and 4 basal antennal segments also have white hairs. **Biology** An active predator, with excellent vision; flies readily and is agile in the air. **Habitat** Sandy areas associated with the coastline.

3 *Chaetodera regalis*
African Riverine Tiger Beetle

Medium-sized (body length 18mm). Head and pronotum copper-coloured, with widely spaced white hairs. Lower face and base of mandibles yellowish. Elytra yellowish, boldly patterned with black. Legs black, with a metallic sheen and white hairs. **Biology** Preys on insects. Most active at midday. **Habitat** Open sandy areas, especially near water, in subtropical forest and savanna.

4 *Cicindela quadriguttata*
Emerald Tiger Beetle

Small (body length 11mm). Dorsal surface of body metallic green, finely sculptured to give a matt finish. Upper lip and base of mandibles white. Legs metallic copper, tending to green on tarsi. White hairs on legs and underside. **Biology** Diurnal. Hunts insects near muddy pools. **Habitat** Succulent karoo and fynbos. **Related species** *Cicindela disjuncta* (body length 10mm) **(4A)**, which is coppery brown, with 4 pale yellow spots on each elytron, and occurs in the warmer eastern parts.

5 *Platychile pallida*
Night Tiger Beetle

Small (body length 13mm), pale yellow. Posterior corners of pronotum produced into acute points. Mandibles conspicuous. The 4 basal antennal segments are shiny brown, the remainder pale yellow. **Biology** Nocturnal predators that rapidly gather around lights or fires on the beach. **Habitat** Beneath cast-up seaweed on west coast beaches.

6 *Manticora*
Monster tiger beetles

Very large (body length 42–57mm), spectacular and unmistakable, dark brown, with enormous mandibles. Elytra flattened, with a dull metallic sheen and fine hairs, each arising from a small tubercle. The sand-dwelling larvae **(6A)** have a massive spade-shaped head and large eyes and jaws. **Biology** Flightless. Voracious nocturnal predators of arthropods and ground-dwelling insects. The larger jaws in males may help to hold the female during mating. **Habitat** Diverse.

1 Subfamily Paussinae

Ants' guest beetles, nest beetles

Small to medium-sized, pale to dark reddish beetles, recognized by their remarkable broad, flattened antennae, which in some species have lost their segmentation. All live in association with ants, e.g. *Paussus lineatus* (1) in association with *Lepisiota capensis*. The ants lick up aromatic secretions produced from glands marked by tufts of yellow hairs on the antennae, pronotum and tip of the beetle's abdomen. Larvae live among the ants' brood, on which both they and the adults feed. Adults leave the host's nest to reproduce, and males may be attracted to lights. Paussids expel a caustic liquid when disturbed, which volatilizes explosively (and audibly), making a puff of 'smoke'. About 85 species occur in the region.

2 *Cerapterus laceratus*

Bombadier Ants' Guest Beetle

Medium-sized (body length 13mm), shiny brown, with smooth surface. Second antennal segment squarish; remaining 9 segments broad and flattened. Elytra dark brown, each with an orange crescent-shaped patch near the end. Pronotum and elytra fringed with hairs. Legs broad and flattened. **Biology** If molested, sprays caustic quinone-containing gas in explosive bursts. Often attracted to lights. **Habitat** Subtropical forest and savanna.

3 *Paussus curtisii*

Small (body length 7mm), brown, with smooth surface. Elytra squared off at tips. Pronotum with central concave depression and small cavity on each side. Antennal segments fused and indistinguishable, except for the 2 basal segments. Tarsi stubby and with indistinct segmentation. **Biology** Restricted to ants' nests, which are left only for mating. **Habitat** Subtropical forest and savanna.

4 Family Dytiscidae

Predaceous water beetles

Small to large (body length 1–45mm), smooth, streamlined and oval, some almost globular. Brownish, dark olive or black, as in the large and widespread Yellow-edge Water Beetle (*Cybister tripunctatus*) (body length 29mm) (4). Some species, such as the small (body length 15mm) *Hydaticus bivittatus* (4A), have distinct colour patterns. Hind legs strongly developed, flattened and fringed with stiff hairs to increase surface area for swimming. Legs row synchronously. Fore tarsi of males may have suckers for gripping females. Larvae have sickle-shaped mandibles and are known as water tigers. They vary in shape, mostly with elongate body tapering posteriorly, as in the large (body length 35mm) *Cybister* (4B). In some species, the larval body is stout and broad and the last few segments of the abdomen narrow as in *Hyphydrus* (4C). There are about 200 species in the region. **Biology** Adults carry a supply of air below their close-fitting elytra, and both adults and larvae surface to replenish their air supply via spiracles at the tip of the abdomen. Both stages are predatory, feeding on a variety of prey, from zooplankton to small fish. Pupation occurs on land, close to the water. Adults are strong fliers, often attracted to lights at night. **Habitat** Most bodies of fresh water.

5 Family Gyrinidae

Whirligig beetles

Small to medium-sized (body length 3–18mm), black, oval and somewhat flattened, with short antennae, typified by the medium-sized (body length 14mm), lead-coloured whirligig beetles *Dineutes* (5). Some are black or metallic, like *Gyrinus*, or have pale borders, as in the smaller (body length 8mm) *Aulonogyrus* (5A). Unusual compound eyes are divided into upper and lower sections, for seeing simultaneously above and below the water. Fore legs long (for holding prey or clinging to the bottom). Mid and hind legs short and flattened, adapted for paddling. Larvae (5B) elongate, 6–25mm long, tapering towards head, with 3 pairs of well-developed thoracic legs, and breathing filaments along each side of the abdomen. About 55 species in the region. **Biology** Adults aquatic and usually observed on water surface, often in large groups. They dart about at great speed if alarmed, often diving below the surface; also fly readily. Eggs are laid on twigs and leaves in water. Larvae fully aquatic, living at the bottom of pools. Adults feed on small insects trapped in water surface film; larvae predatory. **Habitat** Typically in flowing water, rarely vleis.

1 Family Rhysodidae
Wrinkled bark beetles

Small (body length 5–8mm), elongate beetles with a relatively long reddish-brown to black prothorax. Thorax and elytra deeply grooved longitudinally; head also with grooves. There is a short neck between head and thorax. Antennae with 11 segments, resembling a string of beads. Legs robust and short. Larvae pale, soft-bodied and up to 9mm long. **Biology** Adults and larvae found in moist, rotten wood infested with slime moulds on which they probably feed. Adults of some species able to fly. Larvae live in short tunnels, but adults do not burrow, simply squeezing their way through the wood fragments. **Habitat** Cosmopolitan, occurring in moist tropical and temperate forests.

2 Family Hydrophilidae
Water scavenger beetles

Small to very large (body length 1–50mm), shiny black, dark brown or greyish, with oval, convex bodies, resembling some dytiscids. They kick the flattened hind legs alternately when swimming (unlike dytiscids). Adults have short, clubbed antennae, and most are easily recognized by the long maxillary palps and long spine (see arrow) between the legs **(2)**. *Hydrophilus* (body length 14mm) **(2A)** is typical. **Biology** Adults of some smaller species congregate in large numbers on the undersurface of stones lying in water, and surface head-first for air. Some are terrestrial, living in decomposing vegetable matter, including dung. Larvae **(2B)** are predatory. Feeding habits of adults vary, some being predatory, others scavenging on decaying plants. **Habitat** Typically in fresh water; also damp terrestrial habitats.

3 Family Spercheidae
Filter-feeding water scavenger beetles

Small (body length 3.5–5mm), broad, convex, dull brown or yellowish water beetles. Head notched between the eyes. Antennae with 7–9 segments, ending in a club with 3–4 segments. Elytra often with ridges or elongate tubercles. Larvae with large head, the abdomen widest at mid length, with conical projections along each side; long, slender legs and antennae. Only 3 species in the region, *Spercheus cerisyi* **(3)** is typical. **Biology** Larvae and adults aquatic but are poor swimmers. Hang upside down below water surface while filtering food, adults feeding mainly on algae, larvae on organic remains and small invertebrates. Eggs are laid in an egg case carried by the female below the abdomen, supported by the hind legs. Pupation occurs out of water. **Habitat** Cosmopolitan, inhabiting still, shallow waters rich in vegetation such as ponds, dams, stock water troughs and slow-moving streams.

4 Family Histeridae
Steel beetles, hister beetles

Small to medium-sized (body length 1–20mm), hard-bodied, shiny, mostly black but sometimes red, orange, green or even bi-coloured. Elytra shortened, exposing the last or last 2 segments of the abdomen. Recognized by conspicuous mandibles and deep retraction of head into pronotum. Legs short, with broad tibiae. The largest species are *Pactolinus* **(4)**. **Biology** Larvae and adults prey on insects, including dung beetle larvae of genus *Onthophagus*. **Habitat** Near dung or decaying matter such as carrion, in birds' nests, the burrows of small mammals and under bark. A few live in the nests of ants and termites.

5 Family Hydraenidae
Minute moss beetles

Very small (body length 1–7mm). Body outline usually disrupted between pronotum and elytra. Rows of punctures occur on the elytra. About 200 species in the region. Genus *Mesoceration* **(5)** is endemic to South Africa and Lesotho. Larvae very small and elongate. **Biology** Some adults truly aquatic, others semi-aquatic, poor swimmers but able to move around upside down on the underside of the water surface, aided by an air bubble attached to ventral side of the abdomen. Also crawl about on submerged vegetation. Feed on algae, various other microorganisms and detritus occurring on plant material and wet stones. Larvae terrestrial, preferring damp areas close to water, scavenging or feeding on filamentous algae. *Mesoceration* is characteristic of riffle areas and often found together with Elmidae and Dryopidae. Useful indicators of water quality. **Habitat** Cosmopolitan, frequenting wet mosses in benthic zones of fast-flowing rivers and on riverbanks, some in stagnant and saline pools.

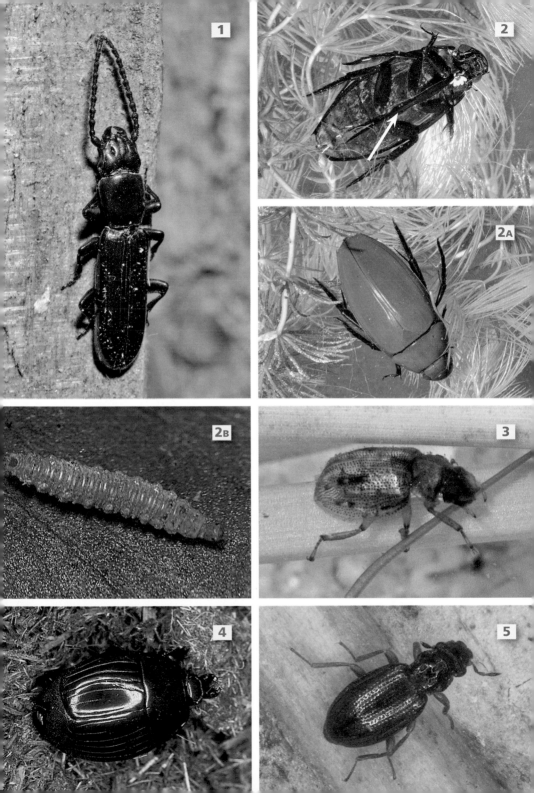

1 Family Silphidae
Carrion beetles

Medium-sized (body length 10–40mm), flattened, dark brown, sometimes with a metallic blue or green sheen. Antennae weakly clubbed. Elytra ribbed and shortened, exposing 1–4 abdominal segments. Pronotum expanded sideways. *Thanatophilus mutilatus* **(1)** shown. Larvae blackish and flattened. Only 3 species in region. **Biology** Larvae and adults feed on carrion, snails or caterpillars. **Habitat** Under corpses of birds and mammals.

2 Family Staphylinidae
Rove beetles

Small to large (body length 1–20mm), narrow and elongate, with very short elytra, exposing at least 5 or 6 abdominal segments. Most species, such as *Zyras* (subfamily *Alaeocharinae*) **(2)** (body length 8mm) and *Dolicaon* (subfamily *Paederinae*) **(2A)** (body length 18mm), are dark. Some other members of the *Paederinae* **(2B)** (body length 6mm) have bright or metallic colours. *Cyparium* (subfamily *Scaphidiinae*) are shorter **(2C)** (body length 5mm), with relatively large elytra. A large and varied family with thousands of regional species, many still undescribed. **Biology** Usually seen running on the ground, with the abdomen curved upwards, as in earwigs. Adults and larvae prey on mites, other insects and small worms. Many feed on decaying plant and animal matter, fungal spores and hyphae, or on algae. Some larvae parasitize fly pupae. *Paederus* species produce caustic liquids that can cause severe skin irritation if the insects are crushed on skin. Adept fliers. **Habitat** Diverse, including flowers, fungi, moss, leaf litter, dung, decaying plant material, carrion, soil beneath stones, sandy areas, the nests of mammals, birds, ants or termites, and some shorelines. Attracted to lights in large numbers.

3 Subfamily Scydmaeninae (Family Staphylinidae)
Ant-like stone beetles

Somewhat atypical Staphylinidae. Small (body length 1–7mm), black or reddish-brown, ant-like with globular abdomen and constriction between pronotum and elytra. Head with distinct neck. Antennae elbowed as in ants, sometimes with rows of long hairs on the 2 basal segments. About 200 regional species; *Mastigus* **(3)** is typical. **Biology** Predatory, feeding mostly on mites. **Habitat** Below bark, in moss on tree trunks, in leaf litter or rock cracks.

4 Family Rhipiceridae
Fan-horn beetles, parasitic comb beetles

Brownish-grey, covered with short, closely set hairs. Males have strongly fan-shaped antennae. The unidentified species shown here (body length 12mm) is typical. Fewer than 10 species in 3 genera known from the region. **Biology** Larvae live in soil and parasitize the larvae of cicadas. Adults may be seen visiting flowers, probably feeding on pollen, and are attracted to lights. **Habitat** Most common in bushveld, but rarely encountered.

5 Family Lucanidae
Stag beetles

Medium-sized to large, smooth, black or reddish-brown. Head projects forwards, males equipped with particularly large toothed mandibles. Antennae elbowed; segments of antennal club cannot be folded together. Pronotum relatively short or, in wingless species such as the Table Mountain stag beetle *Colophon westwoodi* **(5)** (remains of 2 dead beetles shown) (body length 25mm), about half length of body. There are about 22 species in 6 genera of stag beetle in the region. **Biology** Adults do not feed or (possibly) take in nectar, fruit juices or plant sap. Larvae, except those of *Colophon*, bore in decaying wood. *Colophon* larvae probably feed on roots of restios. Specimens of this genus are highly sought after by collectors, and the genus is listed as a Red Data Book genus. **Habitat** Southern and eastern forests. *Colophon* restricted to mountains of the southwestern Cape.

Family Trogidae
Carcass beetles, hide beetles

Small to medium-sized heavily sculptured, greyish-black beetles, with head usually concealed under the pronotum. Legs are retractile. Larvae are typically scarabaeoid, white and C-shaped, with a dark, hardened head and well-developed legs. Adults and larvae feed on animal carcasses, carnivore faeces, feathers, fur, skin, and pellets produced by birds of prey. The long-lived adults occur directly under the food source, the larvae in vertical burrows in the soil and below their food source. Capable of digesting keratin. Adults may be attracted to lights. About 50 species in the region; most common in arid areas and dry savanna.

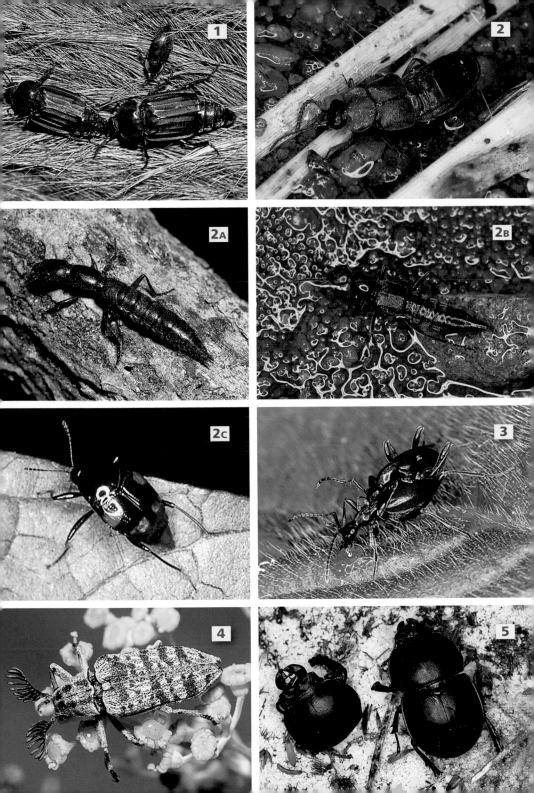

1 *Omorgus asperulatus*

Medium-sized (body length 15mm), with body, especially the abdomen, domed. Head and pronotum coarsely punctured, brownish, with black raised areas and tubercles, forming a pattern. Elytra black, with rows of black tubercles and scattered brownish patches. Legs black, mid and hind tibiae with brownish sections. **Biology** Nocturnal. Produces a noise by rubbing the abdomen against the elytra. **Habitat** Arid savanna and desert. **Related species** *Omorgus foveolatus* (**1A**) (18mm long), which has convex elytra studded with numerous small black tubercles. It occurs in the drier regions of the Kalahari and Namibia, and in Madagascar. There are also numerous similar species in the region.

2 *Trox sulcatus*

Medium-sized (body length 11mm), body black with sparse tufts of stubby reddish-brown hairs and margins of similar hairs. Reddish-brown hairs on the legs. Head and pronotum punctured, with coarse sculpturing. Elytra with serrated longitudinal ridges. **Biology** One of last insect groups to colonize dry old carcasses. Adults live for several years. **Habitat** On or under carcasses, with the larvae in vertical burrows below. **Related species** Many other similar-looking *Trox* species occur in the region.

3 Family Bolboceratidae   Dor beetles

Medium-sized to large (body length 10–25mm), very convex, globular, reddish-brown. Similar in appearance and overall habits to dung beetles (below). Males of some species with conspicuous horns on head and pronotum, the latter punctured. Antennae clubbed, mandibles large and clearly visible. Front tibiae broad, adapted for digging. Elytra with longitudinal grooves. Underside covered with long hairs. *Meridiobolbus faustus* (**3**) is typical. About 40 species in the region. **Biology** Virtually unknown. **Habitat** Bushveld, grassland. Attracted to lights.

Family Scarabaeidae   Scarab beetles, dung beetles

Small to large beetles, stout, with large head and pronotum. Mostly brown or black, sometimes patterned with yellow, white or red; otherwise metallic green or blue. The abdomen is often not completely covered by the elytra. Front tibiae are broad for digging. Shape and appearance highly variable, but members of the family easily recognized by the antennae, which have an apical club of 3–7 flat, expanded, moveable plates that can open out fanwise. Adults and larvae feed on fresh and decaying plant matter, nectar, dung and fungi. They are also found under bark and in the nests and burrows of vertebrates, termites and ants. The family contains about 3,000 regional species, including many serious agricultural and horticultural pests, as well as highly beneficial dung beetle species.

4 *Campulipus limbatus*   Long-legged Flower Chafer

Medium-sized (body length 14mm), relatively soft-bodied and somewhat flattened. Head and pronotum black, with a curved orange band along each side of the pronotum. Elytra orange, with a characteristic black pattern. Legs black, spiny and very long, with especially long tibiae. **Biology** Adults fly about actively in sunshine, visiting flowers on which to feed. In flight they resemble bees. **Habitat** Diverse areas, with flowering plants. **Related species** Several other similar species occur in the genus, some with metallic green on a black body.

5 *Goliathus albosignatus*   Goliath Beetle

Massive (body length 45–70mm), with a velvety back. Top of head white, with a forked extension in the male, rounded in the female. Pronotum black, with white longitudinal lines and margins, scutellum black, with white markings. Elytra white, with irregular black markings. The largest fruit chafer in the region. **Biology** Adults feed on various fruits, also on sap flows and gum oozing from *Senegalia* trees. Males may fight each other, and males of other species, at such sites. Aggregate at dusk on 'sleeping trees', including Marula and Buffalo Thorn. **Habitat** Subtropical savanna and woodland.

6 *Mausoleopsis amabilis*   White-spotted Fruit Chafer

Small (body length 10–13mm), shiny black. Pronotum lightly punctured, with 1 large or 2 smaller white patches down each side. Elytra also punctured, with grooves and 3 large white patches down each side. Some specimens also show smaller white dots. **Biology** Adults attracted to flowers, fruit baits and sap flows. Sometimes found in birds' nests. Larvae develop in horse and goat dung. **Habitat** Subtropical forest and savanna.

1 *Tephraea dichroa*   Wild Potato Fruit Chafer

Small (body length 13mm). Colour variable. Pronotum usually orange, with 4 (sometimes 2) black dots, but may be plain greenish-brown or even yellow. Elytra dark brown or greenish-brown, with longitudinal rows of punctures and a row of 4 small white or yellow dots. **Biology** Adults attracted to fruit and flowers; also feed on fruit and the vegetative parts of poisonous *Solanum* plants. **Habitat** Savanna and bushveld. **Related species** Several similar species occur, but none with a row of white dots on the elytra.

2 *Pachnoda sinuata*   Garden Fruit Chafer,
Brown-and-yellow Fruit Chafer

Medium-sized (body length 24mm), with a smooth surface, colour pattern variable, usually yellow, with a dark brown central area broken by yellow spots and a transverse yellow line across the back of the elytra. **Biology** Adults feed on flowers and fruit. Larvae (**2A**) develop in manure, compost and among plant roots, and pupate in hard, oval clay cells, which are often found in compost heaps. An abundant and familiar garden pest, often damaging fruit on trees. **Habitat** Diverse. Common in gardens.

3 *Rhabdotis aulica*   Emerald Fruit Chafer

Medium-sized (body length 24mm), bright metallic green or blue. Pronotum with a white line down each side. Elytra with longitudinal lines of punctures, white lines and dots around the edges, and a variable number of lines and dots towards the centre. Underside bordered with long brownish hairs, commonly showing copperish-red reflections. Legs metallic green, with brownish hairs. **Biology** Adults feed on fruits and flowers, including proteas. Larvae develop in goat and cattle manure. **Habitat** A range of vegetation types. **Related species** There are 4 other similarly marked species that vary in colour from brown to green to black.

4 *Cyrtothyrea marginalis*   Common Dotted Fruit Chafer

Small (body length 9mm), black, covered with white spots. Pronotum lightly punctured, with reddish sides. Elytra with broad, shiny grooves with lines of punctures between them. **Biology** Adults attracted to fruits and various flowers (especially arum lilies) and may damage ornamental plants. **Habitat** Diverse, including gardens. **Related species** There are several other white-spotted black species, but none with red sides to the pronotum.

5 *Leucocelis amethystina*   Amethyst Fruit Chafer

Small (body length 12mm). Pronotum shiny, reddish-brown, sometimes with a broad black line down the middle. Elytra light metallic blue or green with broad, shiny grooves, sometimes merging to form lines. **Biology** Adults feed on flowers of various shrubs and trees, such as *Senegalia*, and are attracted to fruit baits and cattle dung. **Habitat** A range of vegetation types. **Related species** *L. rubra* (**5A**) (14mm long), which is a uniform dark reddish-brown. There are also about 4 species with similar coloration to *L. amethystina*, but these usually have white markings.

6 *Xeloma tomentosa*   Fluffball Chafer, Gold-haired Fruit Chafer

Small (body length 13mm), brown, with yellowish markings, body characteristically clothed in dense golden hairs. **Biology** Adults feed on composite flowers, including sunflowers. **Habitat** Mountain, highveld and escarpment. **Related species** The 4 other species in the genus are dark brown or black and lack golden hairs. Also, *Atrichelaphinis tigrina* (**6A**) (12mm long), which is a dull copper-coloured species, with black marbling on the elytra, that pollinates proteas, and *Elaphinis irrorata* (**6B**) (11mm long), which is patterned in dull green and white.

1 *Plaesiorrhina plana* — Yellow-belted Fruit Chafer

Large (body length 25mm). Pronotum black, punctured, with yellow margins. Elytra also punctured, without grooves, black, with an irregular, bright yellow transverse band. **Biology** Adults feed on various flowers and fruits, and also on sap flows. Sometimes attracted to beehives. Larvae develop in humus. **Habitat** Eastern savanna areas, excluding highveld. Found in all forests. **Related species** *Pedinorrhina trivittata* (**1A**), which is superficially similar but much smaller (body length 16mm) and more variable; some forms are brown.

2 *Chlorocala africana* — Gleaming Fruit Chafer

Medium-sized (body length 23mm), shiny, iridescent yellow-green to bluish. Front of head with characteristic median ridge ending in a rounded transverse ridge. Pronotum sparsely punctured. Elytra with lines of punctures but not grooved. **Biology** Adults feed on sap flows, the flowers of various trees and shrubs, including proteas, and *Cassia* pods. **Habitat** Subtropical forest, woodland and savanna.

3 *Mecynorhina passerinii* — Orange-spotted Fruit Chafer

Large (body length 30–45mm), with a velvety surface. Pronotum greyish-yellow, with a broad, black central longitudinal line. Elytra black or dark brown, with large orange or red spots. Legs black, except for reddish-brown hind tibiae and tarsi. Large, forward-projecting cream horns in male, but not in female (shown here). **Biology** Adults feed on sap flows and are attracted to fermenting fruit. **Habitat** Moist forests.

4 *Taurhina splendens* — Regal Fruit Chafer

Large (body length 30mm), distinctively coloured. Pronotum metallic green, with a chalk white border. Elytra chalk white, with a metallic green patch around and including the scutellum and, sometimes, a green line on each elytron, from anterior corner to tip. Head oblong, with 2 hooked projections at the rear between the eyes, and a white top in the male (shown here). Legs metallic green. **Biology** Adults feed on sap flows and mangoes. **Habitat** Hot tropical savanna. **Related species** Derby's Flower Beetle (*Dicronorrhina derbyana*) (**4A**), which often feeds alongside *T. splendens*, is a similar, but larger (40–50mm) species, with more green and much less white on the elytra and different ornamentation on the head.

5 *Anisorrhina flavomaculata* — Zigzag Fruit Chafer

Large (body length 25mm), shiny reddish-brown. Pronotum smooth, with 4 black spots that may coalesce to more or less cover it. Scutellum with 2 black spots. Longitudinal rows of punctures on the elytra, most of which may be covered by 2 large characteristically shaped yellow patches outlined in black. **Biology** Adults feed on various fruits, including prickly pears, flowers and sap flows. They also enter beehives. Larvae develop in dung and compost mixtures. **Habitat** Diverse. **Related species** *Pedinorrhina trivittata*, which is very similar but smaller, with 3 black longitudinal lines on the pronotum.

6 *Trichostetha fascicularis* — Green Protea Beetle

Large (body length 25mm). Pronotum smooth, black, with 4 or 5 white longitudinal grooves, scutellum black. Elytra with indistinct grooves and punctures, green in southern forms, becoming reddish with white patches in northern forms. Underside with yellow to orange hairs visible from above as tufts around the elytra. Legs with similar hairs. **Biology** Adults buzz noisily between *Protea* plants or bury themselves head-down in the flowers, feeding on pollen and nectar. Not attracted to fruit or fruit baits. **Habitat** Wherever proteas occur. **Related species** *Odontorrhina pubescens* (21mm long) (**6A**), which is dull green and covered in long, fine hairs, and *T. capensis* (15mm long) (**6B**), which has a white-speckled black pronotum with a brown central stripe and the elytra wholly or partially brown, also with white speckling.

1 *Porphyronota hebreae* — Marbled Fruit Chafer

Medium-sized (body length 21mm), yellowish-brown, patterned with dark brown spots and patches. Pronotum sparsely punctured. Elytra with longitudinal grooves. **Biology** Adults feed on a variety of fruits and sweet exudates from grass stems (as shown here), and are attracted to fruit baits. Sometimes enters beehives. **Habitat** Subtropical bushveld and grassland. **Related species** *P. maculatissima*, which is similar but darker and has an overlapping range; it occurs in subtropical forest, where it feeds on various fruits and flowers.

2 *Diplognatha gagates* — Large Black Nest Chafer

Large (body length 22–32mm), shiny black, smooth; newly emerged adults have reddish-brown tinge. Thorax strongly convex. **Biology** Larvae develop in roofing thatch, large birds' nests and compost heaps. Adults feed on green vegetation, flowers, fruits and sap; larvae feed on bird droppings. **Habitat** Diverse. **Related species** The Hive Chafer (*Oplostomus fuligineus*), which is shiny black, but has a finely punctured head and pronotum and a few broad, shallow grooves on the elytra. It frequents beehives, feeding on the brood and honey, and destroys the nests of social wasps. Also, *Xiphoscelis shuckardi* (**2A**), which is dull grey, with a raised tubercle at the end of each elytron, and with conspicuous tibial spines on the hind legs. Its larvae live in the frass (faecal pellets) of the Southern Harvester termite (*Microhodotermes viator*) (p.64), and adult beetles fly in early spring. Widespread across the Western and Northern Cape.

3 *Scelophysa trimeni* — Blue Monkey Beetle

Small (body length 9mm). Male covered with powder blue to greenish-blue scales. Female green to brown. Both sexes have a sparse covering of black hairs. Hind legs strongly developed in male. Tarsal claws double and uneven. **Biology** Adults feed on flowers, especially blue-flowered flax (*Heliophila*). May embed themselves in the discs of daisies and mesembs when feeding. Larvae probably feed on roots and plant debris. **Habitat** Coastal sandy areas, and sandy parts of the 'Knersvlakte'. **Related species** *Lepisia rupicola* (**3A**), which is a small (8mm long) apple green species found in the Western Cape. Also, *Scelophysa militaris* (**3B**), which is 1 of a group of similar daisy-inhabiting species, with maroon to orange elytra and a white abdomen.

4 *Anisonyx ditus* — Glittering Monkey Beetle

Small (body length 9mm). Pronotum, elytra and exposed terminal segments of abdomen covered with metallic turquoise scales. Body, except for elytra, densely clothed in long, fine hairs. Legs black and hairy, hind legs long and moderately strongly developed. **Biology** Adults feed on flowers, mainly *Leucadendron*. Larvae probably feed on roots and plant debris. **Habitat** Dry fynbos. **Related species** Other *Anisonyx* species, which occur at altitude, except for *A. ignitus* (**4A**), which is found in the Ceres Karoo and has bright red scales on the body. An undescribed *Anisonyx* species occurs in the Kamieskroon uplands (**4B**) and has pink body scales.

5 *Heterochelus chiriagricus* — Striped Monkey Beetle

Small (body length 10mm). Sexually dimorphic. Male black, with yellow or orange elytra that show variable black patches; female smaller, with dull brown elytra with 4 faint bars. Legs reddish-brown; hind legs stout and very strongly developed in male, with spines that form an efficient pincer when the leg is folded. **Biology** Males display aggressively if disturbed and fight over females. **Habitat** Succulent karoo and dry fynbos. **Related species** *H. detritus* (9mm long) (**5A**), which is highly dimorphic: the female is dull brown, the male (shown) larger, with enlarged, reddish hind legs and a black pronotum.

1 *Peritrichia cinerea* Oval Hairy Monkey Beetle

Small (body length 9mm), matt black, clothed in long, soft, greyish-brown and black hairs. Hind legs stout and strongly developed. **Biology** No aggression between males. Feeds mostly on pollen and probably also on the nectar of white, blue and purple flowers, particularly Liliaceae. **Habitat** Strandveld. **Related species** The Shaggy Monkey Beetle (*Anisonyx ursus*) (**1A**), which occurs in the same montane habitat, is easily confused with *P. cinerea*. *Anisonyx* species are more commonly found on fynbos flowers, including proteas and leucodendrons.

2 *Mauromecistoplia nieuwoudtvillensis* Large Barred Monkey Beetle

Small (body length 10mm). Male brownish-black, with white-fringed pronotum and black legs. Chequered black-and-white edge of abdomen not covered by elytra. Female with pronotum and elytra covered in orange scales, elytra with 4 black ribs devoid of scales, and underside with white and orange hairs. Scutellum white in both sexes. **Biology** Adults frequent *Didelta*, *Berkheya* and *Gazania* flowers. **Habitat** Karoo hill slopes.

3 *Pachycnema marginella* Brown Monkey Beetle

Small (body length 10mm). Head and pronotum black, with dark bristles. Pronotum bordered with yellow or white scales, with 2 or 4 patches of scales in the centre. Scutellum black, with yellow or white scales. Elytra brown, with a few white scales. Underside with long white hairs, except on the abdomen, which is covered with white scales. Legs stout, especially the hind pair. **Biology** Adults burrow into discs of daisies and thistles. **Habitat** Succulent karoo. **Related species** *Hoplocnemis crassipes* (**3A**), which is a massive monkey beetle; males have hugely enlarged and armed hind legs. They apparently do not feed on flowers and spend most of their lives underground.

4 *Lepithrix pseudolineata* Spider Monkey Beetle

Small (body length 9mm). Head black, with yellow hairs on the face. Pronotum black, outlined with yellow scales and with 2 yellow stripes in the centre. Elytra brownish, each with a yellow line along the inner margin, and long black hairs, especially around the margins. Underside densely covered in long golden hairs. Legs long and slender. Larvae (**4A**) small and C-shaped. **Biology** A very active flier, seen only in spring, often on mesembs. Larvae feed on organic matter in sandy soil. **Habitat** Succulent karoo. **Related species** *Kobousa plumipes* (**4B**), which is a rare species known only from around the Clanwilliam area. The feathery hind legs of the males are probably used in courtship or male–male competition.

5 *Popillia bipunctata* Yellow Shining Leaf Chafer

Medium-sized (body length 14mm). Head, pronotum, scutellum and legs shiny black. Legs quite stout. Elytra yellowish-brown, with longitudinal rows of punctures. End of abdomen not covered by elytra, black, with 2 patches of white hairs. Bands of similar white hairs on underside. **Biology** Adults active by day, visiting flowers on which they feed. Larvae develop in soil. **Habitat** Diverse, usually in areas with flowers, in summer-rainfall regions.

6 *Adoretus ictericus* Wattle Chafer

Medium-sized (body length 13mm), yellowish-brown, with a darker brown head. Body surface shiny and finely punctured. **Biology** Adults feed nocturnally on foliage and flowers, commonly damaging garden plants, especially roses, and hide by day in leaf litter or soil. Often attracted to lights. The subterranean C-shaped larvae, known as 'white grubs', feed on decaying plant material, sometimes plant roots. **Habitat** Diverse. **Related species** There are many similar species in this and other genera, including the honey-coloured *Anomala* species (18mm long) (**6A**).

1 *Hypopholis sommeri* Rose Chafer, Large Wattle Chafer

Medium-sized (body length 18mm), shiny and finely punctured. Elytra pale brown, with 4 darker longitudinal bands and darker margins. Pronotum dark brown, sometimes with 2 dark patches. **Biology** Adults feed nocturnally on flowers and the foliage of wattle, potato, sugar cane and other plants. Larvae feed on tubers of potatoes and various other plants. Commonly damages garden plants, especially roses. Attracted to lights. Hides by day in leaf litter or soil. **Habitat** Forest, gardens and crop lands.

2 *Schizonycha* Brown flower chafers

Medium-sized (body length 18mm), soft-bodied and bulbous, entirely reddish-brown. Head, pronotum and elytra punctured. Contains several similar species. **Biology** Hide by day in leaf litter or soil. Typical 'white grub' larvae. Adults feed at night on flowers and the foliage of plants such as wattle, sugar cane and potato. Cause extensive damage to garden plants. **Habitat** Diverse, including gardens. **Related species** *Eucamenta transvaalensis* (**2A**) (body length 12mm), which is a smooth reddish-brown species, common in the eastern summer-rainfall region and typically attracted to lights.

3 *Sparrmannia flava* Woolly Leaf Chafer

Large (body length 24mm), with a dense covering of long, fine, fawn-coloured hair on the head, pronotum, legs and underside. **Biology** Larvae live in burrows in soil in and around accumulations of the dung of springbok and, probably, other mammals, and feed on dung pellets collected from the surface. Adults fly at dusk or at night and feed on vegetation. **Habitat** Arid and semi-arid savanna and karoo. **Related species** There are 18 other species in the genus.

4 *Cyphonistes vallatus* Fork-horned Rhino Beetle

Large (body length 29mm), dark brown to black. Pronotum shiny and smooth, with sparse punctures. Elytra with lines of shallow punctures. Forked horn on head and forward-pointing forked projection on pronotum of male (**4**). **Biology** Adults probably do not feed; they are nocturnal and commonly attracted to lights. The typical C-shaped scarabaeoid larvae feed on rotting vegetation and pupate in the soil without making a cocoon. **Habitat** Diverse.

5 *Heteronychus arator* Black Maize Beetle, African Black Beetle

Medium-sized (body length 12–15mm), entirely shiny black. Head with transverse sculpturing. Pronotum very lightly and sparsely punctured. Elytra with longitudinal rows of distinct punctures set in shallow grooves. Front tibiae toothed for digging. **Biology** Eggs are laid below the soil surface. Larvae (**5A**) feed on decomposing plant material in soil. Adults chew into the stems of plants just below the surface, especially maize and pineapple, and also bore into potato tubers. Commonly causes damage to lawns and turf. Adults fly at night and are attracted to lights. Swarming flights common in spring and early autumn. **Habitat** Grassland, cultivated pastures and maize fields. Cosmopolitan, found in Africa, Australia and South America.

6 *Oryctes boas* Rhinoceros Beetle

Large (body length 44mm), familiar, shiny dark brown, with lines of shallow punctures on the elytra. Males with long, backward-curved horn on the head, sometimes equal in length to pronotum, which has a concave depression with elevated sides and 2 projections at the rear. **Biology** Adults burrow into the growing apex of a coconut palm to feed on the juice generated from the young shoot, sometimes killing the plant. Also a pest of bananas and sugar cane. Nocturnal, and commonly attracted to lights. The typical C-shaped larvae feed on rotting vegetation, especially compost and horse manure, and pupate in soil without making a cocoon. **Habitat** Diverse, but most common in savanna. **Related species** *O. monoceros*, which is similar, but darker in colour, and has a shorter horn and smaller concavity in the pronotum.

1 *Garreta nitens*
Green Dung Beetle

Medium-sized (body length 13–18mm), metallic green or copper, finely sculptured. Pronotum strongly domed. Elytra with fine grooves, outer margins with indentations. **Biology** Active by day. Cuts dung from a fresh dung pad and moulds it into a ball, which it rolls away and buries at some distance. The dung ball is then remodelled by the female into a brood ball into which an egg is laid. When mature, the larva pupates within the remains of the brood ball. **Habitat** Savanna and bushveld. **Related species** There are several other similar species with different metallic colours.

2 *Gymnopleurus humanus*
Small Green Dung Beetle

Small (body length 12mm), dark metallic blue or green with a finely punctured surface. Pronotum strongly domed, concealing the head from above. Outer margin of elytra with characteristic indentation. **Biology** Active by day. Shapes fresh dung into a ball, which it rolls away to bury at some distance. The female remoulds the ball to create a brood ball into which an egg can be laid. The mature larva pupates within the remains of the brood ball. Recorded from a wide variety of dung types, but omnivore dung is preferred; also favours carrion. **Habitat** Dry savanna and bushveld. **Related species** *G. ignitus* (**2A**), which is a small (11mm-long) species that occurs in a range of metallic colours from blue to red.

3 *Heteronitis castelnaui*
Grooved Dung Beetle

Large (body length 28mm), shiny black, with oblong shape typical of the genus. Head and pronotum with sculpturing, elytra with conspicuous longitudinal grooves. Pronotum notably large. **Biology** Adults are active mostly around dusk and draw dung down into tunnels dug beside and below the dung source. Recorded entirely from the coarse, fibrous dung of elephant, rhino, horse and zebra. For breeding, dung is packed into the blind ends of the tunnels, forming sausage-shaped brood masses, into each of which a series of eggs is laid. Mature larvae make pupal cases from their own faeces before pupating. **Habitat** Dry to moist open woodland savanna.

4 *Onitis alexis*
Bronze Dung Beetle

Medium-sized (body length 16mm), with oblong shape typical of genus. Head metallic green. Pronotum metallic green and punctured. Elytra brownish, with longitudinal grooves and tinges of green. **Biology** Adults active at dusk and dawn, burying dung in tunnels dug beside and below the dung source. Prefers the dung of large herbivores, such as elephant, buffalo and cattle. For breeding, a nest is built by male–female pairs, with dung packed into the blind ends of the tunnels, forming sausage-shaped brood masses, into which eggs are laid sequentially. Larvae construct pupae from their own faeces, and emerging adults dig their way to the surface. **Habitat** Diverse.

5 *Sisyphus*
Spider dung beetles

Small (body length 3–12mm), dull black, dark grey or brown. Pronotum and elytra covered with hairs, varying in thickness and colour (black, reddish-brown or yellow), sometimes arranged in tufts. Hind legs very long, due to the elongate femora and tibiae. **Biology** Adults congregate to feed on fresh dung, especially that of small mammals, but also of tortoises, toads, birds and, uniquely, large carnivores. They form balls of dung for breeding and roll them away to be buried up to 50mm below the surface. A single egg is laid in the brood ball. Some species leave the brood ball on the ground or attached to grass stems or twigs. **Habitat** Savanna, bushveld and forest.

6 *Copris mesacanthus*
Nursing Dung Beetle

Medium-sized (body length 17mm), dull black, surface strongly punctured, especially on the elytra, which also have shallow grooves. Males have a backward-curving horn on the head, with a small rear projection. Pronotum with central concavity, each side drawn up into a blade-like extension. **Biology** Adults fashion soil-coated brood balls from dung. These are taken down burrows below or beside fresh dung pads. Some dung is retained as food. Females remain in the nest until their progeny mature. Most active by night. **Habitat** Savanna and bushveld. **Related species** *C. caelatus* (**6A**) (17mm long), which is a matt black species with fine punctations on the elytra, and which occurs at high altitudes and is active by day. Also, there are many other similar species, differing in the sculpturing of the pronotum and structure of the horn in males.

1 *Heliocopris neptunus* — Trident Dung Beetle

Large (body length 37mm), dark brown, surface of head wrinkled. Pronotum punctured, elytra shiny with fine grooves. Male **(1)** with fork-like 3-pronged extension to pronotum and backward-curved horn on the head, slightly forked at tip. **Biology** Adults burrow below or beside fresh dung pads, taking quantities of dung down into a chamber, where several soil-coated brood balls are made, each with an egg. Some dung is retained as food. Females remain in the nest until their progeny emerge as adults. **Habitat** Savanna and bushveld.

2 *Catharsius tricornutus* — Three-horned Dung Beetle

Large (body length 26mm), black. Head and pronotum lightly punctured, elytra shiny with fine grooves. Male **(2)** with a more or less straight horn on the head and 2 variably developed, diverging horns on the pronotum. Female with a short, transverse central ridge at front of pronotum. **Biology** Nocturnal, often attracted to lights. Adults burrow below or beside fresh dung pads, taking down quantities of dung. Jointly, males and females pack dung for breeding into hemispherical layers in expanded sections of tunnels, or at the ends of tunnels, forming spherical brood masses that resemble brood balls, each with an egg. No brood care. **Habitat** Diverse, mostly savanna, grassland and pastures.

3 *Anachalcos convexus* — Plum Dung Beetle

Large (body length 24mm), globular, black with a coppery sheen and finely punctured surface. **Biology** Nocturnal. Adults feed on carrion as well as dung. Male–female pairs jointly roll balls of dung away from the source, burying them in shallow burrows. After an egg is laid in the first brood ball, several balls are introduced into the nest. Females, sometimes accompanied by males, care for their progeny until they emerge as adults. **Habitat** Subtropical forest and savanna. Prefers clay-loam soils.

4 *Proagoderus aciculatus* — East Coast Shade Dung Beetle

Medium-sized (body length 12mm), metallic green, body surface strongly punctured. Elytra with widely spaced shallow grooves. Reddish hairs on head and margins of the pronotum and elytra. **Biology** Burrows below a dung source, packing dung for breeding into the blind ends of tunnels, forming a sausage-shaped mass or series of oval masses. A single egg is laid in each sausage or oval mass. **Habitat** Diverse. **Related species** *P. bicallosus* **(4A)**, which is 1 of several similar species, all of which feed on the dung of small antelope.

5 *Proagoderus tersidorsis* — Bi-coloured Dung Beetle

Medium-sized (body length 13mm). Head and pronotum metallic green, male with a long backward-curving horn on the head. Elytra brownish, darker at base. **Biology** Probably tunnels below a dung source, filling the blind end of a tunnel with dung and creating a sausage-shaped mass or series of oval masses, with a single egg laid in each mass. **Habitat** Savanna. In association with the dung of large herbivores, such as elephant, rhino and buffalo.

6 *Circellium bacchus* — Addo Flightless Dung Beetle

Large to very large (body length 30–50mm), black, moderately shiny, globular with a rounded outline. Pronotum very large, almost equal in size to the abdomen. One of a few flightless dung beetles. **Biology** Adults congregate at elephant dung (for feeding) and buffalo dung (for feeding and breeding). Dung is shaped into balls that are rolled away to be buried **(6A)**. Only 1 nesting sequence occurs per year, with a single offspring produced from each brood ball. **Habitat** Fish River scrub and 'spekboomveld'. Conservation status 'Vulnerable'.

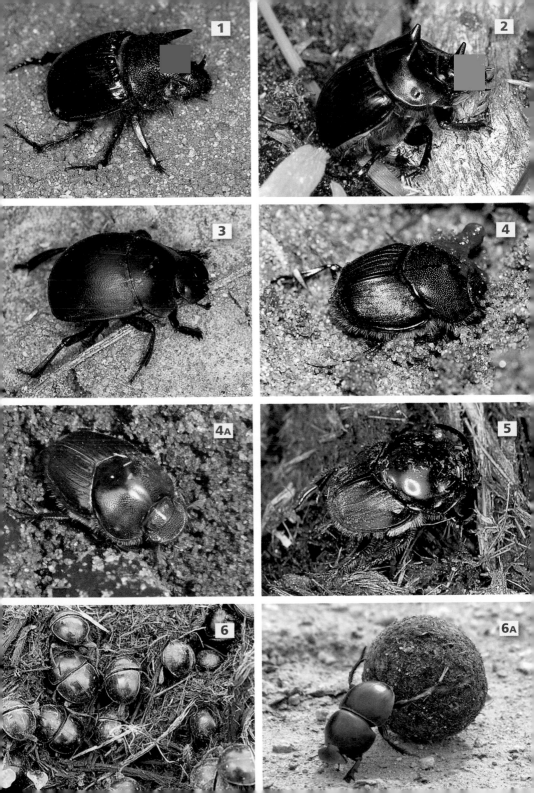

1 *Scarabaeus rugosus* — Green Grooved Dung Beetle

Medium-sized (body length 21mm), shiny black. Head strongly toothed. Pronotum and elytra with both fine and coarse punctures; elytra also with fine grooves. Tarsi end in 2 claws (unlike *S. nigroaeneus*, which has 1). **Biology** Nocturnal, but also active in the early morning. Adults cut portions of dung from fresh dung pads, forming them into balls that are rolled away jointly by male–female pairs to be buried. These are used for feeding and for preparing brood balls, in each of which a single egg is laid. **Habitat** Fynbos and succulent karoo. Prefers deep, sandy soil. **Related species** *S. proboscideus* (**1A**), which is a dull black species, common on the west coast, and *S. costatus*, which is a small (12mm long), matt black species, with a single tarsal claw and equal-sized teeth on the tibia; it occurs in the Western Cape.

2 *Scarabaeus nigroaeneus* — Large Copper Dung Beetle

Large (body length 35mm), black, with a coppery sheen. Head and pronotum punctured. Elytra with fine grooves. Tarsi end in 1 claw. **Biology** Active by day. Adults cut portions from the fresh dung of various mammals, rolling them into balls that are pushed away by both male and female, working as a pair. After a dung ball is buried, it is used for feeding and creating a brood ball ('pear'), in which a single egg is laid. Females care for the brood, and 2 broods may be raised per year. Body temperature of males is raised before fighting over females. **Habitat** Savanna and woodland.

3 *Pachylomera femoralis* — Flattened Giant Dung Beetle

Massive (body length 36–50mm), broad, somewhat flattened, dull black, with raised polished areas. Pronotum wider than abdomen. Fore legs prehensile and powerfully developed, especially the femora, which are equipped with teeth on the leading edge. **Biology** Active by day and commonly seen in flight. Adults burrow beside the fresh dung of various mammals, dumping excavated soil on top of the dung and filling tunnels with dung, probably for feeding only. Tunnels are sealed with soil. Also rolls dung balls away for burial, presumably for making brood balls. Attracted to human faeces. **Habitat** Savanna and bushveld. Prefers deep sand.

4 *Aphodius* — Miniature dung chafers

Very small (body length 4mm), pale brown, resembling miniature chafers. Elytra with longitudinal grooves. **Biology** Adults feed on and reproduce in dung or organic matter in soil. Some breed in dung collected by other species. They never bury dung. **Habitat** Diverse. Abundant in fresh cow pats. Attracted to lights in numbers.

5 *Coptorhina klugii* — Fork-nosed Dung Beetle

Medium-sized (body length 14mm), black, the surface finely or coarsely punctured. Top of head characteristically forked, with 2 broad upturned prongs. Pronotum with a steep frontal face, separated from the hind portion by a more or less sharp ridge that may have a variety of projections. Sides of elytra indented. **Biology** Despite the common name, adults are obligate feeders on wild mushrooms, portions of which are buried, presumably for larval food. **Habitat** Diverse.

6 Family Psephenidae — Water pennies

Adults seldom seen, small (body length 3mm), dark, with soft, squared-off elytra. *Afrobrianax ferdyi* has fine punctures on its elytra (**6**). Larvae small (4–7mm diameter), oval, circular and flat, resembling tan limpets (*A. ferdyi* shown) (**6A**). **Biology** Adults short-lived, eating little or nothing. Nocturnal. Larvae are aquatic and slow-moving, and can pump water out from under their bodies, improving their suction to rocks. Larvae cling to and crawl over stones or submerged wood, feeding on algae attached to the substrate. Females crawl underwater to lay eggs on a rock before dying. Some species pupate under water. There are 4 genera in the region, each with a single species. **Habitat** Larvae occur on rocks in fast-flowing streams.

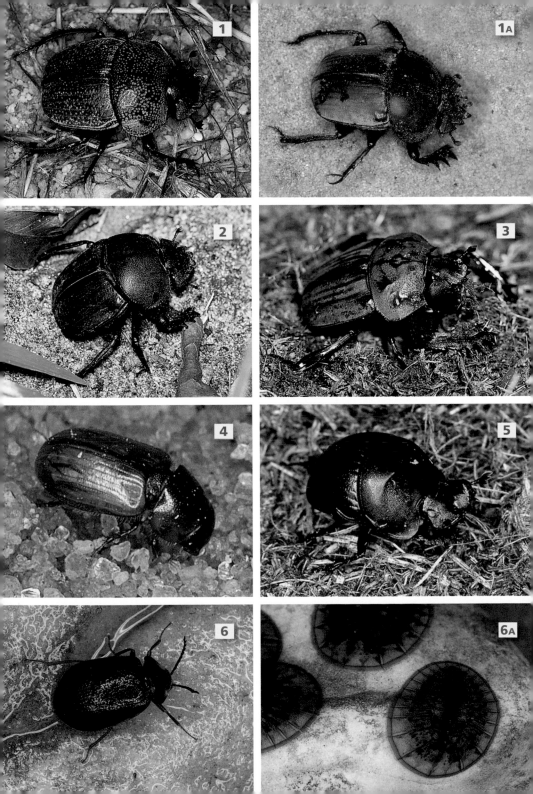

1 Family Buprestidae
Jewel beetles, Flat-headed borers

Very small to very large (body length 2–50mm) torpedo-shaped beetles with extremely hard bodies, usually bright metallic copper, blue or green. In some, these colours are restricted to the upperside of the abdomen beneath the cryptically coloured elytra, and are visible only in flight. Others are brown or black, with or without white or yellow markings. Some have bright yellow and red tufts of wax-coated hairs (setae). The antennae are thread-like, saw-toothed or comb-like, and the eyes are very large. Buprestids are alert, sun-loving and fly readily. Adults feed on nectar, pollen and foliage, and are often seen flying around the tops of flowering bushes and trees. Females lay eggs in crevices in bark. The larvae **(1)** excavate oval tunnels in wood or the stems of herbaceous plants, some even in grasses. Adults are relatively short-lived, but immatures can take up to 35 years to complete their development, longer than for any other insect. There are approximately 1,200 species in the region.

2 *Julodis cirrosa*
Brush Jewel Beetle

Large (body length 27mm), surface metallic blue-green and heavily punctured. Covered with tufts of yellow, white or red wax-coated hairs. **Biology** Larvae tunnel in stems and roots or are free-living root-feeders. Adults are short-lived and feed on the flowers and foliage of karoo shrubs, especially *Lycium*. **Habitat** Semi-arid and arid areas. **Related species** The genus contains 25 other variable and easily confused species, including *J. albomaculata* (27mm long) **(2A)**, which is black, with white waxy tufts.

3 *Neojulodis tomentosa*
Plum-and-Green Brush Jewel Beetle

Medium-sized (body length 16mm). Pronotum black, with a green or purple metallic sheen and long white and yellow hairs. Elytra green, punctured, and bearing tufts of yellowish hairs. Underside black with a green sheen, densely clothed in long, silky white hairs. Legs brown. **Biology** Details not known. Larvae probably free-living root-feeders. **Habitat** Shrubs in areas of montane fynbos. **Related species** *N. picta* **(3A)**, which is smaller (body length 12mm) and has a black pronotum with a metallic sheen, yellow stripes and punctures, brown elytra with longitudinal black stripes and bold white patches, and long silky body hairs.

4 *Sphenoptera*
Protea borers, jewel beetles

Medium-sized (body length 18mm), black. Pronotum punctured. Elytra strongly tapered, with longitudinal grooves and rows of punctures. Legs with a bronze sheen. **Biology** Larvae are wood borers of *Senegalia*, *Rhus*, *Terminalia*, *Afzelia*, Proteaceae and other shrubs and trees. **Habitat** A variety of habitats with woody shrubs and trees.

5 *Acmaeodera viridaenea*
Glittering Jewel Beetle

Small (body length 13mm). Female metallic green to purple, male greenish-brown, with coppery sides. Head and pronotum heavily punctured. Elytra have longitudinal ridges and rows of distinct punctures, and are fused together down the centre. Legs black. **Biology** Adults active during the hottest time of day, tending to hover around flowers. They feed on flower petals, mostly of *Senegalia* or *Grewia*. **Habitat** Diverse. One of the most common jewel beetles in the region. **Related species** There are many similar spotted *Acmaeodera* species (body length around 8mm), which feed on daisy flowers during the hottest time of day; some have larvae that bore into the flowering stems of aloes.

6 *Agelia petelii*
Meloid-mimicking Jewel Beetle

Large (body length 26mm), distinctive, with a black body. Mimics *Mylabris* blister beetles. Pronotum and elytra punctured, pronotum with a pair of backward-facing wing-like extensions, with metallic red bases. Elytra have 4 large yellow patches, the rear patches edged in red. **Biology** Often found in association with blister beetles and on the Sickle Bush (*Dichrostachys cinerea*) and various species of *Grewia*. **Habitat** Arid thorny vegetation.

7 *Lampetis amaurotica*
Eyed Jewel Beetle

Large (body length 24mm), black. Pronotum, except for 4 circular black patches, covered with white punctures. Elytra have white punctured longitudinal grooves, with clumps of white punctures between them, and a broad white lateral line on each elytron. **Biology** Larvae bore in wood. Host preference not known. **Habitat** Bushveld.

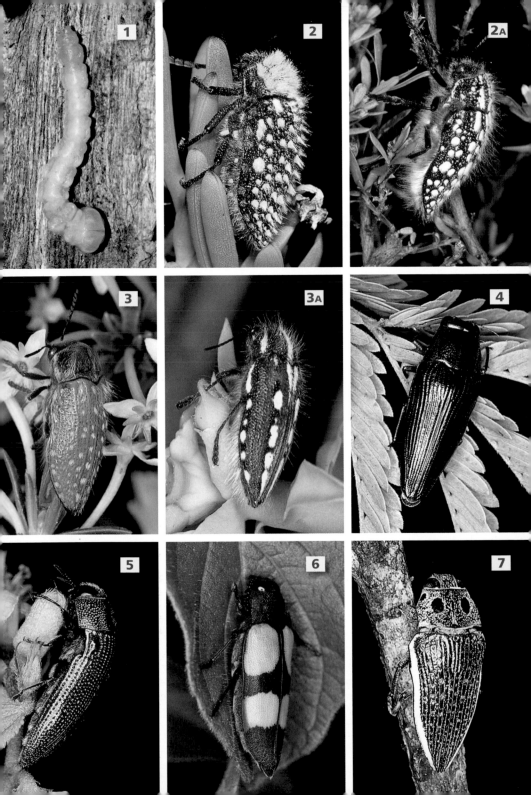

1 *Evides pubiventris* — Emerald Jewel Beetle

Large (body length 25mm), metallic green. Head metallic green and copper, punctured. In flight shows rainbow coloration at top of the abdomen. Elytra with slightly sinuous longitudinal ridges, strongly punctured in between. Underside with short white hairs. Tibiae and tarsi metallic green. **Biology** Associated with Anacardiaceae trees, including Marula (*Sclerocarya birrea*) and Live-long (*Lannea discolor*). **Habitat** Savanna.

2 *Sternocera orissa* — Giant Jewel Beetle

Large (body length 35–45mm), black and hefty. Head white or yellowish on top and between the compound eyes. Pronotum with large white or yellowish punctures and a patch on each side. Elytra each with rows of punctures and white or yellowish patches at base and along sides. **Biology** Larvae free-living root-feeders. Adults clumsy in flight, often seen around flowers of *Senegalia* and *Vachellia*. Females lay very large eggs in soil or on the ground. **Habitat** Moist savanna, bushveld and sand forest.

3 *Meliboeus punctatus* — Painted Jewel Beetle

Medium-sized (body length 12–15mm), black, with heavily punctured surface and metallic blue-green sheen. Pronotum with 4 reddish-orange patches. White patches along sides of abdomen. **Biology** Adults visit flowers. Possible velvet ant mimic. **Habitat** Mixed bushveld.

4 *Phlocteis exasperata* — Metal Jewel Beetle

Small (body length 12mm), body coarsely sculptured, with large tubercles. Head and pronotum metallic bronze. Longitudinally grooved between eyes. Elytra metallic bronze in centre, black laterally and at rear. **Biology** Associated with *Rhus* trees. **Habitat** Diverse.

5 Family Dryopidae — Long-toed water beetles, hairy water beetles

Minute to small (body length 1–8mm). Recognizable by the extended tarsal claws. Distinguished from the similar Elmidae (below) by their short, clubbed antennae. Dense covering of fine hairs on the body allows the beetles to breathe while immersed. Only about 8 species in the region. **Biology** Adults unable to swim, but cling to floating debris when in the water. Feed on aquatic plants and decaying vegetable matter. May be attracted to lights. Some larvae may prey on small animals as well, and most are terrestrial. **Habitat** Cosmopolitan; more common in the tropics. Generally associated with water. Useful indicators of water quality.

6 Family Elmidae — Riffle beetles

Minute to small (body length 1–6mm), with relatively long legs and tarsal claws. Distinguishable from the very similar Dryopidae (above) by their slender antennae, which lack a distinct club. Ventral surface of body densely covered with minute hydrophobic hairs that trap a layer of air used in respiration. Some larvae small and elongate; others larger, resembling trilobites, up to 8–10mm long with retractable, filamentous or feathery gills at tip of the abdomen. About 50 species in the region. **Biology** Larvae and most adults aquatic, feeding on pieces of dead plant and other organic matter and various microorganisms attached to the substrate. Adults attracted to lights. Mature larvae emerge from the water to pupate in soil cavities. On emerging from their pupae, adults may fly once before returning to water, where they remain permanently. **Habitat** Cosmopolitan, occurring mostly in the riffle zones of freshwater streams in temperate and tropical regions, where they crawl about on stones and other solid substrates. Useful indicators of water quality.

7 Family Heteroceridae — Variegated mud-loving beetles

Minute to small (body length 2–6mm), elongate, parallel-sided, convex; generally brownish, covered in small setae of 2 different lengths and displaying a variety of colour patterns. Resemble miniature scarab beetles, with rows of flattened spines on the broadened fore tibiae. Short, clubbed antennae. Some 23 species in the region; *Heterocerus* (7) is typical. **Biology** Distinct colonies live in galleries excavated in damp soil around lakes, rivers and ponds and on intertidal sand flats, feeding on debris and algae. Attracted to lights. **Habitat** Cosmopolitan, mostly in subtropical and tropical areas.

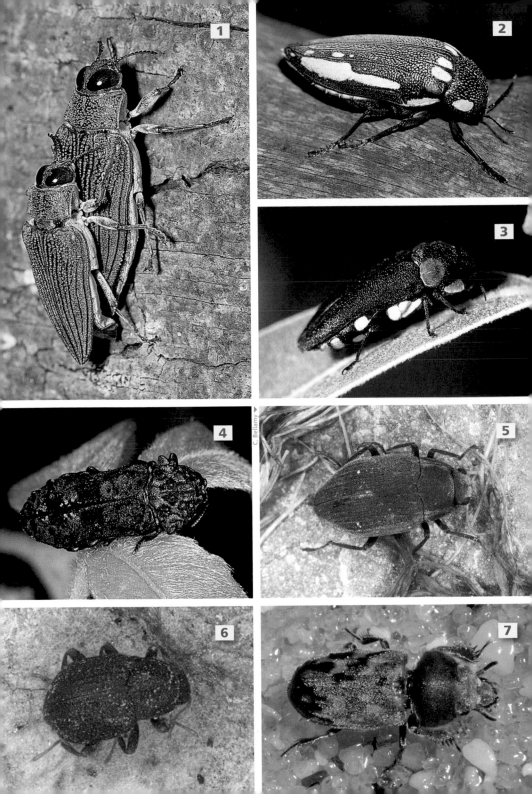

1 Family Elateridae

Click beetles

Small to very large (body length 4–80mm), elongate, parallel-sided beetles. Brownish or black, sometimes with yellow, red or white markings. Hind angles of the large pronotum produced into points, the head deeply inserted into the pronotum. Antennae thread-like, saw-like or with large plates (lamellae). Most characteristic is the ability of adults to leap into the air when stranded on their backs, using a special spine-and-notch mechanism on the underside, between the pro- and mesothorax. This generates a loud click, hence the family's common name. Adults mostly nocturnal, and may be attracted to lights. They feed on foliage, flower petals or pollen. Larvae (wire worms) are usually long, cylindrical and shiny yellow **(1)**. Some inhabit soil, feeding on plant roots, bulbs and tubers, and can cause considerable damage to crops; others, sometimes covered with long hairs, live in rotting wood and prey on other insects. There are some 700 species in the region.

2 *Tetralobus flabellicornis*

Giant Acacia Click Beetle

Very large (body length 60–80mm), smooth, dark brown, covered with a brownish-grey down that rubs off, leaving bare, shiny patches. Antennae with large plates in male, serrate in female. Larvae covered with long hairs. **Biology** Typically associated with large *Senegalia* trees on which it feeds. Larvae live in termites' nests. **Habitat** Subtropical forest and savanna.

3 *Calais tortrix*

Harlequin Click Beetle

Large (body length 30mm), black. Head with punctures and a patch of white hairs between the compound eyes. Pronotum also with punctures and 4 patches of white hair. Elytra with longitudinal rows of punctures, 2 basal spots, and 3 bands of white hairs. Antennae serrate. **Biology** Details not known, but thought to be a nocturnal herbivore. **Habitat** Subtropical forest. **Related species** *Aliteus caffer* **(3A)**, which is a large species (body length 25mm), with white mottling, that occurs in forests.

4 *Cardiotarsus acuminatus*

Common Brown Click Beetle

Small (body length 11mm), brown. Head and pronotum smooth. Elytra with longitudinal punctured grooves. Antennae thread-like. **Biology** Recorded from Australian Black Wattle (*Acacia mearnsii*). **Habitat** Bushveld and forest. Common on foliage.

5 *Cardiophorus obliquemaculatus*

Dwarf Cross Click Beetle

Very small (body length 5mm), with a black head. Pronotum smooth, brown or black. Elytra black, with 2 creamish basal spots and 4 curved creamish patches. Antennae thread-like. **Biology** Larvae of the genus may be predaceous or feed on plant material. Some larvae in the genus are luminous. **Habitat** Under dry cow pats or in grassland. **Related species** *Selatosomus* species **(5A)** (body length 19mm), which are dark reddish-brown and spindle-shaped. The larvae of European species of this genus are known to feed on the stems of various cereal crops.

6 *Selasia pulchra*

False Firefly Beetle

Previously included in a separate family, Drilidae, but now transferred to the family Elateridae. Adults show remarkable sexual dimorphism. Male **(6)** (body length 7mm) is typical, pale yellow to black, parallel-sided, with large, fan-shaped antennae. The larvae-like female **(6A)** is wingless, bristly and much larger than the male. **Biology** Females lay large batches of eggs in the ground. The larvae prey on snails, and pupation tales place within the shell of a consumed snail. Males sometimes attracted to lights. **Habitat** Arid subtropical parts. **Related species** There are 11 other similar regional species of *Selasia*.

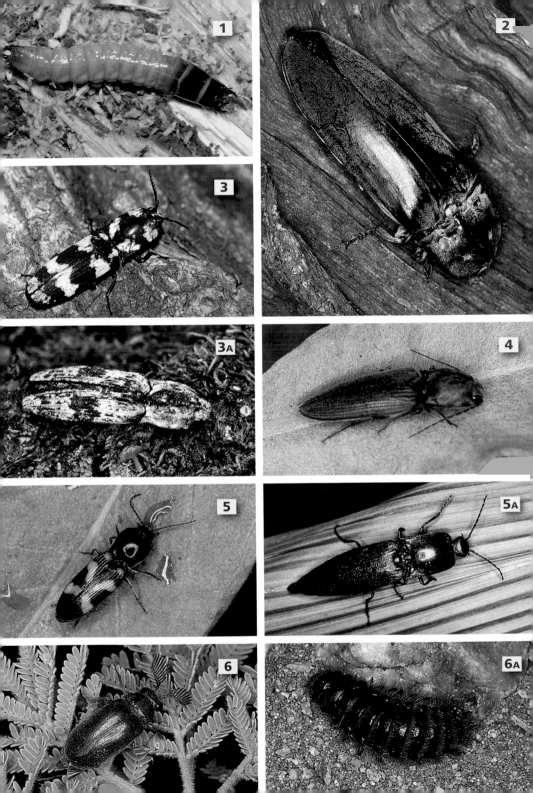

1 Family Lampyridae
Fireflies, glow worms

Small to large (3–30mm), parallel-sided, soft-bodied, mostly black or brown beetles; males with very large eyes. In some species, adult females resemble larvae and never develop wings **(1)**. Both wingless females and larvae of all species are called glow worms. In these species male beetles are large **(1A)** (body length 10mm), with the head concealed under the pronotum. All adults, whether beetle or larviform females, have light-producing organs, usually (in females) on the penultimate or (in males) **(1B)** on the last 2 segments of the abdomen. *Luciola* **(1C)** comprises small species (body length 7mm) in which both sexes are winged. The family includes about 30 regional species. **Biology** Nocturnal. Male fireflies are more active fliers than females. In those species with larviform females, these rest on vegetation and under logs and stones, and glow to attract males. Adults do not feed. Larvae predatory, especially on snails. **Habitat** Diverse, including vegetation in woodland and moist grassland. Some KwaZulu-Natal species have aquatic larvae that glow underwater in streams.

2 Family Lycidae
Net-winged beetles

Small to medium-sized (body length 6–25mm), soft-bodied and flattened. Elytra are often considerably expanded to the side, with characteristic longitudinal ridges and a network of minor ridges, hence the common name. Head elongate, almost entirely hidden under the pronotum, which is much smaller than the elytra. Conspicuously coloured in orange and black and contain cantharadin, making them distasteful to predators, especially birds. Antennae conspicuous and serrate. Adults are active by day and are often found on grasses and other plants, sometimes in large numbers, feeding on nectar from a wide variety of flowers. Larvae **(2)** live in decaying wood, under loose bark or in leaf litter, probably feeding on fungal material. The warning coloration of adults is mimicked by various other insects. About 50 species occur in the region.

3 *Lycus trabeatus*
Tailed Net-winged Beetle

Medium-sized (body length 18–22mm). Pronotum black, with broad orange borders. Elytra vary, some widely expanded with a constriction three-quarters along the length, some slender, others intermediate; black at base and tip but not on expanded portion. Antennae mildly serrate, black. Femora orange, remainder of legs black. **Biology** Feeds on flowers, including those of trees. **Habitat** Subtropical forest, savanna and grassland. **Related species** *L. ampliatus* **(3A)** (22mm long), which can be distinguished by the absence of black spots on the anterior part of the elytra, and *L. dentipes* **(3B)** (20mm long), which has an M-shaped black marking on the anterior margin of the elytra, and a pronounced ridge in this region.

4 *Lycus melanurus*
Hook-winged Net-winged Beetle

Small to medium-sized (body length 11–25mm). Antennae moderately serrate. Pronotum black, with broad orange lateral borders. Elytra with anterior expansion, produced into backward-pointing hooks; base black at centre, frontal half orange, rear half black. **Biology** As for family. **Habitat** Forest, savanna and grassland.

5 Family Scirtidae
Marsh beetles

Adults (not depicted) unremarkable, minute to small (body length 2–8mm), with soft elytra. Mostly brownish to black, sometimes with yellow or red patches. Body slightly flattened or almost subglobular, covered with semi-erect hairs. Some species have enlarged hind femora and can jump like flea beetles. Larvae **(5)** unique among pupal-stage insects in having long, multi-segmented antennae. **Biology** Adults attracted to lights. Often found feeding on flowers. Larvae are aquatic filter feeders, ingesting detritus from leaf and stone surfaces. Pupation occurs in wet soil or among dead leaves above the water surface. **Habitat** Cosmopolitan. Adults occur on vegetation and in rotting plant material in wet areas and along shorelines, the larvae in flowing or stagnant water rich in decomposing plant matter, also in water-filled tree hollows or in wet moss.

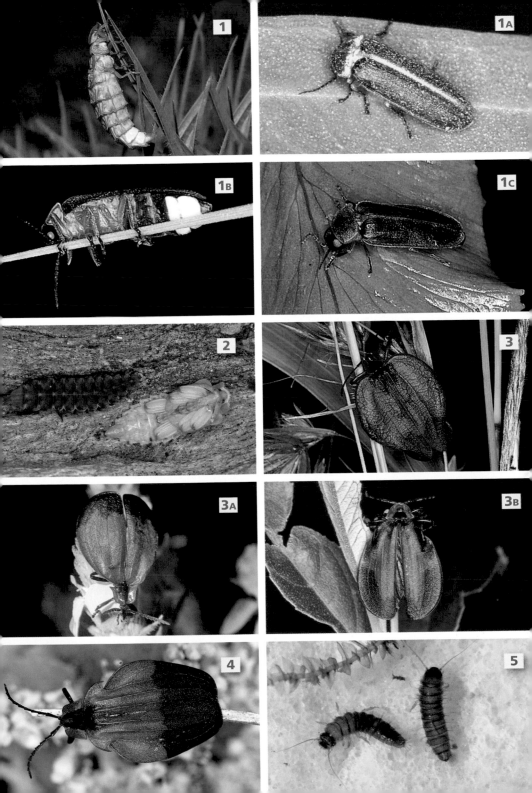

1 Family Ptinidae
Spider beetles

Small (body length 2–5mm), spider-like, rounded, with long legs and thread-like antennae. Head and pronotum narrow. Elytra globular. The American Spider Beetle (*Mezium americanum*) **(1)** (body length 3mm) is typical and has smooth, dark elytra. There are about 30 species in the region. **Biology** Adults are scavengers; larvae feed on dry vegetable and animal matter. Some species infest grain, dried fruit, fabrics, and other stored materials. **Habitat** Nests of birds and other animals, dry carcasses and cave floors. *M. americanum* is an alien species found in hollow walls or under floors.

2 Family Anobiidae
Furniture beetles, death watch beetles

Small (2–5mm), with a cylindrical body and downward-pointing head. The introduced Tobacco Beetle or Cigarette Beetle (*Lasioderma serricorne*) **(2)**, body length 2–3mm, has a rounded oval shape and is reddish-brown. Antennae uniformly serrated, with 11 segments. Elytra weakly punctured. The introduced Drugstore Beetle (*Stegobium paniceum*) and common Furniture Beetle (*Anobium punctatum*) are similar, the latter being covered in a fine pile of hairs and having a distinctive cowl-like prothorax **(2A)**. The family is poorly studied in the region, and the number of species is not accurately known. **Biology** Most species bore small, round holes in dry wood, and some can be serious pests of furniture and wooden structures. The common name is derived from the tapping sound made by some species. *A. punctatum* is a particularly common household pest, the larvae feeding on dry wood. Adults leave small piles of sawdust when they emerge from infested wood, this often being the first sign of an infestation. Adults do not feed. Eggs of *L. serricorne* are laid on the chosen substrate, especially tobacco, but also oilseeds, cereals, dried fruit, flour, and certain animal products on which the larvae feed. **Habitat** Cosmopolitan, introduced species that are associated with human dwellings; *L. serricorne* is found wherever various dried foodstuffs are available, such as in kitchens, supermarkets and food-processing factories.

3 Family Dermestidae
Hide beetles, museum beetles

Small to medium-sized (body length 2–10mm), oval, blackish or dull brown, often with patterns of white, yellow, brown or red scales or hairs. Antennae clubbed and can be folded (with the legs) against the body. The Varied Carpet Beetle (*Anthrenus verbasci*) **(3)** is small (body length 3mm), oval and patterned with white scales; its larvae are small and cylindrical. The Hide Beetle (*Dermestes maculatus*) **(3A)** is larger (body length 8mm), black, with white on the sides of the pronotum and abdomen, and has cylindrical, hairy larvae **(3B)**. There are about 85 regional species. **Biology** Adults feed on pollen and nectar, larvae on a wide variety of dry animal material rich in protein. Larvae can be serious pests, damaging insect collections, but are also of use in cleaning delicate vertebrate skeletons in museums. *D. maculatus* attacks carcasses and fish, the larvae also infesting dry poultry manure. **Habitat** Diverse, some inhabiting birds', bees' and wasps' nests. *D. maculatus* is cosmopolitan.

4 Family Bostrichidae
Auger borers, shot-hole borers

Small to large (body length 2 32mm), cylindrical, dark brown or black, with body characteristically squared-off at the rear. Pronotum strongly domed (concealing the downward-pointing head), sometimes equipped with 2 forward-pointing spikes. Elytra smooth, ribbed or sculptured, and can be shortened; they end in spines or other projections. Tibiae with saw-like teeth along outer edge. *Apate femoralis* **(4)** (body length 12mm) is typical. About 80 species in the region. **Biology** Adults and larvae are borers, the latter mostly in dry wood, the former in live trees. Species in the subfamily Dinoderinae are very small and include pests of bamboo, stored grain, dried fruit and nuts. The subfamily Bostrichinae comprises larger species, some boring into live or weakened trees, edible roots and tubers such as sweet potatoes. Adult *Apate* ring-bark branches to create suitable conditions for the development of their larvae. Other *Apate* species attack date palms. The Powderpost Beetle (*Bostrychoplites cornutus*) **(4A)** (body length 16mm) is commonly intercepted in wood carvings and other artifacts from Africa and is considered an invasive species in Europe. Often attracted to lights. **Habitat** The wood of host plants and timber products.

Family Cleridae
Chequered beetles

Small to large (body length 2–30mm), elongate, slightly flattened and parallel-sided beetles, best recognized by their densely hairy bodies and large heads. Most are brightly coloured, some banded brown, black and white on the elytra. The antennae are serrate or clubbed. One or more of the tarsal segments have membranous flaps. Adults are found on flowers, tree trunks and the foliage of woody plants, and are sometimes attracted to lights. Some species eat pollen and nectar, others are predaceous. The slender, cylindrical larvae, some of which are brightly coloured, commonly occur in dead wood. They prey on the larvae of wood-boring beetles and grasshopper egg pods, or parasitize bees and wasps. Others feed on stored animal products or carrion. There are about 200 regional species.

1 *Tenerus variabilis*
Orange-and-Black Clerid

Medium-sized (body length 14mm). Body and legs pinkish-red. Antennae black, except for 3 pinkish-red basal segments. Pronotum finely punctured, with a black anterior patch. Elytra each with a large black anterior patch and a black posterior band. **Biology** As for the family. **Habitat** On shrubs and trees in subtropical bushveld.

2 *Exkorynetes analis*
Green Bacon Clerid

Small (body length 6mm), metallic blue-green, body with hairs. Head and pronotum heavily punctured, elytra with rows of distinct punctures. Eyes separated by no more than the width of an eye. Legs and antennae pale brown. **Biology** Predaceous, but also feeds on old cheese and cured meats. **Habitat** Diverse. **Related species** The very similar Red-legged Ham Beetle (*Necrobia rufipes*), which has more widely spaced eyes and a dark brown or black antennal club.

3 *Phloeocopus ferreti*

Medium-sized (body length 14mm). Body, antennae and legs dark brown, entire body thinly covered with long, erect hairs. Head and pronotum punctured. Elytra more strongly punctured, each with 4 or 5 reddish or yellowish streaks and spots. **Biology** Nocturnal, commonly attracted to lights. **Habitat** Subtropical forest and savanna.

4 *Gyponyx signifier*
Velvet Ant Clerid

Medium-sized (body length 14mm), an apparent mimic of some velvet ants (Mutillidae p.470), entire body covered with long, erect white hairs. Head and pronotum black. Elytra with longitudinal rows of large punctures, reddish-brown at base, remainder black with 2 transverse white bands that distinguish it from other similar species. Legs black with brown tarsi. **Biology** Nocturnal, attracted to lights. Adults and larvae may be carnivorous. **Habitat** On tree trunks or under bark in subtropical forest and savanna.

5 Family Trogossitidae
Bark-gnawing beetles

Small to medium-sized, elongate and flattened, body in some species clothed in scales. Head and mandibles directed forwards and fully visible from above. Pronotum with distinct lateral margins. Antennae bead-like, ending in a loosely segmented club with 3 segments. Last tarsal segment with lobe between claws. *Gymnochila varia* (body length 19mm), sometimes attracted to lights, is shown here **(5)**. **Biology** Larvae live under bark or in the tunnels of wood-boring insects and are probably predators. *Gymnochila varia* is found on White Ironwood (*Vepris undulata*). The Cadelle Beetle (*Tenebroides mauritanicus*) is an economic pest, infesting stored grain and cereal products, but also preying on the larvae of other pest species. **Habitat** Under bark or on bracket fungi on tree trunks in subtropical forest; also in granaries and food stores.

6 Family Prionoceridae

A small family formerly included within the Family Melyridae (p.268), and comprising about 150 global species, most of which belong to the genus *Idgia*. *Idgia dimidiata* **(6)** is medium-sized (body length 14mm), dull orange and black, with soft elytra. Eyes large and protruding. **Biology** Adults are pollen-feeders and may be seen on flowers, while the larvae are predatory or feed on dead insects or decomposing wood. Rarely seen; adults seasonal and short-lived. **Habitat** In open woodland and grassland in the warmer summer-rainfall regions.

Family Melyridae

Soft-winged flower beetles

Small (body length 4–11mm), soft-bodied, elongate, oval, and dorsally flattened. Often brightly coloured, with covering of long, erect hairs. Distinct margins to pronotum. Elytra smooth; may be shorter than abdomen. Antennae variable, with 10 or 11 segments, the basal segments modified in males for courtship and mating. Adults predacious in some species; others often found on flowers and feed mostly on pollen. Larvae are mostly carnivorous. There are at least 350 regional species.

1 *Astylus atromaculatus*

Spotted Maize Beetle

Small (body length 10mm), body covered with scattered, erect black hairs. Head and antennae black. Pronotum with flat-lying white hairs and 2 black patches. Elytra yellow, with large black patches. **Biology** Eggs are laid in clusters under dry leaves. Adults eat pollen, clustering on various flowers, including those of grasses, and often on maize tassels. Larvae live in soil, feeding on decaying vegetable material and sometimes damage germinating maize seedlings; also prey on insects. **Habitat** Introduced in 1916 from South America, now a major pest in gardens and on agricultural land. Ingestion of adults has caused cattle deaths.

2 *Melyris*

Metallic groove-winged flower beetles

Small (body length 7mm). Head and pronotum metallic green, finely punctured. Elytra metallic copper with strong longitudinal ridges and sculpturing in between. Legs metallic green. **Biology** Adults feed on pollen and visit carcasses and cow dung. Larvae omnivorous, feeding on dry decaying kelp and other vegetable matter, cow dung and carcasses; also prey on insects. **Habitat** Typically on daisies and mesembs in strandveld and dune vegetation. Most abundant in spring and early summer.

3 *Apalochrus*

Small (body length 6mm), body covered with short whitish hairs that are interspersed with longer black hairs. Head and pronotum metallic purple, finely punctured. Elytra more heavily punctured, metallic green and shorter than abdomen. Antennae black or brownish. **Biology** Adults and larvae prey on other insects. **Habitat** Grassland and bushveld.

Family Erotylidae

Pleasing fungus beetles

Typically small (body length 2–6mm), but with a few large species (body length 28mm). Body oval and typically shiny, patterned in red, yellow and black in the larger species. The head is sunk into the broad, straight prothorax, and the elytra cover the abdomen entirely. There are 5 tarsal segments for each leg, but the fourth segment may be very small and indistinct, especially in the smaller species. The short, 11-segmented antennae generally end in a flattened club, comprising 3 segments. The spined larvae of some species feed on the outside of fungi, whereas the paler and warty larvae of others feed within fungi. There are approximately 32 species in the region.

4 *Amblyopus*

Small, reddish species (body length 5mm) with finely punctate elytra. **Biology** Found on fruiting bodies of fungi, especially bracket fungi growing on tree stumps and logs. **Habitat** Coastal and montane forest, and bushveld in the moister parts of the region. Adults of some species attracted to lights and sap fluxes.

5 *Promecolanguria*

Lizard beetles

Shiny, mostly black and orange, elongate and cylindrical. Pronotum relatively large (body length 6mm). **Biology** Larvae are stem or wood borers, or feed in leaf litter or stored produce. **Habitat** Diverse, often forest. Rare. **Related species** At least 20 other lizard beetle species in the region.

6 Family Lymexylidae

Ship-timber beetles

Medium-sized to large (body length 15–45mm), brown, exceptionally long and narrow. Elytra shortened, exposing entire abdomen and folded hind wings in *Atractocerus brevicornis* (6) and the last 2 abdominal segments in *Melittomma*. Antennae conspicuous and saw-like in *Melittomma*, relatively short and pointed in *Atractocerus*. Eyes large, almost meeting on the face. Larvae with very elongate body segments. A small family; only 3 species in the region. **Biology** Adults attracted to lights. Larvae bore into hardwood and stems of plants, including those of palms. **Habitat** Subtropical forest and savanna.

1 Family Nitidulidae

Sap beetles

Minute to small (body length 1–12mm), ovoid, mostly dull-coloured beetles, sometimes with red or yellowish patches. Antennae have 11 segments and end in a distinct knob or club. Elytra sometimes shortened, exposing up to 3 abdominal segments. Larvae pale, elongate and somewhat flattened. **Biology** Diverse feeding habits; most feed on decaying vegetable matter, overripe fruit, fungi and sap exuding from tree wounds, but also associated with flowers (pollen), carrion, the nests of ants and bees and stored products. The larvae of some species prey on bark beetle larvae (Curculionidae, Scolytinae, p.302, 306) or various coccids. Some species are pests of agricultural crops and dried grains and fruits. **Habitat** Cosmopolitan and widespread.

2 Family Silvanidae

Sylvan flat grain beetles

Small (body length 1.5–5mm), elongate, flattened, parallel-sided, with a conspicuous head. Long 11-segmented antennae with more or less conspicuous 3-segmented club. Edges of pronotum toothed in many species. **Biology** Feed primarily on fungi. Adults and larvae probably also prey on small arthropods. Some, like the Saw-toothed Grain Beetle (*Oryzaephilus surinamensis*) **(2)**, are pests of stored produce. **Habitat** Diverse. Under loose bark or in rotting vegetable matter or dry foodstuffs.

Family Coccinellidae

Ladybeetles, ladybird beetles

Easily recognized by their rounded hemispherical shape, short legs and bright colours, mostly black, red, yellow or orange, often with distinct spots or stripes. This warning coloration is backed up by the bleeding of alkaloid-rich blood from the leg joints. Some are shiny and smooth; others are covered in short hairs that give a dull appearance. The head is concealed below the pronotum. Distinguished from similar-looking beetles by apparently 3-segmented tarsi (in fact there are 4, the third very small). Larvae commonly black, with conspicuous yellow or white markings, some with tufts of white wax. Adults and larvae are usually carnivorous, feeding on various homopteran bugs, scale insects, small insects and mites, but members of subfamily Epilachninae are herbivorous.

3 *Chilocorus distigma*

Black Two-spot Ladybeetle

Small (body length 6mm), smooth, black with a red central spot on each elytron. **Biology** Adults and larvae predatory, commonly feeding on Red Scale (*Aonidiella aurantii*) (p.194) on citrus. **Habitat** On plant stems and foliage. **Related species** *C. solitus*, which feeds on White Aloe Scale (*Duplachionaspis exalbida*) (p.194).

4 *Oenopia cinctella*

Black-ringed Ladybeetle

Small (body length 5mm), smooth. Head and pronotum black, with pale yellow markings. Elytra orange-yellow, with black margins. **Biology** Adults and larvae feed on psyllids and other small insects, leaf beetle eggs and larvae, and are important in biological pest control. **Habitat** Plant stems and foliage. **Related species** *Cheilomenes propinqua* **(4A)** (5mm long), which has yellow elytra, with 6 black stripes, and is widespread in Africa and Europe.

5 *Declivitata hamata*

Humbug Ladybeetle

Small (body length 4mm), smooth. Head and pronotum yellow, with brown markings. Elytra each with 1 brown and 2 red longitudinal lines and with brown margins. **Biology** Adults and larvae prey on small insects, such as thrips. **Habitat** Common on plant stems and foliage in grassland.

6 *Hippodamia variegata*

Variegated Ladybeetle

Small (body length 4mm), smooth. Head and pronotum black, with pale yellow markings. Elytra red, with 9 black spots, 1 split by the midline. Pale yellow line directly behind the pronotum. **Biology** Adults and larvae predatory, specializing in aphids. Introduced from Europe and first reported in Cape Town around 1965. Now widespread and common across the region. Great potential for control of pests such as Russian Wheat Aphid. **Habitat** On plant stems and foliage. **Related species** *Psyllobora variegata* **(6A)** (4mm long), which is usually yellow, but can have white patches on the elytra or a white pronotum. Both pronotum and elytra have bands of large black spots running across them. Larva (below beetle) and pupa (to right of beetle) are banded yellow and white and have similar black spots. Both adults and larvae feed on fungus and downy mildew, especially on grapevines and poplar trees.

1 *Harmonia vigintiduomaculata*　　　　Chequered Ladybeetle

Medium-sized (body length 8–9mm), smooth. Colour pattern highly variable: either black, with yellow or orange on each side of the pronotum and 10 symmetrically placed yellow patches on each elytron, or yellow to orange, with a few black spots on the pronotum and elytra **(1A)**. **Biology** Adults and larvae consume vast numbers of soft-bodied pest insects. **Habitat** On plant stems and foliage.

2 *Cheilomenes lunata*　　　　Lunate Ladybeetle

Medium-sized (body length 7mm), shiny. Head and pronotum black, with yellow markings. Elytra black, with a pattern of red or yellow patches. Larvae **(2A)** black, with yellow markings and a transverse row of 6 short tubercles on each abdominal segment. **Biology** Adults and larvae predatory, especially on aphids, including wheat aphids. Freshly hatched larvae remain on their egg batches for a while, eating unhatched eggs. **Habitat** Diverse. Common in gardens on aphid-infested plants, especially sow thistle (*Sonchus*).

3 *Exochomus flavipes*　　　　Black Mealy Bug Predator

Small (body length 4mm), very round, shiny, black, with a pale yellow patch on each side of the pronotum. **Biology** Adults and larvae feed on aphids, including wheat aphids, mealy bugs, soft scales and cochineal insects. **Habitat** Diverse. On plant stems and foliage, often on prickly pears infested with cochineal insects.

4 *Henosepilachna bifasciata*　　　　Curcubit Ladybeetle

Medium-sized (body length 8mm), covered with fine down. Head and pronotum orange-red. Elytra each with 6 large black patches. Larvae yellowish, with black spines. **Biology** Adults and larvae feed on the leaves of various cucurbits, including pumpkin, watermelon and cucumber, usually from the underside, and occasionally on potato leaves. **Habitat** Diverse. On plant stems and foliage.

5 *Epilachna infirma*　　　　Red Herbivorous Ladybeetle

Medium-sized (body length 8mm), body covered with short whitish down. Head and pronotum black, the latter with coarse sculpturing. Elytra brownish-red, with black margins, and each with 7 yellowish patches. **Biology** Both adults and larvae are herbivorous. **Habitat** Subtropical weedy areas.

6 *Epilachna paykulli*　　　　Nightshade Ladybeetle

Medium-sized (body length 6mm), pinkish-red, covered with short whitish down. Elytra each with 9 black spots. **Biology** A pest of vegetables. Feeds on the leaf surfaces of potato, tomato and other solenaceous plants, but avoiding the veins, which results in a characteristic net-like appearance to the leaves. **Habitat** Low vegetation.

7 *Epilachna dregei*　　　　Potato Ladybeetle

Medium-sized (body length 8mm), covered with short whitish down. Head and pronotum orange. Elytra black, each with 10 large red patches. Larvae yellow, covered with long, spiny, black-tipped projections **(7A)**. **Biology** Feeds and lays clusters of yellow eggs on the leaves of potato, tomato and, sometimes, cucurbit plants. Larvae feed on the underside, adults on the upper surface of leaves, skeletonizing them. Pupae are attached to the plant. Adults congregate in large numbers on hilltops, where they spend the dry season. **Habitat** Vegetation, including crops.

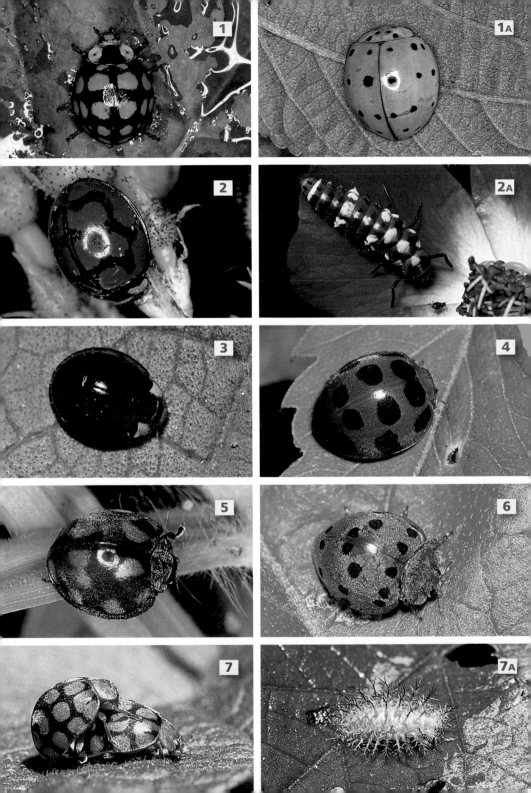

1 Family Discolomatidae — Discolomatid beetles

Small (body length 2–8mm), mostly oval, evenly convex, densely covered with short, stiff, flat-lying hairs. Head recessed into front of pronotum. Antennae 9- or 10-segmented, ending in a single large clubbed segment. *Notiophygus* **(1)** are grey or black, often ornamented with white or grey spots and patches of minute flattened hairs. Larvae dark, with light yellow spots, or lighter, with dark markings, and with hair-like extensions along the edge of the body. About 140 species in the region, of which 138 are from the diverse genus *Notiophygus*. **Biology** More active in wet weather, congregating on or below stones or crawling on grasses. They feed on bracket fungi or crustose lichens. **Habitat** On or under the bark of tree trunks or in plant litter in forested areas. Also in large numbers on isolated koppies in the Karoo and savanna regions. Occasionally in thatch roofs.

Family Tenebrionidae — Darkling beetles

A large and very diverse family. Shape highly variable. Predominantly black, smooth and shiny, or dull and sometimes strongly sculptured. Many species lack hind wings and are flightless. Most live on the ground and many are nocturnal. The larvae, known as mealworms or false wireworms, are elongate, cylindrical and tough-bodied. Most species are scavengers, feeding on dead plant or animal remains; some are pests of dried foodstuffs. Widespread, especially in arid and desert areas. There are some 3,200 species in the region.

2 *Lagria vulnerata* — Hairy Darkling Beetle

Smallish (body length 11mm), soft-bodied, dull metallic brown, green or violet, covered with whitish hairs, surface strongly punctured. Head and pronotum narrower than bulbous abdomen. Antennae bead-like, thickening towards tip. **Biology** Adults and larvae are herbivorous. A pest on beans. **Habitat** Common on grass in bushveld and savanna. Many similar species, including *Chrysolagria* **(2A)** (11mm long), which has a coppery cast on the elytra.

3 *Zophosis testudinaria* — Frantic Tortoise Beetle, Koffie-pit

Medium-sized (body length 14mm), oval, slightly flattened, developing a pinkish waxy bloom under hot and dry conditions. Elytra with small flattened tubercles. Antennae long. Legs long and slender. **Biology** Adults are active in the heat of the day and scuttle about rapidly, almost swimming across the sand. Feed on plant and animal detritus, also subterranean stems of plants. **Habitat** Coarse, gravelly sand, hard silt pans or the base of small shrubs in arid areas. **Related species** There are many other common but smaller species, with a bluish waxy bloom.

4 *Stips dohrni* — Ridged Seed Beetle

Smallish (body length 11mm), oval, flattened, creamish-white overall. Head with 1 longitudinal ridge; pronotum with 2; elytra with 1 ridge down the midline and 2 semicircular ridges that join posteriorly. Legs fairly short. **Biology** Nocturnal, slow-moving. **Habitat** Inhabits vegetated dunes in sandy, arid areas, such as the Kalahari.

5 *Eurychora* — Mouldy beetles

Medium-sized (body length 13mm). Pronotum expanded into flanges on either side of head. Upper surface flat and often covered with soil and vegetable debris, held in place by long, stout hairs and waxy filaments. Over 20 species in the region. **Biology** Adults nocturnal. Rest under stones and logs during the day. **Habitat** Rocky overhangs and aardvark or porcupine burrows.

6 *Psammodes bertolonii* — Spindle Toktokkie, Giant Toktokkie

Very large (body length 53mm), unusually elongate, elytra tapering. Head and pronotum black, the latter with buff-coloured down on each side. Elytra with low longitudinal ridges and buff-coloured down along the sides. **Biology** Adults tap out rhythms on the ground to locate and attract mates. They feed on plant and animal material. Flightless and unusually fast-running. **Habitat** Subtropical forest and woodland. **Related species** *P. virago* **(6A)**, which is large (body length 28mm), stout, globular and black. The pronotum has scaly sculpturing and the elytra are sparsely punctured. Biology and habitat are similar to those of *P. bertolonii*. Occurs in the Kalahari and arid parts of Limpopo.

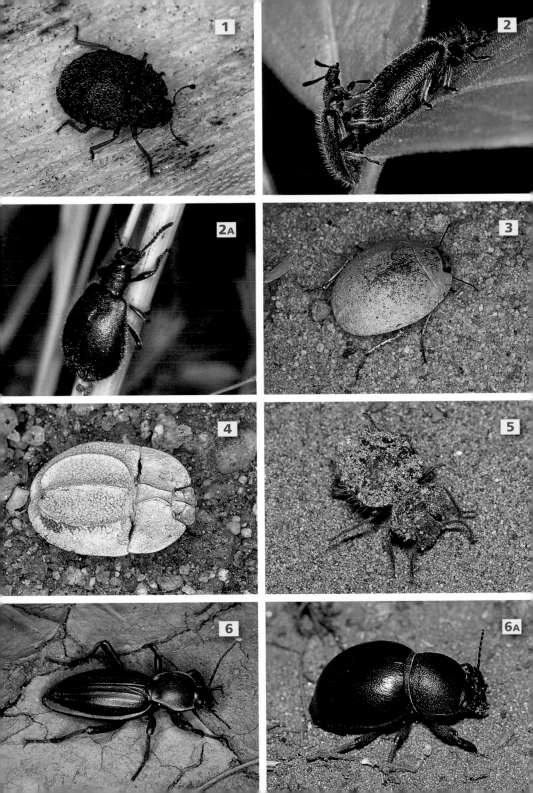

1 *Psammodes striatus* — Striped Toktokkie

Large (body length 24mm), stout, globular, black. Pronotum smooth, with some puncturing at the sides. Elytra smooth, with reddish-brown longitudinal lines. Male has pale yellow felt-like patch under the abdomen. **Biology** Adults locate and attract mates by tapping out rhythms on the ground. Feed on plant and animal material. Flightless and slow-moving. The tough, elongate yellow larvae (**1A**) are found under rocks. **Habitat** Ground-dwelling, often on koppies. **Related species** *P. hirtipes*, which is very similar, black, with punctured femora and whitish pubescence on the tibiae and tarsi. Also, *P. vialis* (**1B**), which is large (body length 26mm), stout, globular, black or brownish, with prominent tubercles at the sides and rear of the smooth elytra, and occurs in the arid northwest, and *P. fartus* (**1c**) (32mm long), which is black, with a broad, dull red stripe along the centre of the elytra, and occurs in the summer-rainfall parts of the region.

2 *Dichtha incantatoris* — White-legged Toktokkie

Medium-sized (body length 22mm), stout, globular, black. Pronotum sparsely punctured in the centre, more heavily at the sides. Elytra smooth, with 3 longitudinal ridges. Antennae and legs pale tawny, due to dense covering of pubescence. **Biology** Feeds on both plant and animal remains. Communicates by tapping on the ground. **Habitat** Ground-dwelling.

3 *Moluris pseudonitida* — Rounded Toktokkie

Medium-sized (body length 15mm), smooth, globular, black. Pronotum unusually globular. Legs strongly developed, black with some reddish down on the tibiae and tarsi. **Biology** Most active at dusk and at night, when it emerges from a retreat in a rock crack. **Habitat** Ground-dwelling in arid savanna. **Related species** Other *Moluris* species, some of which are black, with a greyish bloom, the elytra somewhat flattened on top, with vertical sides, mild sculpturing and 3 low longitudinal ridges.

4 *Somaticus aeneus* — Tar Darkling Beetle

Medium-sized (body length 22mm). Head and pronotum matt black. Elytra shiny black with reticulate sculpturing in between longditudinal ridges. Legs black, long and strongly developed, with punctures. **Biology** Adults scavenge on plant and animal material, while larvae feed on roots, sometimes damaging maize seedlings. **Habitat** Semi-desert to savanna and succulent karoo. **Related species** There are several other similar species, 7 of which are injurious to maize. Many are black and covered with silver, golden and reddish-brown hairs, with silvery hairs on the head, and longitudinal ridges on the pronotum and elytra.

5 *Epiphysa flavicollis* — Tortoise Darkling Beetle

Medium-sized (body length 22mm), black, with a transverse band of golden hairs on the back of the head and front of the pronotum, and swellings in front of the eyes. Mandibles short but thick. Pronotum smooth, with side ridges. Elytra very rounded, covered with fine tubercles. **Biology** Adults sluggish, living for years and scavenging plant and animal material at night. **Habitat** Arid savanna and semi-desert. Found in rodent burrows in sandy areas.

6 *Alogenius cavifrons* — Pitted Darkling Beetle

Medium-sized (body length 14mm), black. Head and pronotum coarsely punctured. Elytra globular, very coarsely sculptured, the raised parts making a reticulate pattern. Legs punctured. **Biology** Adults sluggish, scavenging plant and animal material. Especially active at sunset. **Habitat** Rocky hills in arid savanna.

7 *Renatiella*

Medium-sized (body length about 12mm), black, with strongly pitted femora, head and pronotum. Elytra dull, covered with a network of prominent polished ridges forming a reticulate pattern. Legs relatively long. **Biology** Adults and larvae feed on dead or decaying plant material under or on the soil surface. Adults will climb stumps or stems of shrubs to escape immersion in water. **Habitat** Widespread from South Africa to Tanzania. Found on various sandy substrates.

1 *Physodesmia globosa*   **Globular Darkling Beetle**

Medium-sized (body length 12mm), black and moderately shiny, with rows of prominent backward-pointing tubercles on the elytra. Body very rounded, with a smooth pronotum and fairly long legs. **Biology** A relatively slow-moving diurnal species, often seen with the male following the female closely. Feeds on diverse plant detritus, including flowers that have dropped from trees onto the ground. **Habitat** Arid, sandy areas in the Northern Cape and especially Namibia.

2 *Stenocara gracilipes*   **Striped Fog-collecting Beetle**

Medium-sized (body length 14mm), black. In very hot and dry conditions body becomes covered with a waxy white bloom that forms 4 longitudinal stripes. Head finely punctured, antennae brown. Pronotum punctured. Elytra with longitudinal rows of tubercles. Legs very long, slender, brown and punctured. **Biology** Adults scavenge plant and animal material. Well known for their habit of drinking by adopting a head-down posture to direct dew and ocean fog condensation from the body surface into the mouth. **Habitat** Arid vegetation types, especially arid savanna.

3 *Stenocara dentata*   **Long-legged Darkling Beetle**

Medium-sized (body length 17mm), black, body covered with varying amounts of waxy white bloom. Head separated by a fine yellow line from the lightly punctured pronotum. Elytra with longitudinal ridges separated by a row of tubercles. Antennae brownish. Legs moderately long. **Biology** Fast-running. Adults scavenge plant and animal material. **Habitat** Succulent karoo and Nama karoo, often below dry cow pats. **Related species** *S. longipes* (**3A**) (body length 20mm), which is shiny black and very common in Namaqualand and dry parts of the Western Cape. Scuttles to the centre of bushes when approached.

4 *Cryptochile assimilis*   **Streaked Darkling Beetle**

Smallish (body length 12mm), covered with tiny scales. Head and pronotum dark brown, with brown and creamish-white markings, pronotum with streaky sculpturing. Elytra dark brown, with longitudinal ridges and paler brown and creamish-white streaks. Antennae dark brown. Legs stout, creamish-white. **Biology** Adults scavenge plant and animal material and are active in spring. **Habitat** Succulent karoo and Nama karoo.

5 *Psorodes gratilla*   **Warty Darkling Beetle**

Medium-sized (body length 16mm), brownish-black and moderately globular. Pronotum covered with small blunt tubercles. Elytra with more or less longitudinal rows of large blunt tubercles alternating with rows of very small tubercles. **Biology** Adults scavenge plant and animal material. **Habitat** Usually found crawling on the ground.

6 *Psorodes tuberculata*   **Tuberculate Darkling Beetle**

Medium-sized (body length 15mm), black, not particularly globular. Pronotum sculptured with tubercles and streaks. Elytra with longitudinal rows of tubercles. Antennae brown, long and thin. Legs black, becoming brown on the tibiae and tarsi; femora of fore legs strongly developed, each with a stout forward-pointing spine. **Biology** Adults probably scavenge plant and animal material. **Habitat** Common under rocks overlying larger rocks.

7 *Eupezus natalensis*   **Tree Darkling Beetle**

Large (body length 26mm), dull greyish-black, with long antennae and long, strongly developed legs. Elytra have longitudinal rows of well-defined punctures, with a felt-like black bloom between them. Undersides of tibiae with a conspicuous brush of golden hairs. **Biology** Details unknown, but probably feeds on scavenged plant material. **Habitat** Subtropical forest, usually on tree trunks. **Related species** *Hoplonyx* species, which are similar in appearance, but shiny black.

1 *Anomalipus elephas* — Large Armoured Darkling Beetle

Large (body length 28–38mm), heavily built, entirely black. Head and pronotum smooth and lightly punctured, the pronotum expanded sideways and wider at the middle than the abdomen. Elytra strongly sculptured, with several ribs. **Biology** Adults lay a single large egg in a shallow hole. They are active by day, feeding on plant litter. Larvae live in soil and feed on roots and detritus. **Habitat** Shaded areas within clumps of bushes. **Related species** The smaller *A. sculpturatus* **(1A)** (body length 29mm), which is similar in shape, but with the pronotum less expanded laterally, and fewer, more widely spaced longitudinal ridges on the elytra (3 on each and 1 down the midline), with buff-coloured granular bands in between them. A highly variable species occurring in Mpumalanga and Limpopo.

2 *Gonopus tibialis* — Armoured Darkling Beetle

Medium-sized (body length 19mm), shiny black. Pronotum flattened, squarish, with a beaded margin and a longitudinal central groove. Elytra moderately rounded, with longitudinal rows of tubercles. Legs stout and strongly developed, each of the front femora armed with a stout forward-pointing spine, and the front tibiae armed with teeth. Antennae fairly short, the last 4 segments with short golden hairs. **Biology** Larvae live in soil, feeding on roots or plant detritus. **Habitat** Among plant debris in loose dune sand in arid savanna and semi-desert.

3 *Gonocephalum simplex* — Dusty Maize Beetle

Small (body length 10mm), grey or brownish, somewhat flattened, parallel-sided, with the pronotum extending on either side of the head. Head and pronotum covered with flat-lying scale-like hairs. Elytra with low longitudinal ridges covered with similar hairs, alternating with rows of punctures. **Biology** Gregarious adults feed on bark, ring-barking larger plants and felling small plants and seedlings. A common pest of many crops. Larvae, known as False Wireworms, are attracted to germinating seeds, scraping away the seed coat and damaging the developing roots and shoots. Can be a significant pest of maize. **Habitat** Diverse. Sometimes found in dry cow pats. Common in gardens, where lawns grow against rocks or buildings.

4 *Trigonopus*

Small to medium-sized (body length 9–13mm), greyish-black. Head and pronotum densely punctured, the latter more or less flattened on top, squarish in outline, with curved sides and a beaded margin. Elytra densely covered with short tubercles and slightly rounded, with indistinct longitudinal ridges. Legs moderately stout. **Biology** Larvae live in soil, feeding on roots or plant detritus. **Habitat** Diverse.

5 *Neoeutrapela* — Brown-striped flower beetles

Small (7mm) flower-feeding beetles, belonging to the subfamily Alleculinae (comb-clawed beetles). Elongate body and legs and long antennae. Head, pronotum and legs black. Elytra dark centrally and marginally, but with a broad brown longitudinal stripe running parallel to the lateral margin. Body sparsely covered with long hairs. **Biology** Commonly seen feeding on daisies in Namaqualand. **Habitat** Flowers in fynbos and succulent karoo.

6 *Strongylium purpuripenne* — Metallic Tree Darkling Beetle

Large (body length 26mm). Head, pronotum, legs and antennae striking metallic green and punctured. Elytra metallic purple, with well-defined, punctured longitudinal grooves. **Biology** Details unknown, but appears to feed on trees. Sometimes attracted to lights. **Habitat** Often found climbing tree trunks in subtropical forest and woodland.

7 *Tenebrio molitor* — Yellow Mealworm

Medium-sized (body length 15mm). Adults somewhat shiny, dark brown to black. Head and pronotum smooth. Pronotum rectangular, with beaded margins and curved sides. Elytra with longitudinal punctured grooves. **Biology** Often reared to provide larvae **(7A)** as food for pets and laboratory animals. Adults and larvae feed on diverse plant and animal matter, especially grain and grain products. **Habitat** Usually in stored produce; also in birds' nests, particularly those of pigeons. Cosmopolitan.

1 *Asthenochirus*
Mouldy tree darkling beetles

Small (body length 10mm), oval, with long antennae and legs. Dark brown, often with a bloom of white wax. Head and pronotum punctured. Elytra with longitudinal punctured grooves. **Biology** Details unknown. **Habitat** Tree trunks in wooded savanna.

2 *Metallonotus aerugineus*
Royal Tree Darkling Beetle

Medium-sized (body length 17mm) and spectacular species. Head and pronotum metallic copper, punctured. Elytra metallic purple, coarsely punctured. Abdomen somewhat bulbous. Antennae and legs metallic greenish-bronze. **Biology** Probably a scavenger of dead plant material. Occurs in small groups. **Habitat** Tree trunks and branches in subtropical forest and woodland.

3 *Emmallus australis*
Fuzzy Darkling Beetle

Small (body length 10mm), brown, with a globular body covered in long, erect, reddish-brown hairs. Pronotum punctured, domed to conceal the head from above. Elytra with longitudinal rows of large, shallow punctures. **Biology** Nocturnal. Species-level details unknown. **Habitat** Under *Senegalia* bark in arid savanna.

4 *Pachyphaleria capensis*
Beach Darkling Beetle

Small (body length 7mm), oval, pale yellowish to orange. Antennae and legs also yellowish. Pronotum smooth. Elytra with fine longitudinal grooves. **Biology** Feeds on decaying seaweed and confined to the west coast. **Habitat** Under cast-up seaweed, especially kelp, on the drift line of sandy beaches.

5 *Himatismus*
Tapering darkling beetles

Smallish (body length 12mm), elongate to oval, dark brown, with a pointed abdomen and protruding eyes. Head and pronotum covered with golden hairs. Elytra and pronotum with longitudinal bands and patches of golden hairs. **Biology** Sluggish scavengers. **Habitat** A range of vegetation types, often under rocks or logs.

6 Family Anthicidae
Ant beetles

Small (body length 2–9mm, mostly under 4mm), with head, pronotum and abdomen clearly separated, giving an ant-like appearance. Resemble miniature ground beetles (Carabidae) but have ungrooved elytra and 4 tarsal segments on the hind legs. Black or metallic blue, sometimes with brown-and-yellow patterns, often with a forward-projecting horn on the pronotum. *Formicomus caeruleus* (6) (body length 4mm) is dark metallic blue. Some 93 species in the region. **Biology** Adults omnivorous. Larvae feed on insect eggs or are scavengers. Adults attracted to lights. *Anthicus* feed on the larvae of blister beetles (Meloidae p.284) and are attracted by dead adult blister beetles. **Habitat** Decaying plant material and leaf litter, under cow pats and bark or on dry carcasses. Abundant on living vegetation.

7 Family Mordellidae
Tumbling flower beetles, pintail beetles

Small (body length 2–10mm), with a characteristic body shape: broad at the front, with the abdomen tapering to a spine-like tip that projects well beyond the elytra. In profile, body appears hunched, with the head bent strongly downwards. Brown or black, sometimes with patterned patches of short white, red or yellowish hairs (7A). Hind legs much longer than other legs and with expanded femora. Some 36 species in the region. **Biology** Adults feed on nectar. Larvae live in rotting wood, plant stems or in the fruiting bodies of fungi. Some species are predatory. Adults prefer various daisies, and will kick and tumble if disturbed. **Habitat** Adults on flowers, mainly daisies and members of the carrot family, or on dead wood in the sun.

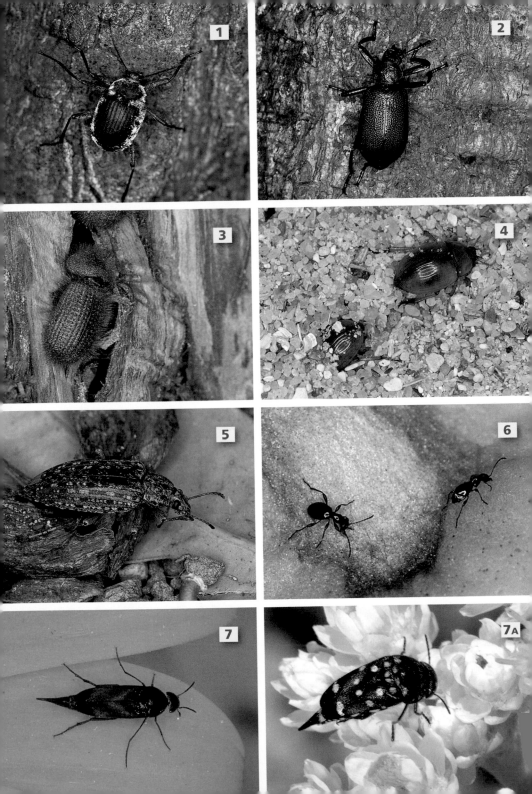

Family Meloidae

Blister beetles, oil beetles, CMR beetles

Mostly elongate and soft-bodied, with a large head separated by a distinct neck from the narrower pronotum. Pronotum is always narrower than the elytra, and its shape varies. Elytra are fully developed or, in ground-living species, shortened to expose part of the abdomen. Coloration uniformly black, grey, brown, red, yellow, metallic green or blue; or black with red, yellow or orange bands, warning of toxicity or distastefulness. Some display the colours of the old Cape Mounted Rifles (CMR), hence the local name. Adults secrete liquids containing the poison cantharadin from the leg joints. May blister human skin and, if eaten, can prove fatal. Some lay eggs in holes in soil and the first-stage larvae (triungulins) seek out grasshopper egg pods on which to feed. Others lay eggs on flowers, and their first-stage larvae attach themselves to visiting bees, which carry them back to their nests. The beetle larvae then occupy cells and feed on the hosts' stores of pollen and nectar. After reaching their hosts, they moult into fat, inactive grubs. Adults feed on flowers, foliage or nectar. About 350 species in the region.

1 *Lydomorphus bisignatus*

Slender Grey Blister Beetle

Medium-sized (body length 16mm), elongate. Head black, with flat-lying greyish hairs. Antennae long and thin. Pronotum orange. Elytra black, also covered with greyish hairs. **Biology** Attracted to lights. **Habitat** Flowers in a range of vegetation types. **Related species** *L. mesembryanthemum*, which feeds on mesemb flowers.

2 *Ceroctis capensis*

Spotted Blister Beetle

Medium-sized (body length 12mm). Head, pronotum, antennae and legs black. Elytra black, each with 3 yellow patches along the inner and 3 more along the outer margin. **Biology** Adults feed on a variety of flowers, e.g. *Watsonia*, peas and beans. **Habitat** Diverse. Rarely in gardens.

3 *Actenodia curtula*

Red-banded Blister Beetle

Small (body length 11mm), body covered with erect black hairs. Head and pronotum metallic green, punctured. Antennae and legs black. Elytra black, each with 3 large red patches along the inner and 3 more along the outer margin, sometimes merging to form bands. **Biology** Adults often feed on groundnut flowers. **Habitat** Flowers in a range of vegetation types.

4 *Mylabris oculata*

CMR Bean Beetle

Large (body length 27mm). Head and pronotum black. Antennae orange, except for the 2 basal segments, which are black. Elytra black, with 2 yellow basal spots and 2 broad yellow transverse bands, the posterior band sometimes red. One colour variant has red bands on the elytra and all-black antennae **(4A)**. **Biology** Adults feed on flowers, often damaging ornamental garden plants and cotton, beans, peaches, citrus and other crops. The larvae, however, regulate the numbers of grasshoppers (including plague locusts) by parasitizing their egg pods. The dried beetles are still used as an aphrodisiac across Africa, but the cantharidin toxin can cause damage to internal organs. **Habitat** Diverse. Often seen swarming on flowers, particularly on *Senegalia*.

5 *Mylabris burmeisteri*

Felt Blister Beetle

Medium-sized (body length 13mm). Head, pronotum, elytra, legs and underside covered with greyish-bronze hairs. Antennae brownish-red, except for the 2 basal segments, which are black. Elytra have black spots and irregular yellow markings among the greyish-bronze hairs. **Biology** Adults feed on the flowers, leaves and buds of citrus, cotton and vegetables, including beans. **Habitat** Diverse. Common in the Nama karoo.

6 *Hycleus lunata*

Lunate Blister Beetle

Medium-sized (body length 14mm), black, body covered with erect black hairs. Pronotum and elytra punctured, elytra with 3 yellow transverse bands. Antennae black, the last 5 segments yellow. **Biology** Adults feed on the flowers of cotton, fruit trees and vegetables, including peas and beans. **Habitat** Diverse. **Related species** *H. transvaalica* **(6A)**, which is medium-sized (body length 12mm), with a black head, antennae, pronotum and legs. The elytra are beige, punctured, and outlined in red, with several scattered black patches. It occurs in a band across the northern parts of the region, and feeds on the flowers of groundnuts and cowpeas.

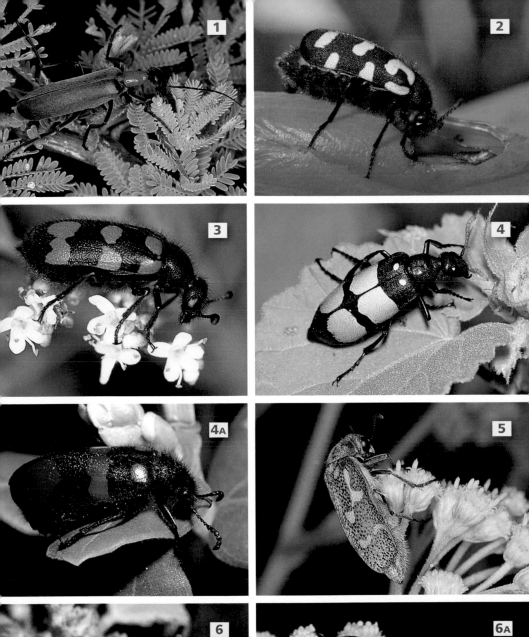

1 *Lytta nitidula* — Green-headed Blister Beetle

Medium-sized (body length 14mm). Head and pronotum metallic green, smooth. Antennae and legs black. Elytra brown. Underside with a thin covering of white hairs. **Biology** Adults feed in the fresh flowerheads of various proteas, as well as the flowers of certain irises, including *Bobardia spathaca*. **Habitat** Fynbos.

2 *Epicauta velata* — Little Grey Blister Beetle

Small (body length 11mm), body and legs covered with a grey pubescence. Head, pronotum and elytra with small black spots. Antennae relatively long, brownish at the base, darker at the tips. **Biology** Adults eat the flowers and leaves of cotton, millet, potato, lucerne and maize. **Habitat** Subtropical forest and savanna. Often near water.

3 *Meloe angulatus* — Oil Beetle

Large (body length 26mm), shiny bluish-black, with very short elytra and no hind wings. Head punctured. Pronotum squarish, coarsely punctured. Antennae bead-like. **Biology** Sluggish. Freshly hatched larvae (triungulins) climb up plants onto flowers and wait for bees, to which they attach themselves. They are carried to the bees' nests, where they feed on the eggs, nectar and pollen. **Habitat** Low open vegetation on the ground.

4 *Synhoria testacea* — Carpenter Bee Blister Beetle

Large (body length 22–35mm). Head, thorax and abdomen sealing-wax red. Antennae and legs black, except for reddish-brown basal half of femora. Males have huge mandibles. **Biology** Adults probably do not feed. Larvae parasitize nests of carpenter bees (genus *Xylocopa*). **Habitat** Diverse, usually near carpenter bee nests in logs, timber posts or dead wood. Fairly common around Cape Town.

Family Cerambycidae — Longhorn beetles, timber beetles

Small to very large (body length 3–100mm), with elongate cylindrical or flattened bodies. Antennae very long – two-thirds to 4 times body length. Usually held forwards, but can be swept back at rest. Chewing mandibles are large and strong, especially in males. Elytra usually cover the abdomen, but in some species are greatly shortened. Nocturnal species are often brownish; those that live on the ground, bark or dead wood are cryptic. Diurnal species are often bright, with black, orange or yellow warning coloration. Adults occur on flowers, feeding on pollen and nectar, or leaves, wood or roots. Nocturnal species often attracted to lights, but found under logs or bark by day. Eggs are laid in cracks in the stems or roots of woody plants, or in living or dead timber. The larvae feed as they burrow into wood, and many species are consequently pests of timber. There are about 659 species in the region.

5 *Chorotyse* — Red stump-winged longhorns

Medium-sized (body length 12mm), wasp-like, easily recognized by very short red elytra and exposed hind wings (folded over abdomen at rest). Head and pronotum black. Antennae long, brown and thread-like. **Biology** An adult has been found in the cell of a solitary bee, suggesting a parasitic association. Adults observed pollinating ground orchids. **Habitat** Succulent karoo and fynbos.

6 *Zographus niveisparsus* — Zebra Longhorn

Large (body length 30mm), black. Pronotum with fine white transverse lines and a pointed projection on each side. Elytra with white patches and dotted white longitudinal lines. A white line encircles the base of antennae. **Biology** Like similar *Zographus* species, larvae may feed on *Xymalos monospora* and *Strychnos*. **Habitat** Subtropical forest and savanna. **Related species** The Orange-eyed Longhorn (*Z. oculator*) **(6A)**, which is large (body length 33mm), black, with large, orange-centred yellow spots and bars on elytra. On flowers, especially *Senegalia*, on the south and east coasts and in Namaqualand.

7 *Aristogitus cylindricus*

Medium-sized (body length 18mm), black. Pronotum cylindrical, with 4 dorsal tubercles and a blunt thorn-like protrusion on each side. Base of the elytra is wider than the pronotum, tapering posteriorly to a blunt point. Body and legs have white hairs; on the elytra white hairs are limited to the centre line and to 2 small patches near each anterior corner. Antennae quite thick and longer than the body. **Biology** Larvae burrow into alien Monterey Pine (*Pinus radiata*). **Habitat** Fynbos and renosterveld.

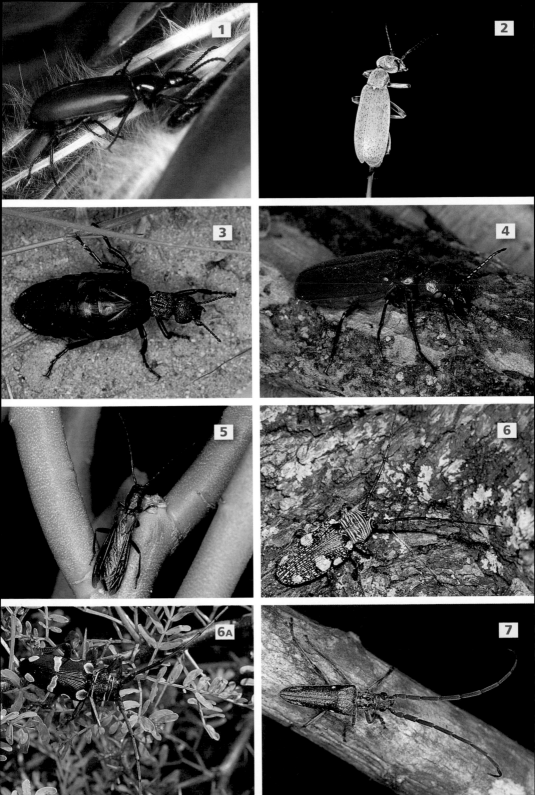

1 *Ceroplesis thunbergii* — Pondo-Pondo Longhorn

Large (body length 24mm). Head, antennae and legs black. Pronotum brown, punctured. Elytra black, with 3 red bands, brown at the base and strongly punctured, especially at the base. **Biology** Eggs are laid in Australian *Acacia* and native *Senegalia*, the female making a slit for this purpose in the bark of a dying branch. *Ceroplesis* larvae develop within woody stems of the host plant. Adults fly readily in hot sunshine, and feed on succulent bark. Known to damage wattle trees. **Habitat** Diverse. Usually on branches and trunks of trees. **Related species** There are several similar species, including *C. capensis* (**1A**), which is large (body length 37mm), with several blunt tubercles on the pronotum, and small tubercles and 4 red transverse bands on the elytra. Its larvae bore into Rooibos Tea Bush (*Aspalathus linearis*) and probably Australian Black Wattle (*Acacia mearnsii*) and *Cassia*. Found on tree branches and shrubs in the southwestern Cape. Also, *Pycnopsis brachyptera* (**1B**) (26mm long), which is similar but has cream bars on the elytra and occurs in the warmer summer-rainfall region.

2 *Tragocephala formosa* — Orange-tree Borer

Large (body length 30mm), black. Sides of pronotum yellow to orange, with a pointed lateral projection. Elytra each with 3 broad yellow to orange patches, the last 2 with a series of small white dots between them. Legs and antennae black, the latter about as long as the body. **Biology** Larvae develop in the stems of citrus trees, regularly causing serious damage in citrus groves in KwaZulu-Natal and the Cape provinces. **Habitat** Savanna and bushveld. **Related species** The genus includes 34 other similar-looking species, including this unidentified *Tragocephala* species (**2A**) from northern KwaZulu-Natal.

3 *Phoracantha recurva* — Lesser Eucalyptus Borer

Large (body length 25mm), brownish-black, with a broad yellowish band across base of elytra, interrupted by 2 black spots on each elytron. **Biology** Females lay eggs in crevices in the bark of eucalyptus trees (usually sickly, felled or dead) and the newly hatched larvae burrow down into the cambium layer beneath the bark and then disperse outwards, creating a radiating pattern of burrows, each of which is packed with frass (sawdust-like faeces) and becomes progressively wider as the larvae (**3A**) feed and grow. The burrows run parallel to one another and can extend for at least 30cm in all directions, effectively ring-barking the tree and killing infected branches. Larval development normally takes 3–4 months. The attractive pattern of feeding scars that remains on the surface of dead logs (**3B**) is a characteristic feature of eucalyptus forests. Pupation occurs in cells excavated by the mature larvae. An economically important pest of eucalyptus plantations. **Habitat** Diverse, wherever eucalypt trees are planted. Introduced from Australia in about 1906, probably in freshly cut railway sleepers. **Related species** The closely related Eucalyptus Longhorn Borer (*P. semipunctata*) (**3C**), which is also a pest of plantations and differs in that it is larger, with a black zigzag line across the yellow base of the elytra.

4 *Dirphya nigricornis* — Orange Coffee Borer

Large (body length 26mm), narrow, elongate, with a distinctive bright orange head and pronotum and black antennae. Elytra with basal third bright orange, remainder silvery grey, the inward-curving of the margins towards the rear giving a waisted appearance. Legs orange. **Biology** Flies by day. Coloration probably aposematic. Larvae develop in the stems of plants of the coffee family Rubiaceae and are sometimes pests in coffee plantations. Larvae eject excrement through a series of holes as they bore along a stem. **Habitat** Coastal forests in KwaZulu-Natal and forested parts of Limpopo.

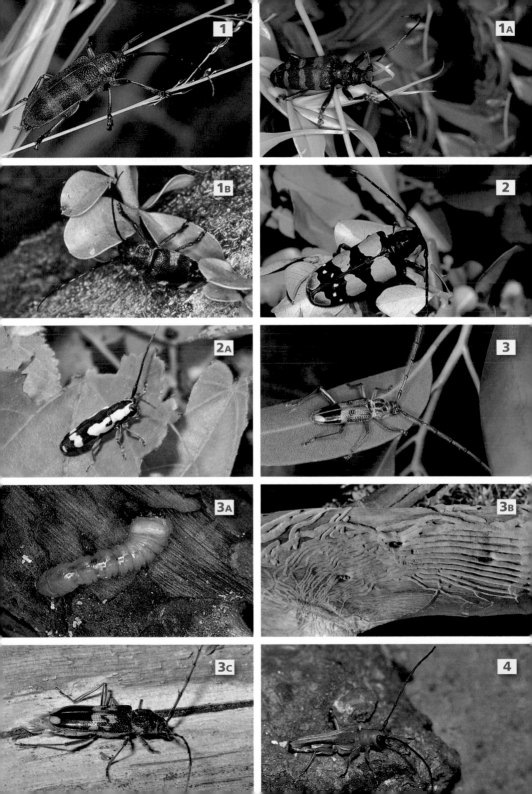

1 *Merionoeda africana* — Wasp Longhorn

Small (body length 9mm). Head, pronotum and antennae black, antennae somewhat serrate. Elytra shortened, pointed, yellow to orange, with black tips. Fore and mid legs yellow, last part of femora swollen; femora of hind legs with basal half yellow, and last part swollen and black. **Biology** A mimic of Braconid wasps. **Habitat** Subtropical savanna. **Related species** The Sombre Twig Pruner (*Cloniocerus krausii*) **(1A)** (body length 25mm), which is mottled black and yellow and has conspicuous tufts of black hair along the base of the antennae, which are strongly toothed distally. Fairly common in moist areas from the Western Cape to KwaZulu-Natal; also occurs in Madagascar.

2 *Tetradia lophoptera* — Dumpy Longhorn

Medium-sized (body length 13mm), short and stocky, coarsely sculptured on the pronotum and especially at base of the elytra. Black, with a covering of grey down and a greyish-white patch on each side of the pronotum. Near base of each elytron is a raised area with a U-shaped tuft of brown hairs. Antennae and legs blackish-brown, with a covering of greyish-white hairs. **Biology** Larvae recorded boring into acacias. **Habitat** Arid savanna.

3 *Phryneta spinator* — Fig-tree Borer

Large (body length 35mm), with prominent spines on the sides of the pronotum. Body and legs dark, with pale mottling on elytra. **Biology** Adults very robust and strong, gnawing the bark of figs and various other fruit trees; squeak loudly when handled. Larvae **(3A)** burrow into the base of a fig tree trunk, ejecting pelleted wood shavings from the holes and ultimately killing the tree. **Habitat** Indigenous and particularly common in Cape Town, where it is a serious pest of fig trees. **Related species** *Monochamus* species, which are the most important pests of coffee trees in South Africa, including *M. spectabilis* **(3B)** (20mm long), which is a similarly coloured Afrotropical species with characteristic M-shaped, white-edged black markings on the elytra and extremely long antennae; occurs in the warmer eastern parts of the region.

4 *Prosopocera lactator* — Turquoise Longhorn

Large (body length 33mm), pale turquoise, with a striking pattern of broad brown lines on the head, pronotum and elytra. Antennae brown, becoming turquoise towards tips. Legs brown, with some turquoise markings. **Biology** Adults cut through branches and young stems of the Buffalo Thorn (*Ziziphus mucronata*) in preparation for laying eggs. If disturbed, they drop to the ground. **Habitat** Subtropical forest and savanna. Prefers higher altitudes.

5 *Tragiscoschema bertolonii* — Ocellate Longhorn

Medium-sized (body length 13mm), narrow and elongate, with antennae longer than the body. Head and pronotum black down the centre, yellow on the sides. Elytra powdery blue, with 14 yellow patches outlined in black. **Biology** Females ring-bark stems before laying eggs. Larvae bore into the stems and twigs of cotton, *Hibiscus* and other plants. **Habitat** Savanna.

6 *Promeces longipes* — Common Metallic Longhorn

Medium-sized (body length 14mm), narrow and elongate. Entire body, including antennae and legs, a striking metallic bluish-green. Head and pronotum finely punctured. Elytra punctured, with indistinct longitudinal ridges. **Biology** Larvae bore into the stems of *Rhus*. Common and conspicuous; adults usually seen visiting flowers, often those of Melkbos (*Euphorbia mauritanica*). **Habitat** Diverse. Releases a foul-smelling odour when handled. **Related species** There are several other species in this and related genera that look similar.

7 *Philematium virens* — Large Green Longhorn

Large (body length 25–29mm) and broader than *Promeces longipes* (above). Head, pronotum and elytra metallic green and punctured. Elytra wider than the pronotum at their base, tapering to the rear. Antennae black. Legs black, with red femora. Pronotum cylindrical, with pointed tubercles. **Biology** Flies conspicuously by day, often settling on vegetation. Larvae bore into the sapwood and heartwood of Red Milkwood (*Mimusops africana*) trees. **Habitat** Subtropical, often coastal, forest and savanna.

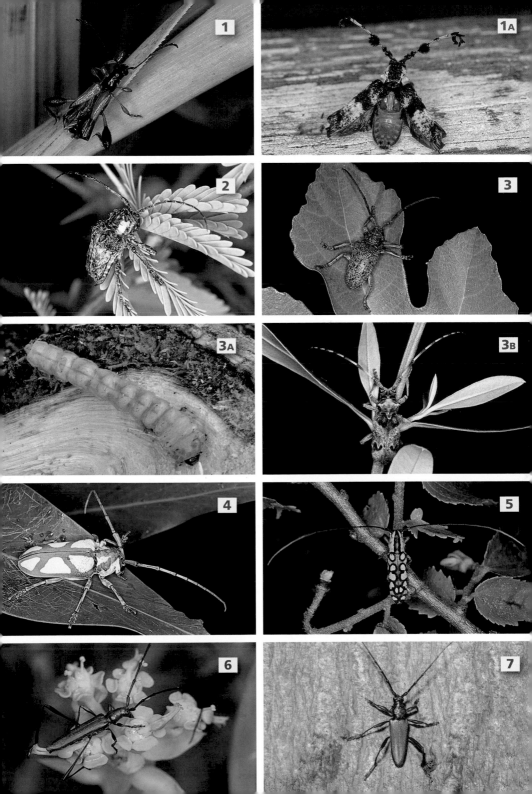

1 *Anubis scalaris* — Skunk Longhorn

Medium-sized (body length 17mm), narrow, elongate, antennae longer than body. Head, pronotum and 2 basal segments of antennae metallic greenish-bronze, remainder of antennae black. Elytra shortened (exposing last 2 abdominal segments), metallic green, with 3 orange transverse bands. Legs metallic green or bronze. **Biology** Has warning coloration and releases a musky scent if grasped. **Habitat** Diverse. Prefers open bushveld and savanna. **Related species** *A. clavicornis*, which have yellow rather than orange transverse bands in the adults and prefer yellow composite flowers.

2 *Tithoes confinis* — Giant Longhorn

Massive (body length 55–90mm), entire body dark brown and covered in backward-lying fawn-coloured hairs that rub off and leave patches of brown. Pronotum with 2 or 3 spines on the sides. Mandibles enlarged and strongly developed. **Biology** Adults nocturnal. The huge larvae **(2A)** bore into Cashew Nut trees; probably also Marula, Wild Plum and others in the mango family (Anacardiaceae). Attracted to lights. **Habitat** Subtropical forest.

3 *Macrotoma palmata* — Large Brown Longhorn

Very large (body length 50mm), with flattish, dark reddish-brown body, legs and antennae. Pronotum squarish, with several sharp spines along each side. Elytra punctured. Femora and tibiae of fore legs have short spines (less well developed on the tibiae); femora and tibiae of mid and hind legs also spiny. **Biology** Adults nocturnal, often attracted to lights. Larvae have an enzyme for digesting cellulose and bore into various hosts, including *Senegalia*, *Citrus* and *Ficus*. **Habitat** Subtropical forest.

4 *Erioderus hirtus*

Large (body length 40mm), brown, with a furry underside and pronotum. **Biology** Adults nocturnal, often attracted to lights. Larvae possess an enzyme for digesting cellulose and bore into a variety of host trees. **Habitat** Eastern forested parts of South Africa. **Related species** *Delocheilus prionoides* **(4A)** (body length 28mm), which is similar in overall appearance, but has pale brown ridges on the elytra and is found from the southern Cape to KwaZulu-Natal.

5 *Cacosceles* — Giant Spined Longhorn

Large (body length 25–70mm), robust beetles. Very dark brown, with massive mandibles. Antennae long and prominent. Pronotum with a sharp lateral spine on each side. **Biology** Adults nocturnal, attracted to lights. **Habitat** Larvae found in rotten wood, often in contact with the soil, such as stumps and roots.

Family Chrysomelidae — Leaf beetles

Small to medium-sized (body length 1–35mm) beetles showing great diversity in shape and colour. Some resemble ladybeetles (Coccinellidae), but have 4, not 3, clearly visible tarsal segments. Adults and larvae of all species are herbivorous and many are pests. Occur on all types of plants and in all terrestrial habitats. Adults eat flowers and foliage; larvae feed on or bore into or mine leaves, stems and roots. The number of species in the region is unknown, but well over 1,000.

6 *Sagra bicolor* — Swollen-legged Leaf Beetle

Medium-sized (body length 18mm). Head, antennae and pronotum black, with a bluish-green sheen. Elytra metallic purple, smooth, with longitudinal rows of punctures. Eye margins indented. Legs strongly developed, hind pair in male (shown) longer than body and with swollen femora. Elliptical cavity close to base of lower side of hind legs filled with dense reddish hairs; tarsal segments expanded and flattened, with reddish velvety hairs on undersides. **Biology** Larvae live within, and can cause swellings in, the stems of woody plants. **Habitat** Savanna and woodland.

7 *Lema*

Small (body length 8mm). Head and pronotum red and punctured, elytra red, with longitudinal rows of large, distinct punctures. Antennae fairly thick. **Biology** Adults alert and active, flying off when approached. Adults and larvae of the group to which *Lema* belongs feed together on their foodplant leaves, the larvae usually covered in their own moist frass, which they place on their backs. Foodplants not known. **Habitat** Subtropical forest, bushveld and grassland. Recorded from Pretoria.

1 *Lema trilineata* Tobacco Slug, Three-lined Potato Beetle

Small (body length 7mm), moderately elongate, elytra wider than pronotum. Head black, pronotum orange, with 2 circular black patches, slightly punctured. Elytra pale yellow, with 3 black longitudinal bands and longitudinal rows of punctures. **Biology** Feeds on nightshades (e.g. Cape Gooseberry); a pest of tobacco. Larvae cover themselves in faeces (**1A**). Pupate in soil. **Habitat** Gardens, subtropical forest and savanna. Originate in South America. **Related species** *L. bilineata*, which has broader black bands on the elytra and is also a pest of tobacco.

2 *Pseudorupilia ruficollis* Swollen Restio Beetle

Small (body length 5–9mm), with a grossly inflated abdomen and rudimentary elytra. Yellow-brown body, with black legs and antennae. **Biology** Feeds on pollen, especially that of restios. **Habitat** Restio veld and grassland. **Related species** Other *Pseudorupilia* species, most of which feed on restio flowers.

3 *Cryptocephalus decemnotatus* Ten-spotted Leaf Beetle

Small (body length 6mm), short and compact, colour variable – from uniform yellow with a few dark spots on thorax and elytra, to yellow with dark blotches on prothorax and elytra, 2 on prothorax, and 4 on each elytron, to yellow with dots fusing to form 2 transverse bands. **Biology** Feeds on leaves and daisy florets. If disturbed, drops to the ground from foliage and stems of host plants. Larvae live on host plants in sac-shaped cases made of faeces and plant debris. **Habitat** Grasslands and flowering meadows.

4 *Chrysolina clarkii* Ladybird Leaf Beetle

Medium-sized (body length 9mm), rounded and globular, like a ladybeetle (p.270). Head and pronotum shiny brown, sparsely punctured. Elytra black, with large orange patches. Antennae bead-like. **Biology** Adults and free-living larvae feed on various plants. **Habitat** Subtropical forest and savanna. **Related species** Other *Chrysomela* (body length 9mm) species, some of which have bright red spots on black elytra (**4A**), probably mimicking noxious ladybeetles.

5 *Chrysolina progressa* Gold-spotted Leaf Beetle

Small (body length 7mm), rounded, resembling ladybeetles (Coccinellidae). Head and pronotum black. Elytra black, with 12 gold patches. **Biology** As for family (Chrysomelidae). **Habitat** Low succulent vegetation. **Related species** There are many other similarly shaped beetles, most of which feed on *Gymnosporia* species.

6 *Monolepta bioculata* Two-eyed Flower Beetle

Small (body length 5mm), rotund. Head and pronotum shiny brown. Elytra soft, shiny brown, each with 2 large black-edged yellow patches. Legs brown; antennae brown, thread-like and much longer than head and pronotum together. **Biology** Can jump. Feeds on flower petals, including those of *Pelargonium capitatum* and *Monopsis*. **Habitat** Common in grassland, often on flowers.

7 *Sonchia sternalis* Banded Cucurbit Leaf Beetle

Small (body length 7mm). Head and pronotum shiny orange, the latter with 2 black patches. Elytra soft, shorter than abdomen, black, with a broad orange transverse band and 2 orange patches at the rear. Legs and antennae yellowish. **Biology** A minor pest of cucumbers, pumpkins and other squashes, feeding on their leaves and flowers. **Habitat** On foliage of various plants.

8 *Megaleruca* Celtis leaf beetles

Medium-sized (body length 10mm), soft-bodied, with dull, matt, pale yellow surface. Head and pronotum with black markings. Larvae cylindrical, short-legged, yellow and black, with a black head. **Biology** Adults and larvae feed on foliage, often in large numbers. **Habitat** Leaves and stems almost exclusively of Stinkwood (*Ocotea bullata*) and White Stinkwood (*Celtis africana*) trees. **Related species** There are several other very similar species, including *Palaeophylia tricolor* (**8A**), which is small (body length 6mm), dull metallic green, with a black-spotted pronotum and occurs in the Western Cape.

1 *Blepharida*   Rhus flea beetles

Small (body length 6mm) and compact. Head and pronotum brown, the latter with creamish-white markings and lateral margins. Elytra with longitudinal rows of brown punctures and covered with small, tile-like, creamish-white and brown patches. Hind femora strongly developed for jumping. Antennae brown, thread-like, not much longer than head and pronotum together. Larvae **(1A)** have arched bodies that conceal the head and legs when feeding. A sucking disc at the rear end of the body helps them attach to the host plant. Faeces are retained on their backs as a deterrent to predators. **Biology** Pupation occurs in the ground within a cocoon of sand grains. Adults may aggregate on *Rhus*. **Habitat** Savanna. **Related species** Other species in this genus, and in the related *Diamphidia* and *Polyclada*, including some species in which the larvae feed on corkwood (*Commiphora*) plants; the San (Bushmen) use *Polyclada* larvae for poisoning their arrowheads **(1B)**.

2 *Dicladispa*   Spiny leaf beetles

Small (body length 5mm), oblong, brown, with conspicuous long, pointed spines on the pronotum and elytra. **Biology** Larvae are leaf miners, feeding while hidden within the leaves of the host plant. **Habitat** Usually on foliage. **Related species** There are many similar species in *Dorcathispa* and *Dactylispa* in the region.

3 *Oncocephala*   Flat spiny leaf beetles

Small (body length 4mm), oblong and flattened. Elytra are squared off at the rear, dark brown in colour, with paler markings, and have longitudinal ridges interspersed with large punctures. Pronotum heavily punctured. Antennae brown, clubbed, and about as long as the head and pronotum together. **Biology** Some are leaf miners, feeding on *Ipomoea*, sweet potato and other plants of the family Convolvulaceae. **Habitat** On foliage.

4 *Basipta stolida*   Green Tortoise Beetle

Medium-sized (body length 11mm), broad and flattened. Elytra widely expanded sideways, the corners extended on each side of the pronotum, giving the body an oval outline. Head hidden below pronotum. Elytra metallic green over abdomen, translucent over pronotum and remaining areas, with short white hairs. **Biology** Larvae and adults feed on Coast Silver Oak (*Brachylaena discolor*). **Habitat** Subtropical forest and savanna. **Related species** Species in the related genus *Cassida* (body length 5mm) **(4A)**, which are bright green and circular in shape. The spiny larvae protect themselves by carrying their faeces attached to spines on their dorsal surface.

5 *Conchyloctenia punctata*   Spotted Tortoise Beetle

Medium-sized (body length 11mm), moderately flattened, body with oval outline. Pronotum semicircular, orange. Head hidden below pronotum. Elytra pinkish-red, with several black spots. **Biology** Eggs are laid in a flattened, papery, brownish egg case attached to the host plant. Feeds on morning glory (*Ipomoea*). **Habitat** Diverse. **Related species** *C. hybrida* **(5A)** (body length 10mm), which has mother-of-pearl colouring on the elytra. Also, *Cassida virgintimaculata* **(5B)** (body length 9mm), which is dull yellow, with irregular black blotches on the body and occurs in the warmer eastern parts of South Africa, and *Trachymela tincticollis* **(5C)** (body length 7mm), which is a dark brown Australian tortoise beetle introduced inadvertently to the Western Cape, where it feeds on *Eucalyptus*.

6 *Plagiodera caffra*   Red-headed Leaf Beetle

Small (body length 7mm), rounded, resembling a ladybeetle (Coccinellidae). Head, pronotum and legs reddish-brown. Elytra metallic blue-green. **Biology** Foreign members of this cosmopolitan genus often feed on willows and poplars. **Habitat** Diverse. On tree foliage in bushveld.

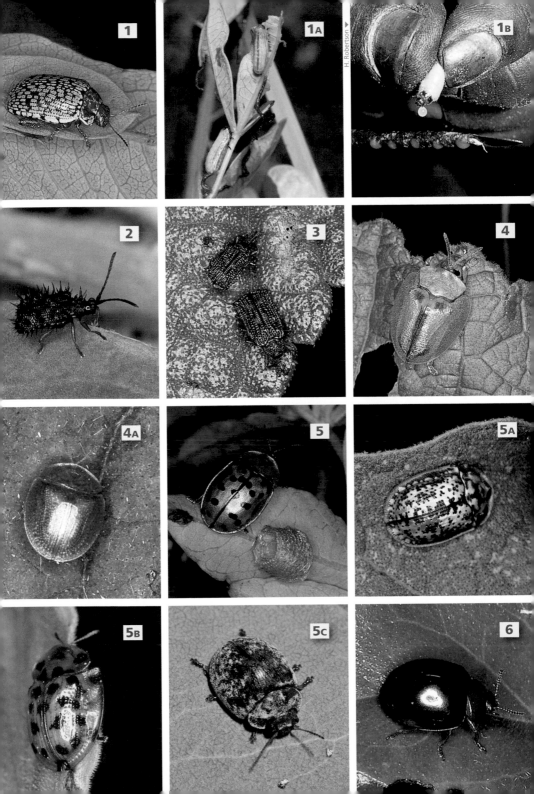

1 *Aspidimorpha areata*
Red Tortoise Beetle

Small (body length 7mm), broad, moderately flattened, oval in outline. Pronotum red, with 4 black spots. Expansions of pronotum and elytra translucent, with an orange tinge and a network of veins. Elytra red, with a characteristic pattern of black lines and longitudinal rows of punctures. **Biology** Known to reproduce on 3 host plants of the Convolvulaceae, *Ipomoea batatas*, *I. cairica* and *I. plebeia*. Produce 3 generations a year. **Habitat** Savanna. On foliage.

2 *Aspidimorpha tecta*
Fool's Gold Tortoise Beetle

Small (body length 9mm), broad, flattened. Pronotum and elytra widely expanded, transparent, with bright gold tortoise-shaped marking; pronotum smooth, elytra with sparse black punctures. Head hidden below pronotum. **Biology** Retention of faecal filaments and cast-off skins of larvae on abdominal spikes is restricted to first 2 instars. Feeds on the leaves of solanaceous plants (potato, tomato) and Morning Glory (*Ipomoea*). Egg case is papery, as in *Conchyloctenia punctata* (p.296). **Habitat** Foliage in diverse habitats. Common in gardens. **Related species** The small Circular Tortoise Beetle (*A. submutata*) **(2A)** (body length 7mm), which is flat, with the pronotum and elytra describing an almost perfect circle. Its metallic colours fade after death, as in all tortoise beetles. Found in foliage in subtropical forest and savanna.

3 *Aspidimorpha puncticosta*
Dune Tortoise Beetle

Medium-sized (body length 13mm), broad, flattened, oval in outline. Head hidden below pronotum. Expansions of pronotum and elytra almost transparent, with fine white veins extending to margins. Centre of pronotum yellow, with a black spot behind the head and margins transparent, elytra transparent, with 5 black patches around the borders. **Biology** Feeds on Dune Morning Glory (*Ipomoea pescaprae*) and other Convolvulaceae species. Cast-off larval skins of all instars (but not faecal filaments) accumulate on the end of the abdomen, which is held over the spiny body for protection **(3A)**, especially from ants. **Habitat** The foliage of *I. pescaprae* on dunes adjacent to beaches.

4 *Macrocoma aureovillosa*
Furry Grassland Beetle

Small (body length 6mm) and compact, head and pronotum large in relation to abdomen. Pronotum metallic green, punctured. Elytra metallic copper, punctured. Legs long, metallic copper. Entire body clothed in fine silvery white hairs. **Biology** Feeds on grasses, notably the anthers of *Trachypogon plumosus* and turpentine grass (*Cymbopogon*). **Habitat** Typically associated with grasses in open grassland.

5 *Antipus rufus*

Small (body length 5–6mm), cylindrical with enlarged fore legs, legs yellow and black, head and thorax reddish-brown, elytra yellow, with variable amount of dark brown to black patches, which may fuse to form black transverse bands. **Biology** Leaf feeder. **Habitat** Grassland and flowering meadows. **Related species** *Crabronites equestris* **(5A)**, which is small (body length 10mm), with the head, pronotum and elytra metallic green, elytra with red margins and bands, very long fore legs, and large, strong jaws. Found on *Lebeckia meieriana* and other leguminous plants and on Sea Guarri (*Euclea racemosa*) in the southwestern Cape.

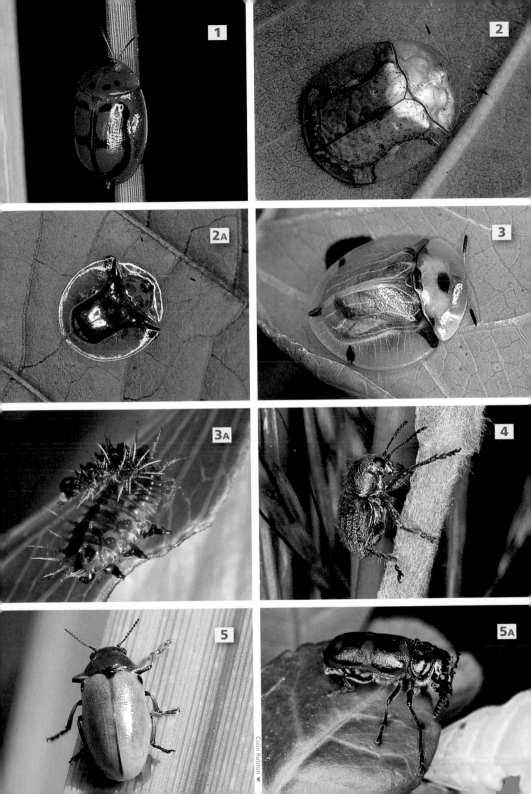

1 *Platycorynus dejeani* — Milkweed Leaf Beetle

Medium-sized (body length 11mm), rotund and elongate. Head and pronotum with metallic copper or purple sheen. Elytra metallic green. **Biology** Adults feed on various plants, including cotton, but are typically found in groups on Milkweed (*Asclepias fructicosa*). **Habitat** Vegetation in natural and disturbed areas.

2 *Lysathia* — Myriophyllum leaf beetles

Small (adult 3.8mm long, mature larva 8mm). Adults metallic green, with pale yellow legs and dark tarsi. **Biology** Only feeds on Parrot's Feather (*Myriophyllum aquaticum*), an invasive aquatic weed. Beetles introduced from Brazil, released since 1994. Adults and larvae feed on growth tips and leaflets, the larvae causing most damage. Large beetle populations severely retard growth, but the weed may recover cyclically. Booster releases may be required, but control of the weed is currently regarded as satisfactory. **Related species** There are some other superficially similar indigenous leaf beetles in shoreline habitats, but none are found on Parrot's Feather.

3 Subfamily Bruchinae — Pea weevils, bean weevils, seed weevils

Small (body length 2–5mm), compact, oval, with shortened, grooved elytra that expose the end of the abdomen. Small head, often not visible from above, with mandibles on short snout. Hind legs may be thickened and have tooth-like projections. Dull brown or black, often mottled, with pale hairs or scale prongs. About 150 species in the region. **Biology** Eggs are laid on seeds, into which larvae burrow. Emerging adults leave characteristic round holes in the seeds (**3A**). *Neltumius arizonensis* (body length 4mm) was imported as a biocontrol agent for mesquite (*Prosopis*). Adults may feign death and drop off a plant when disturbed. **Habitat** On foliage, especially that of leguminous plants, and in stored pulses.

Family Anthribidae — Fungus weevils

Small to large (body length 3–30mm), oblong beetles, darkly coloured and usually densely patterned with white or dark scales or hairs. They have a broad, short to long snout (rostrum). Antennae are fairly long and slender, sometimes ending in a 3-segmented club. Adults feed on pollen, fungi or bark. Larvae develop in rotten wood, fungi or plant tissue. Poorly studied, but about 120 known species in the region.

4 *Chirotenon longimanus* — Long-legged Fungus Weevil

Large (body length 21mm), moderately elongate, with short, broad snout and well-developed mandibles. Antennae long, thread-like, with a club of enlarged segments. Body covered with a pile of short hairs. Head and pronotum with brown and creamish-white markings. Elytra mottled, with large creamish-white hexagonal patch towards rear. Legs long and thin, the fore legs of male twice body length. **Biology** Nocturnal. **Habitat** Subtropical forest and bushveld.

5 *Xylinades rugicollis* — Fungus Weevil

Medium-sized (body length 19mm), moderately elongate, parallel-sided, with short, broad snout and large mandibles. Antennae fairly short, thick, and somewhat bead-like. Body covered in short, flat-lying hairs, mostly black, but with small, scattered reddish-brown patches. Elytra with longitudinal rows of shiny, rounded tubercles. **Biology** Nocturnal. **Habitat** Trunks of forest trees in moister eastern areas.

Family Brentidae — Straight-snouted weevils

Small to medium-sized (body length 4–20mm), parallel-sided, elongate beetles, with a narrow body, more or less elongate snout on head, and swollen femora. Antennae are bead-like and never clubbed. There are about 100 species in the region.

6 *Orfilaia vulsellata* — Common Brentid

Medium-sized (body length 11mm), body dark brown, male with distinct mandibles. Head and pronotum shiny and smooth. Elytra with longitudinal rows of punctures and yellowish markings. **Biology** Females lay eggs in holes, which they drill in wood with their snouts. Adults feed on fungi, sap or wood-eating insects, larvae on decaying wood or fungi. Adults often attracted to lights. **Habitat** Under bark or in dead wood in wooded areas. **Related species** *Amorphocephala imitator* (**6A**) (body length 10mm), which is uniformly reddish-brown, with grooved elytra, and occurs across most of the region, apart from the Western Cape.

1 *Nothogaster picipes*   Ridge-winged Brentid

Small (body length 12mm), brown, with reddish- or yellowish-brown head, rostrum, legs and apexes of the elytra. Striae 3 and 4 deeply punctate. **Biology** Details unknown. **Habitat** Common throughout sub-Saharan Africa**. Related species** *N. laevicollis*, which is similar, but reddish-brown; the rear parts of the elytra have a dark spot and are less brightly coloured than the prothorax. Also striae 3 and 4 are only slightly punctate.

2 *Antliarhis zamiae*   Hose-nosed Cycad Weevil

Small (body length 9mm, excluding snout), shiny brown and flattened. Females have an extraordinarily long, thin snout (up to 3 times as long as the body), with mandibles at the tip; snout much shorter in male. Elytra with longitudinal grooves. **Biology** Using their mandibles, females drill holes through cycad cone scales to reach the seeds, in which they lay batches of eggs. Larvae feed on and hollow out the inside of the seed, within which they pupate **(2A)**. **Habitat** Associated with *Encephalartos* cycads; usually found on the cones. **Related species** *A. signatus*, which is similar but smaller, and the female has a shorter snout than that of female *A. zamiae*.

3 *Cylas puncticollis*   Sweet Potato Weevil

Small (body length 5–14mm), convex, with dome-shaped elytra. Snout mostly down-curved and length varies. Antennae straight or elbowed. Body smooth or hairy, black, reddish-brown or pale. *C. puncticollis* is an imported cosmopolitan pest of sweet potato. **Biology** Eggs laid in stems hatch into grubs that bore into sweet potato tubers, packing their burrows with faecal matter, with resultant crop spoilage. Adults cryptic, emerging at night, and capable of short flights. **Habitat** Agricultural fields where sweet potatoes are grown. **Related species** *C. formicarius*, which has an orange thorax and black abdomen, and is restricted to KwaZulu-Natal.

4 **Family Attelabidae**   Leaf-rolling weevils

Small (body length 2–8mm), stout, recognizable by the square elytra, which do not cover the final abdominal segment. Snout not more than half body length. Antennae bead-like and never elbowed, ending in a 3–4-segmented club. Body sometimes covered with sharp spines. Mostly black or reddish-brown, as in *Parapoderus submarginatus* **(4)** (body length 6mm). Some species, such as *P. nigripennis* **(4A)** (body length 6mm), may be red and black; others have a metallic sheen. There are about 45 regional species. **Biology** Adults feed on flower buds and leaves, in which they cut square windows. Females lay eggs in rolled-up leaves or a slit in a leaf, which then wilts or falls off the plant. *P. submarginatus* rolls the leaves of Water Berry (*Syzygium cordatum*). Mature larvae pupate in soil. **Habitat** Flowers and foliage.

Family Apionidae

Minute to small (2–8mm) weevils, with dome-shaped elytra and a down-curved snout of variable length. Antennae straight or elbowed. Body smooth or (often) hairy. Adults feed mostly on leaves. The larvae develop within roots, stems, flowers or fruits of their host plants. Poorly studied in the region, and the number of species is hard to estimate.

5 *Setapion quantillum*   Wild Almond Weevil

Very small (body length 1.5mm), uniformly matt black with hairy, grooved, parallel-sided elytra. **Biology** Occurs year-round in large numbers only on the leaves of the Wild Almond (*Brabejum stellatifolium*) on which it feeds. **Habitat** On the leaves of various riverine plants in the southwestern Cape. **Related species** *S. provinciale*, which occurs seasonally and specializes in feeding on the terminal buds of Wild Almond trees, which will develop into flowerheads.

Family Curculionidae   Weevils, snout beetles

Highly diverse in size (body length 1–60mm), habits, shape and colour. All have a snout (with mandibles at the end), ranging from short and broad to long and thin. Antennae characteristically elbowed and clubbed, and attached to sides of snout. Elytra tough, often sculptured, and sometimes fused. Legs short and stout, with strong claws and adhesive pads on the tarsi to grip smooth leaf surfaces. Adults and larvae are herbivorous. Females bore into any plant part to lay their eggs (or may lay them in soil). Widespread and includes pests of crops and stored products; most plants are host to 1 or more species. Probably the largest family in the animal kingdom, with some 86,000 species globally, 3,000 from the region.

1 *Ellimenistes laesicollis* — Grey Coffee Snout Beetle

Small (body length 5mm), mottled grey, with broadly grooved elytra and long, elbowed antennae. **Biology** Originally described as a pest of coffee, but has catholic tastes and feeds on a very wide range of unrelated plants, usually seeking out young and tender plant tissue, including leaf stalks and leaf buds. Also damages banana, macadamia and tea plants, young Eucalyptus and Wattle seedlings. **Habitat** This native weevil is a ubiquitous garden pest, but also occurs in plantations and undisturbed veld.

2 *Timola*

Small (body length 8mm), stout, rounded, with a narrow, curved snout almost half the length of the body. Antennae elbowed, thread-like, with a distinct 3-segmented club. Body and legs densely clothed in hairs, buff-coloured except for large brown anterior area on elytra. **Biology** As for family (Curculionidae). **Habitat** Subtropical forest and savanna.

3 *Curculio hessei* — Slender-snouted Weevil

Small (body length 8mm), stout and rounded, with a long, narrow, curved snout, almost half the length of the body. Antennae elbowed, thread-like, with distinct 3-segmented club. Body densely clothed in hairs. Pronotum dark brown, elytra buff, with faint dark brown spots. Last section of femora swollen. **Biology** Feeds on various wild figs and on Buffalo Thorn (*Zizyphus mucronata*). **Habitat** Subtropical forest and savanna.

4 *Prionorhinus canus*

Medium-sized (body length 16mm), elongate and oval, greyish, with a stout snout. Pronotum heavily punctured, with 4 pale longitudinal stripes. Elytra with longitudinal rows of punctures and 4 raised, elongate white patches on each side. Body clothed in short hairs. **Biology** As for family (Curculionidae). **Habitat** Ground-dwelling in arid vegetation.

5 *Hypolixus* — Elegant weevils

Medium-sized (body length 12mm), elongate and oval, with a thick, curved snout equal in length to the head and pronotum. Elytra with longitudinal rows of punctures, 2 diagonal pinkish stripes, and markings at the rear. Body clothed in short black hairs, with pinkish hairs on the sides of the pronotum. **Biology** The Pigweed Weevil (*H. haerens*) is known to damage commercial amaranth ('morog') crops. **Habitat** Subtropical forest and bushveld. **Related species** *Lixus* species (**5A**) (body length 12mm), which are dusted with a yellow bloom and occur in grassland in the summer-rainfall region. Some *Lixus* species are known to breed in the stems and crowns of the Iceplant (*Mesembryanthemum crystallinum*).

6 *Polyclaeis equestris* — Pink-banded Weevil

Medium-sized (body length 18mm), dark blue-green, with a short, broad snout. Elytra each with a broad pink or red diagonal band, the coloration being due to a clothing of red hairs. **Biology** Adults eaten by people in some areas. **Habitat** Bushveld, often on *Senegalia* bushes.

7 *Hipporrhinus furvus* — Prong-tailed Weevil

Large (body length 30mm), elongate and heavily built, with a thick, curved snout, about equal in length to the pronotum. Antennae elbowed, thread-like, with a distinct 3-segmented club. Head with pinkish scales, mostly at the base. Pronotum and elytra with very pronounced, rounded black tubercles, interspersed with white or pinkish-buff scales. Elytra end in 2 prongs. **Biology** Flightless. Some species feed on Pig's Ears (*Cotyledon orbiculata*). **Habitat** Ground-dwelling in diverse habitats.

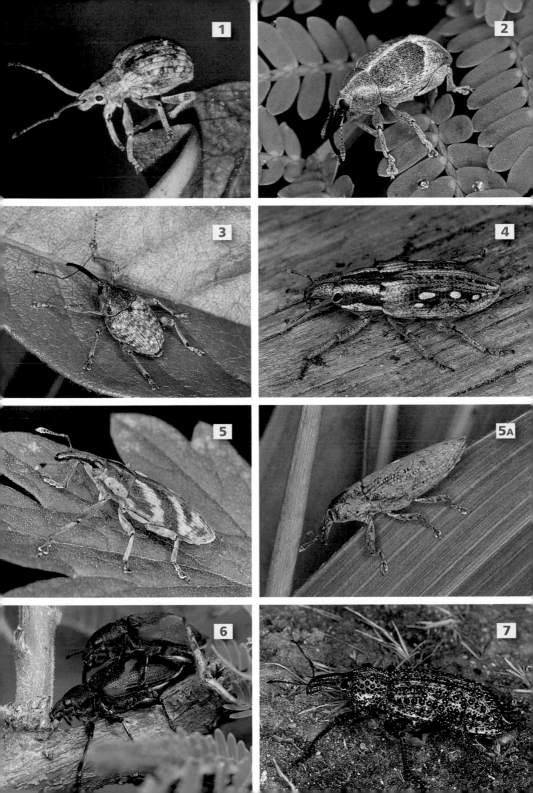

1 *Sciobius*

Small to medium-sized (body length 10mm), fairly stout, with a short, thick snout about equal in length to remainder of head. Pronotum with small, rounded tubercles; elytra with longitudinal rows of small, closely spaced tubercles. Coloration dark brown, with buff hairs. Antennae slender, more than half the length of the body. Legs reddish-brown, with a thin covering of whitish hairs. Femora swollen towards tips. Endemic. **Biology** Larvae of some species feed on the roots of citrus. Adults feed on the leaves of litchi and other trees. **Habitat** Foliage in various vegetation types. **Related species** *Bicodes vittatus* (**1A**), which is a small (body length 4mm) green beetle found in large numbers in savanna vegetation.

2 *Protostrophus* Beaded weevils

Medium-sized (body length 12mm), stout, with a well-rounded abdomen and broad snout about equal in length to the remainder of the head. Body clothed in brown, grey, silver, bronze or iridescent green scales. Elytra with longitudinal rows of punctures. Antennae shorter than head and pronotum. At least 133 species in the region. **Biology** Flightless. Feed on a wide variety of plants, including ornamentals, cotton and tobacco seedlings, young wheat, sweet potato and groundnut plants. Larvae free-living in soil. **Habitat** Ground-dwelling or in foliage.

3 *Brachycerus ornatus* Red-spotted Lily Weevil

Massive (body length 25–45mm), with a rounded abdomen. Head strongly punctured and ridged. Pronotum with robust irregular ridges in the centre and rounded tubercles on each side. Black with red markings. **Biology** Adults feed on the foliage of the toxic Karoo Lily (*Ammocharis coranica*) and Tumbleweed (*Boopone disticha*), the females laying eggs in burrows adjacent to the bulbs. Larvae live in the bulbs and pupate in soil. **Habitat** Ground-dwelling in areas of clay soil. **Related species** There are many similar species, including *B. severus* (**3A**), which is grey, with 2 lines of raised black welts running along the elytra, and occurs in warmer summer-rainfall parts of the region.

4 *Brachycerus tauriculus* Aloe Weevil

Large (body length 19–26mm), heavily built and strongly sculptured. Pale brown head with short, very broad snout. Irregularly sculptured pronotum with large pits, all lined with brown scales. Elytra flattened on top, with 4 coarsely knobbed ridges, dark brown in the centre and white on either side. **Biology** Typically feigns death when disturbed. Adults feed on foliage of various plants and can be destructive when feeding on aloe leaves. **Habitat** Ground-dwelling.

5 *Brachycerus obesus* Obese Lily Weevil

Large (body length 21mm), with a black head. Pronotum with 4 reddish-brown blister-like swellings. Elytra smooth and black with a reticulate pattern of reddish-brown markings. **Biology** Adults feed on young leaves, and larvae bore into and destroy the bulbs of various ammarylid lilies and gladioli. **Habitat** Ground-dwelling.

6 *Sitophilus* Maize and rice weevils

Small (body length 5mm), with moderately long and thick snout. Pronotum almost equal in length to abdomen. Elytra with closely spaced rows of punctures. Dark brown with 4 reddish patches on elytra. Winged. There are 3 species in the genus. **Biology** Drills holes in whole grains for feeding and laying eggs. Larvae feed and pupate within the kernel. **Habitat** Buildings where maize is stored, also maize fields.

7 Subfamily Scolytinae Bark and ambrosia beetles

Mostly small, cylindrical, brown or black beetles, a few millimetres long. **Biology** Adults usually burrow between the bark and wood of various trees. The egg tunnels and larval galleries make patterns indicative of the beetle species (**7**). Some species bore deep into wood, others attack various seeds, such as maturing coffee beans. Ambrosia beetles carry and innoculate the sapwood and xylem with a fungus, which breaks down the wood, nourishing the larvae in their tunnels. The larvae, like those of other curculionids, are apodous (**7A**). Many species are serious pests of timber, killing both healthy and sick trees. **Habitat** Cosmopolitan, where suitable breeding sites occur.

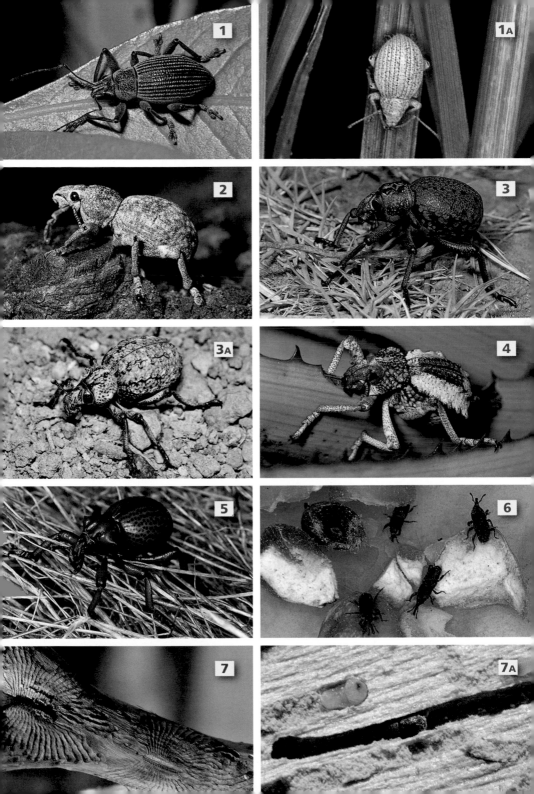

ORDER STREPSIPTERA • STYLOPIDS

A highly specialized, sexually dimorphic group of small insects, living as parasites within other insects. Adult females remain within the host throughout life, with only the reduced head and thorax protruding from the host's abdomen. Adult males are free-living and easily recognized by their bulging eyes, branched antennae, reduced, club-like fore wings and enlarged hind wings.

Stylopids are unusual and seldom encountered, and have a bizarre life history. They were once thought to be highly specialized beetles, but are now recognized as a separate order. Sexes differ greatly in appearance and habits. Males superficially resemble small, dark flies with branched antennae and unusual, bulging, berry-like eyes. Their first 2 thoracic segments are tiny, the third greatly enlarged. The fore wings are uniquely reduced to short, club-like structures, resembling the halteres of flies (but in flies it is the hind wings that are reduced). Hind wings are large and fan-like, with reduced venation. Females lack wings, legs and antennae, and spend their entire lives as internal parasites of other insects, most commonly Hemiptera and Hymenoptera. The reduced head and thorax of females, which protrude from between the host's abdominal segments, are fused.

Males live only a few hours and do not feed. They are strong fliers and seek out hosts infected by pheromone-releasing females. The female is inseminated through a unique, slit-like opening between her head and thorax, and thousands of tiny, mobile larvae move around freely within her haemocoel, effectively consuming her from within (a bizarre system unique to these animals). The active, legged 'triungulin larvae' emerge through the brood canal, then actively seek out and infect new hosts by softening and penetrating their cuticle. They then moult into legless maggots that develop further within the body cavity of the host, absorbing nutrients from its blood. After pupation, males emerge, but females remain within the host. About 600 species are known globally, of which 17 genera and approximately 90 species are known from sub-Saharan Africa. The South African fauna is poorly known, with only about 12 described species.

1 Family Stylopidae   Twisted-wing parasites

Males **(1)** distinguished by 4-segmented tarsi and the form of the antennae, the last 2 segments forming a parallel pair of broad, flat plates. Fore wings are reduced and spoon-shaped. The most diverse of 9 families in the order. **Biology** Insects parasitized by this family are referred to as 'stylopized'. Flattened prothorax and head of females protrude from between abdominal segments of host. Shown here are **(1A)** a stylopized *Ammophila* wasp (p.492) (family Sphecidae), infected by a *Xenos* stylopid, and **(1B)** a stylopized bee of the family Andrenidae. Both are infected by more than 1 stylopid. Parasitic infection is not fatal, and hosts have been observed to carry up to 4 parasites. Microscopic observation reveals hosts carrying tiny infective triungulin larvae **(1C)** as well as empty puparia **(1D)**, from which adult male stylopids have emerged. **Habitat** Rarely encountered; parasitic on bees and Vespidae and Sphecidae wasps. Males free-living, sometimes attracted to lights and flowers.

ORDER MECOPTERA •
HANGINGFLIES & SCORPION FLIES

Slender, predatory insects with elongate bodies, long, grasping legs, narrow wings and mouthparts produced into an unusual 'beak' (rostrum) tipped by biting jaws.

Hangingflies are a minor insect order related to true flies, and easily confused with flies of the Family Tipulidae (p.312), but are distinguished by their beak-like rostrum, and by having 2 pairs of wings. They are sometimes termed scorpion flies, since males of the largest family, Panorpidae, have swollen tips to their abdomens and hold these curled over their backs like scorpions. However, as this family is absent from South Africa, the common name is not appropriate for South African species. All species in the region instead belong to another family, Bittacidae, the hangingflies.

All hangingflies are predatory and mating must, thus, be carried out with caution. The male uses pheromones secreted by special scent glands to attract a female and distracts her by presenting her with a food item. While she eats, he is able to mate in safety. Females select males with larger prey gifts (sexual selection). Eggs are dropped at random in damp habitats and develop into predatory larvae resembling caterpillars, but with 8, rather than 5, pairs of abdominal legs. Larvae conceal themselves by covering their dorsal surfaces with faeces or soil particles, and move about with a looping motion, reminiscent of inchworms. Pupation takes place in a chamber just below the soil surface, and the newly hatched adult emerges via a specially constructed lid at ground level. Of about 600 mecopteran species globally, only 1 family and about 35 species occur in South Africa.

Family Bittacidae
Hangingflies

Slender-bodied, with elongate, prehensile legs and long thread-like antennae, the front of the head extended into a curious elongate rostrum with mandibles at its tip. There are usually 2 equal pairs of narrow membranous wings, but these are greatly reduced in 1 rare genus. Hangingflies are predatory, and adults dangle from vegetation by their front legs, using their specialized prehensile hind legs to intercept passing insects. The hind legs are not used for walking. Once captured, the victim is bitten with the mandibles and injected with saliva to dissolve its tissues. After it is sucked dry, its corpse is dropped to the ground.

1 *Bittacus*
Hangingflies

Medium-sized to large (wingspan 30–40 mm). Wings fully developed. By far the most diverse genus in the family, accounting for about 75% of all species globally and all but 1 of the over 30 species regionally. Species are distinguishable by minor variations in wing venation, colour, leg spination, or structure of male genitalia. **Biology** All species are predatory, hanging from vegetation with the powerful raptorial hind legs extended and using these to intercept flying insects such as flies **(1A)**. Adults may be present for only a few weeks each year, usually during summer. Attracted to lights. **Habitat** Each species has a well-defined habitat preference, ranging from grassland to fynbos or forest.

2 *Anomalobittacus gracilipes*
Wingless Hangingfly

A remarkable species and the only representative of this distinctive genus. Medium-sized (body length 19mm), easily identified by the greatly reduced wings, which consist of tiny non-functional filaments. **Biology** Hangs from restios or other fynbos plants, and preys on flying insects, which are grasped by the enlarged hind legs that dangle beneath the body. Details of life history not known. **Habitat** Restricted to moist fynbos vegetation in the southwestern Cape and known only from a few records, almost all of them males collected in spring.

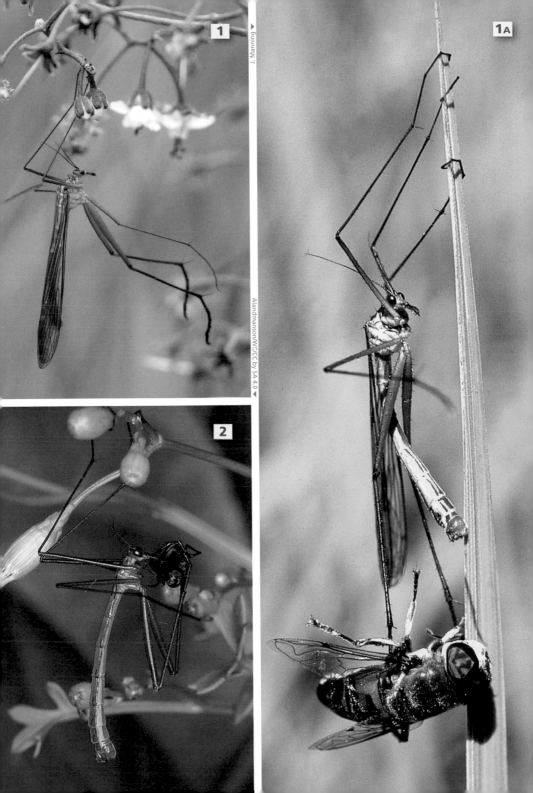

1

1A

2

ORDER DIPTERA • FLIES

Generally small insects, easily recognized in that only the first pair of wings is membranous and used for flying; the hind wings are reduced to a pair of small club-shaped structures called halteres, used for balancing. Some species are wingless.

Flies are among the most diverse of insect orders and include numerous ecologically and economically important species. They occur in all habitats, and have very diverse feeding habits. The mouthparts are usually adapted for piercing and sucking or for lapping and sucking, but chewing mouthparts occur in some bloodsucking species. Flies undergo full metamorphosis and the eggs are usually laid on the larval food source. Larvae are legless, but otherwise variable in appearance, habitat and diet, and larval development can occur within the body of the female to varying degrees. In the more primitive flies larvae have a distinct head capsule and usually live in waterbodies. Fly pupation occurs in a tough capsule. Many species are important vectors for diseases, most notably malaria, dengue fever and sleeping sickness. The larvae of some species feed on live mammals (including humans) and are thus of veterinary and medical importance; many others damage fruit and crop plants. Adult flies, by contrast, can be valuable pollinators. About 16,000 species are known from the Afrotropical region.

The following 11 families (up to, but not including family Stratiomyidae, p.320) are primitive flies, easily recognized by their very long antennae, made up more than 3 segments. The majority have aquatic larvae.

Family Tipulidae
Craneflies, daddy longlegs

A diverse family of medium-sized to large (body length 10–25mm) flies with a V-shaped groove on the thorax. Easily recognized by the very long legs. Some species are wingless. Larvae (known as 'leatherjackets') live in water, moist soil and lawns, and may feed destructively on the roots of garden plants. Adults never feed. Regionally, Tipulidae includes 250 known species.

1 *Conosia irrorata*   Spider Cranefly

Small (wingspan 13mm), yellow-brown, with short, thick legs often held together when at rest. Wings clear brown, with dark mottling at the fore margin. Short beak-like mouthparts. **Biology** Larvae live in rich organic gravel and silt. **Habitat** Low vegetation and rank grass along streams and vleis. Cosmopolitan.

2 *Nephrotoma*

Medium-sized (wingspan 28mm), with 'polished' jet-black-and-yellow body, and thin palps that are longer than the antennae. *N. antennata* has a plain deep orange abdomen. Numerous species in the region. **Biology** *N. antennata* is common in autumn. **Habitat** Very widespread in grass and shrubs in areas with moist soil.

3 *Tipula jocosa*   Giant Cranefly

Very large (wingspan 64mm), unmistakable. Wings light brown with 2 cream spots on margin and dark brown shading over veins. Abdomen and thorax striped. Has a green sheen on thorax and wings when freshly emerged. Larvae (**3A**) very large, with tubercles around the head and long breathing horns. **Biology** Larvae live under mats of vegetation in shallow streams. Adults rest flat against rocks in shaded gorges. **Habitat** Riverine forest. **Related species** Other *Tipula* species, including *T. soror*, which is a very common, usually winged, cranefly occurring in the Western Cape. Also, *T. pomposa* (**3B**), which is among the largest species in the genus, occurring from the Western Cape to KwaZulu-Natal. This genus contains the largest craneflies in the region.

4 *Limonia sexocellata*   Six-spotted Cranefly

Small (wingspan 12mm), easily recognized by 6 dark spots on the fore margin of the wings, a brown thoracic stripe, and very long beak. **Biology** Larvae live inside jelly-like tubes that they construct in very wet moss. Adults bob up and down when alighting. **Habitat** Shaded dank overhangs and grottoes in ravines. **Related species** There are many other small species in this genus.

Family Psychodidae

Moth flies

Small flies easily recognized by the very broad, hairy wings that are held open and flat, thus resembling small moths. Males of native species may have ornamental tufts that are used in courtship displays on leaf surfaces. Very common, adults and larvae both occurring in the vicinity of water. Most indigenous species found in forests, breeding in tree holes. Larvae feed on organic matter and aquatic fungi. Alien *Clogmia* species occur in very large numbers at sewerage plants, where they are beneficial as they assist in breaking down fungal mats that clog filtration systems. Likewise, they keep domestic drains from clogging with fungi. Members of the sandfly group take blood meals and, in the more tropical parts of the world, are important transmitters of disease, including leishmaniasis (which occurs in Namibia, but not in South Africa). Approximately 30 regional species.

1 *Clogmia albipunctata*

Drain Fly

Small (wingspan 6mm), body and broad wings covered in dense hairs. Wings each with 2 small black spots and with a series of white dots along their margins. **Biology** Larvae **(1A)** elongate, grey and hairy, with a distinct head capsule and a conical siphon at the rear. **Habitat** A widespread, cosmopolitan species of the tropics and subtropics, nearly always associated with humans; in a range of foul places, including septic tanks, drains and margins of polluted streams.

2 Family Blepharoceridae

Net-winged midges

Easily recognized by creases in the wings from being folded in the pupal case. Adults usually with grey wings, resembling small craneflies. Bizarre larvae, shown left and middle in **(2)**, have 6 or 7 constricted body segments. A ventral sucker (see arrow) on the abdominal appendage of each segment enables the larva to attach itself to rocks in fast-flowing streams. Pupae black and slipper-shaped, shown right, also attached to rocks. There are 19 species in the region.

3 *Elporia*

South African net-winged midges

Small (wingspan 12mm), delicate, dark grey flies with spindly legs and smoky grey wings. **Biology** Wings of adults are already expanded in the pupal case, which floats to the surface and ruptures, shooting the insect into the air. They gather in the lee of rocks in the fastest-flowing parts of streams, where they skim across the water without getting wet. **Habitat** Larvae on and under smooth rocks in the fastest-flowing parts of cool, unpolluted mountain streams. Adults of the various species of *Elporia* differ only in eye shape and size.

4 Family Chironomidae

Midges, gnats, bloodworm midges

Extremely abundant in all habitats, even those with small temporary bodies of water. Humped thorax and long front legs, males **(4)** with very hairy antennae. Easily confused with mosquitoes, but mouthparts are poorly developed and adults of many species do not feed. At rest, the rather short wings are held apart over the back. Often found in large numbers and may form mating swarms. Larvae are generally green, red or yellow and occur in almost any freshwater or even intertidal habitat, feeding on decaying organic matter. At least 175 species are known from the region.

5 *Telmatogeton*

Seaweed midges

Small (wingspan 8mm), uniformly grey, nondescript, best identified by their association with live seaweeds on the seashore. Cylindrical larvae small (body length 7mm), with hardened head capsule from which a single proleg projects. There are 2 species in the genus. **Biology** Adults may be seen fluttering over the surface of seaweed during low tide, mating and laying eggs. Larvae feed on live intertidal seaweed and are an important component of the diet of intertidal fish. **Habitat** Found among live seaweed on rocky intertidal reefs.

1 *Chironomus formosipennis* — Bloodworm Midge

Adults large (wingspan 12mm), with thorax striped in dull green and brown and a dull green abdomen. Wings have stripes of long brown hairs. Larvae **(1A)** bright red and C-shaped. **Biology** Adults short-lived, attracted to lights. A high concentration of haemoglobin in the blood gives the larvae their red colour and allows them to live in mud at the bottom of stagnant water, extracting what little oxygen there is in these habitats. **Habitat** Larvae found in mud and rotting leaves in any kind of stagnant waterbody. Adults perch on vegetation near the slow-flowing backwaters of heavily polluted streams.

2 Family Culicidae — Mosquitoes

One of the most important groups of flies, as almost all are bloodsucking, and some transmit malaria, as well as yellow fever, elephantiasis, dengue fever and encephalitis. Only females take a blood meal. Males feed on nectar and plant juices, and have bushy antennae **(2)** to detect the whine of females in flight. Wings are folded neatly over the back when at rest (unlike Chironomidae). Eggs, usually laid in a floating raft, are deposited in flowing or still water, or cavities such as tree holes. Larvae and pupae are totally aquatic, swimming actively and breathing through a siphon tube. Adults stay within a few hundred metres of the breeding site. Culicines have only grey scales on the wings and rest with the abdomen parallel to the substrate; their larvae hang at an angle from the surface of the water. Anophelines have wings with patches of black and grey scales, rest with the abdomen tilted at a steep angle to the substrate, and larvae that rest horizontally from the surface of the water. The malaria parasite is taken up by feeding anopheline mosquitoes. Malaria occurs in wet summer months in the northern subtropical parts of the region, but due to more effective management and control the number of cases has declined from over 60,000 in 2000 to less than 10,000 in 2015. About 113 species are known from the region.

3 *Toxorhynchites brevipalpis* — Elephant Mosquito

Very large (wingspan 18mm), with robust, recurved proboscis, metallic green-blue scales on white-striped legs and body, and white stripe on side of body. **Biology** Adults feed on nectar and do not take blood. Larvae are predaceous on other mosquito larvae (especially *Aedes*) that breed in water-filled cavities in tree trunks. **Habitat** Larvae occur in tree holes (often of wild figs) and the leaf axils of large-leaved plants like Wild Banana (*Strelitzia nicolai*).

4 *Aedes* — Bush mosquitoes

Small (wingspan 6–8mm), with characteristic black-and-white-banded legs and abdomen. **Biology** Bite during the day. Eggs are laid singly, often in water that has collected in the leaf axils of *Strelitzia* plants. Yellow Fever Mosquito (*A. aegyptii*) transmits microorganisms that cause yellow fever (not present in the region) as well as filariasis (elephantiasis). **Habitat** Indoors or in forests. The cosmopolitan *A. aegyptii* occurs only in homes and is common in the region. The invasive Asian Tiger Mosquito (*A. albopictus*) **(4A)**, recorded from Cape Town in 1990, is an important vector for a number of viruses and pathogens, and is therefore of medical concern.

5 *Culex* — House mosquitoes

Small (wingspan 8mm), uniformly brown, abdomen striped in yellow. **Biology** The common *C. pipiens* and *C. quinquefasciatus* breed in any small container of stagnant water, and can over-winter as larvae **(5A)**, pupae or adults. The larvae feed on pond microorganisms and detritus. Both species are vectors for filariasis (elephantiasis). Transmits avian malaria, affecting various bird species, including African Penguin. **Habitat** Indoors or in forests.

6 *Anopheles* — Malaria mosquitoes

Small (wingspan 10mm), thin and dark, with white-banded legs and white-flecked wings. **Biology** Characteristically rest at 45° to the horizontal. Eggs are laid singly. Members of the *A. funestus* species group are the main vector of malaria in South Africa and breed all year round; resistant to pyrethroid sprays. The *A. gambiae* complex comprises a number of very similar-looking species **(6A)**, not all of which transmit malaria. *A. arabiensis*, a summer-breeding species, is also malarial. **Habitat** A habit of entering huts contributes to the vector status of some species. Larvae occur in a wide variety of freshwater and brackish habitats such as pools, ditches, streams and seepages.

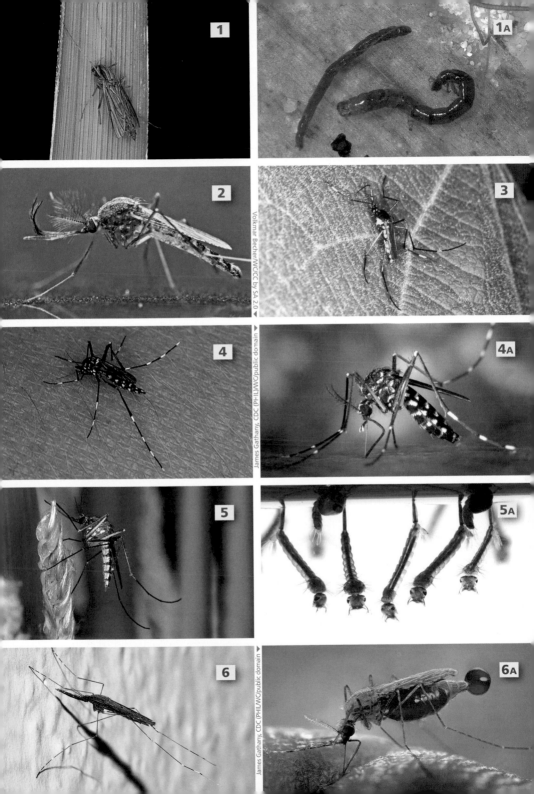

1 Family Mycetophilidae

Fungus gnats

Small, fairly distinctive, delicate, with long, thin antennae and legs, and humped thorax that conceals head from above. Strong legs with enlarged coxae held close together when sitting **(1A)**. Well-developed tibial spines. Wings dark in many species and short relative to abdomen. Eyes do not meet above antennae. About 26 species known from the region. **Biology** Feed mainly on fungus, hence the common name. **Habitat** Larvae in mushrooms, adults in forests; sometimes found in old thatch roofs.

2 Family Ceratopogonidae

Biting midges

Very small (body length under 3mm), easily detected by the bloodsucking habits of females of certain genera. Wings folded over the back when at rest (unlike Chironomidae). Over 126 species known from the region. **Biology** Bite is painless, but intense irritation and itchiness follow. *Culicoides* are thought to be vectors for 'blue tongue' disease in sheep and for African horse sickness. Some species suck blood from insect wing veins, and a few are nectar-feeders. Species that bite humans may transmit filarial worms and various viruses. **Habitat** Larvae of most species develop in waterbodies with a high organic content, but some live in moist or terrestrial locations.

Family Simuliidae

Blackflies

Small biting midges, with characteristic humped thorax and rounded clear or milky wings. In tropical Africa, *Simulium damnosum* transmits the nematode worm that causes river blindness; the fly, but not the disease, is present in South Africa. Many species settle on, but do not bite, humans. Some are pests of poultry and, along the Orange River, clouds of biting blackflies harass livestock. At least 44 species in the region.

3 *Simulium*

Common blackflies

Small (wingspan 5–6mm), stocky, grey-black, with thin, short, somewhat bent legs. There are 36 regional species in the genus. **Biology** Females suck blood from birds, reptiles and other vertebrates, are vectors for a variety of parasites, and transmit filarial disease (elephantiasis) and river blindness in humans. The very characteristic aquatic larvae **(3A)** attach in groups, by means of hooks and silk, to rocks and vegetation in fast-flowing streams or rivers, using an elaborate head fan to filter food particles from the water. The slipper-shaped black pupae are fastened to rocks and underwater vegetation. **Habitat** Widespread in the region, but less common in arid parts. Adults very common, often at some distance from the water. Larvae occur in the fastest-flowing parts of rivers or streams.

Family Bibionidae

March flies, harlequin flies

Smallish, robust, black, often with an orange thorax. In males, the very large eyes take up the entire elongated head. Many of the 18 species in the region belong to the genus *Bibio* and have swollen femora and a shiny thorax.

4 *Plecia ruficollis*

Black-winged Bibionid

Medium-sized (wingspan 12mm), black, with an orange thorax, black wings and thin legs. **Biology** Emerge in large numbers in spring and hence are known as 'March flies' in the northern hemisphere. Sluggish, appearing a little uncoordinated in movement. Larvae soil-inhabiting, probably feeding on roots and tubers. Adults are very likely important pollinators. **Habitat** Adults frequently seen on flowers (especially daisies, onions and their relatives).

5 *Dilophus*

Dwarf bibionids

Small (wingspan 6mm), black and shiny, with a humped thorax and long, thin head. Wings milky white with dark brown veins. **Biology** Adults emerge in large numbers in spring, occupying flowers and have even been recorded from an oily whale skeleton on the shoreline. Males hover above ground to mate with emerging females. Larvae live in soil and feed in large aggregations on plant roots, occasionally causing crop damage. **Habitat** Fynbos, gardens and other habitats, typically on flowers.

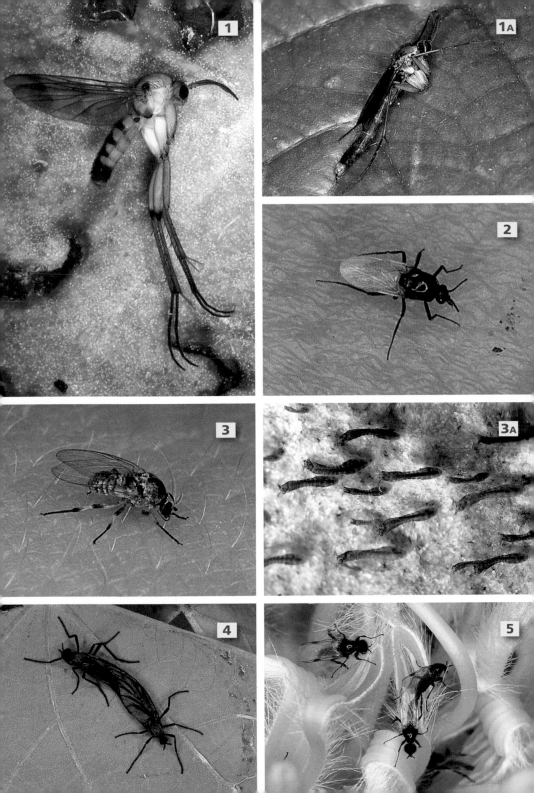

Family Cecidomyiidae
Gall midges

Very small (body length under 4mm), weakly built, with thin antennae and legs. Rounded wings, generally hairy. Long ,bead-like antennae with whorls of hair at each joint. A very large family, but only 25 species known in the region (cf. 600 in Britain). Most are host-specific and cause galls in host plants. A few gall midges feed on fungi and a variety of other foods. Adults are short-lived. Larvae may be agricultural pests, feeding on millet, sorghum and rice.

1 *Dasyneura dielsi*
Acacia Gall Midge

Small (body length 4mm), with fan-shaped wings and long antennae. **Biology** Introduced in the Western Cape in 2001 for the biological control of Australian Rooikrans (*Acacia cyclops*). The eggs are laid on flower buds and the larvae develop in the chambers of the round, fluted gall **(1A)**, inhibiting the formation of seed pods and so reducing the seed banks of the invasive plants. **Habitat** On and around Rooikrans trees. **Related species** *D. rubiformis*, which is another introduced gall-forming species. It was brought to South Africa to control the invasive Black Wattle (*Acacia mearnsii*).

2 Family Sciaridae
Dark-winged fungus gnats

Small (wingspan 7mm), dark, with the eyes forming a bridge above the antennae. **Biology** Larvae not restricted to fungi, but feed in the soil on rotting plant matter and animal faeces. Some species damage mushroom crops, and others are ubiquitous in greenhouses where their larvae damage seedlings by feeding directly on roots and shoots, as well as by transmitting pathogenic fungi. In South Africa *Bradysia* fungus gnats transmit *Fusarium* fungus to pine seedlings. **Habitat** Damp areas with decomposing vegetation, and on the moist soil surface of pot plants. Regionally, 26 species are known.

The following families (up to, but not including Family Phoridae, p.336) have 3-segmented antennae that are shorter than the thorax, with an elongated last segment (the thin arista).

Family Stratiomyidae
Soldier flies

Medium-sized to large flies, fairly easily identified by their flat abdomen (often parallel-sided and very round at the end) and rounded or long third antennal segment. Many have bright or metallic colours and may mimic wasps. Slow-moving, some found on flowers. Larvae are scavengers in a variety of habitats, most frequently in leaf litter, rotten wood or near water. Regionally 112 species are known.

3 *Ptecticus elongatus*
Soldier Fly

Large (wingspan 24mm), elongate, orange, with smoky wings. Hind legs banded, with white tarsi. Larvae broad, flat, yellowish-brown and segmented. **Biology** Adults rub the outstretched fore legs together when at rest, mimicking parasitic wasps, and are attracted to ripe fruit. Larvae have been found in the decaying fruit of *Conopharyngia johnstoni* (Apocynaceae). **Habitat** Widespread in humid areas, especially in forest and scrub. Adults may enter homes.

4 *Hermetia illucens*
Window-waisted Fly

Large (wingspan 16mm), robust, elongate wasp mimic, with smoky wings. Legs banded in white and black. Body with a peculiar clear panel at the junction of the abdomen and thorax. Larvae **(4A)** large, with hardened segments lightly covered in short spines. **Biology** Larvae grow in compost heaps, lying on the surface if the compost gets too hot. They are used commercially to digest blood and animal tissue from abattoir waste and are then ground and incorporated into animal feeds. **Habitat** Near towns. First introduced to KwaZulu-Natal from the New World. Cosmopolitan.

5 *Oplodontha*
Green soldier flies

Medium-sized (wingspan 12mm), brown. Abdomen flattened, with a green edge. Larvae flattened, with a fine honeycomb pattern on the exoskeleton caused by deposition of excreted calcium carbonate. **Biology** Adults perch in exposed positions on leaves. **Habitat** Larvae are aquatic, occurring at the moist edges of dams and streams, with some found at the edges of brackish pans and estuaries. A largely African genus.

Family Rhagionidae

Snipe and wormlion flies

Long-legged, with a thin, hair-like appendage (the arista) on the third antennal segment. Adults have an elongate proboscis and feed on nectar. In 1 group ('wormlions'), the larvae dig a funnel-shaped pit, like that of antlions (Myrmeleontidae), but with a much steeper end to the funnel. Pits are often excavated in rain shadows, such as rock overhangs and the mouths of caves. Rhagionidae includes 24 species known from the region.

1 *Leptynoma sericea*

Wormlion Fly

Medium-sized (wingspan 18mm), with a long, thin proboscis, brown wings with smoky markings, long legs and an arched abdomen. **Biology** Adults are short-lived nectar-feeders. The remarkable grey larvae **(1A)** live gregariously, like antlions (Myrmeleontidae), in steep-walled conical pits **(1B)**. They subdue prey in the pit by whipping around their victims and sucking them dry. **Habitat** Larvae common and widespread in very fine sand beneath rock overhangs and cave entrances.

2 *Chrysopilus tuckeri*

Snipe Fly

Small (wingspan 12mm), delicate, with large, clear, iridescent wings. Body brown, covered in shining golden scales. Head very large. **Biology** Adults visit herbaceous plants, perching on broad leaves; larvae live in leaf mould. **Habitat** Moist shady forests. **Related species** There are 2 other regional species in this cosmopolitan genus.

Family Athericidae

Watersnipe flies, phantom midges

Members of this small, rather primitive family resemble small horseflies, with large, often smoky, wings. Adults are found on rocks or vegetation alongside fast-flowing streams, where the aquatic larvae are predators. The region has the richest athericid fauna in the world, with at least 16 species in 3 genera.

3 *Trichacantha atranupta*

Long-legged Watersnipe Fly

Medium-sized (wingspan 16mm), with small darkish-brown females, shown left in **(3)**, and remarkable males **(3A)**, which are predominantly yellow, with a black patch on each large, broad yellow wing. Hind legs of males greatly enlarged and elongated and are held splayed outwards. **Biology** Adults of both sexes 'subsocial' or gregarious. **Habitat** Cling to underside of logs and causeways above fast-flowing streams, where they guard communal egg masses.

4 *Atherix barnardi*

Small (wingspan 12mm), slender, dark brown, with broad smoky wings and long thin legs. **Biology** Uses its fore legs to gently stroke the substrate. Adults gregarious, congregating in bushes near water. **Habitat** Forest undergrowth. **Related species** *A. adamastor* **(4A)**, which is a small (wingspan 12mm) black species with very broad smoky wings and elongate legs; sucks blood from frogs and occurs on rocks in mountain streams. Also, *Suragina bivittata* **(4B)**, which is small (wingspan 14mm), dull orange with a single grey thoracic stripe, broad clear wings, long fore legs, and a stabbing proboscis; recorded sucking blood from the eyelid of a Giant Eagle Owl at Skukuza.

Family Tabanidae

Horse flies, clegs, deer flies, perdebye

Medium-sized to large, stoutly built flies, often with iridescent eyes. The third antennal segment bears a characteristic thick bristle with rings along its length. Wing veins diverge at the wing tip to form an open 'V'. Both sexes feed on nectar and other juices, but females also suck blood voraciously, attacking a range of vertebrates from frogs to mammals (rarely birds). Females are responsible for spreading diseases, including 'surra' in horses, 'nagana' in cattle and 'loiasis' in primates – the latter transmitted by the worm loa loa (a parasite on the eyeball) to humans and monkeys. The larvae develop in moist places, feeding on decaying vegetable matter (although a few are predaceous), often in mud at the edges of waterbodies. The region has a rich fauna of 227 described species.

5 *Chrysops*

Deer flies

Medium-sized (wingspan 25mm), with emerald green eyes and brown-banded wings. **Biology** Bite humans (causing headaches and joint pain), and a range of domestic and wild mammals. Transmit loiasis and the trypanosome parasite that causes sleeping sickness in tropical Africa. The maggot-like brown larvae have rings marking each segment of the body, and feed on decomposing organic matter in mud and boggy soil. **Habitat** Adults prefer the shade of forests and scrub.

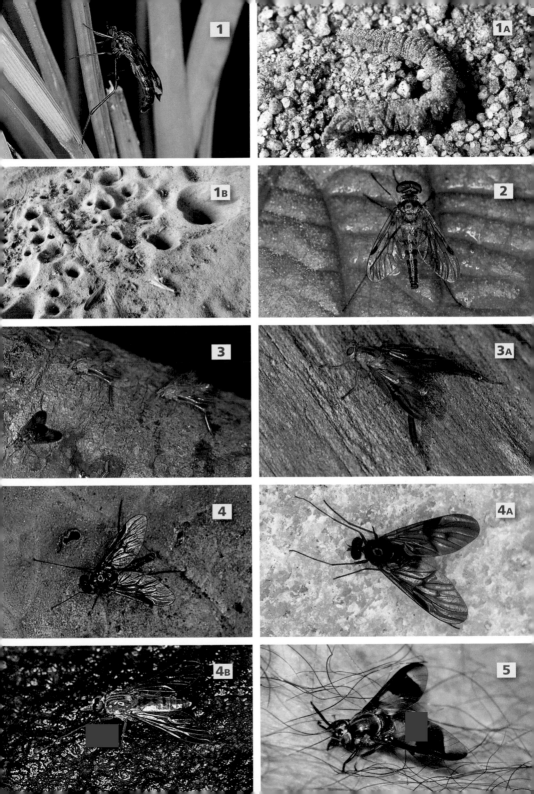

1 *Rhigioglossa* — Horse flies

Medium-sized (wingspan 20mm), with iridescent green eyes, clear wings, and a grey-and-black-striped abdomen. Easily confused with related genera. **Biology** A few are bloodsuckers of humans and especially bothersome in the Western Cape in spring. Both sexes are frequently dusted with pollen after feeding on daisies. **Habitat** Very common in arid succulent karoo vegetation.

2 *Tabanus taeniatus* — Striped Horse Fly

Medium-sized (wingspan 20mm), with clear wings, and 2 pale stripes along length of abdomen. **Biology** A bloodsucker of large mammals, as are others in the genus. **Habitat** Open bushy veld, but common also in open grassland, near streams at high altitude. Most species in the genus occur widely across Africa, but not in forests. **Related species** *T. leucostomus* **(2A)**, which is a dull grey species with white-banded legs that occurs throughout the region, except in the central Karoo.

3 *Tabanus biguttatus* — Hippo Fly

Very large (wingspan 46mm), with a black body, thorax white to pale yellow with a bare black central patch, and black wings with clear tips. One of the region's biggest and best-known horse flies. **Biology** Attacks large mammals such as cattle and hippos, obliging hippos to spend most of the day underwater. Larvae live in mud in pans, constructing mud pillars above the surface to avoid dehydration and in which to pupate; they feed on insect larvae and tadpoles. **Habitat** Widespread in sub-Saharan Africa.

4 *Philoliche rostrata* — Needle-nose Fly

Large to very large (wingspan 34–44mm), with black-marked grey thorax, black-and-white-banded abdomen (orange towards thorax) and white undersides. **Biology** Flies very rapidly and, like *P. aethiopica* (below), often found taking nectar from *Pelargonium cucullatum*. **Habitat** In forest or at high altitude. **Related species** *P. zonata* **(4A)**, which is large (wingspan 34mm), with a black-and-white-banded abdomen, and is fairly common and widespread in the western parts of the region. Also, *P. spiloptera* (body length 12mm) **(4B)**, which has mottled brown wings and pollinates *Wachendorfia*. Most species in the genus are endemic to the region.

5 *Philoliche aethiopica* — Needle-nose Fly

Large (wingspan 30–44mm), with broad brown stripe along thorax and orange abdomen. Geographic variation underlies a complex of 6 closely related and very similar-looking species. Proboscis length varies within each of these species. **Biology** Usually heard before it is seen, as it produces a loud booming sound. Hovers in the areas of shade under horses and cattle, causing them to become restless. Can hover while attached to host. **Habitat** Very common in open grassland near forest.

6 *Braunsiomyia* — Beach tabanids

Medium-sized (wingspan 20mm), light coloured, with opaque milky wings. Abdomen sandy grey in the south, banded in KwaZulu-Natal. **Biology** Always rest on dry sand, above the shoreline. Occur in winter in KwaZulu-Natal. **Habitat** On beaches beyond the high-tide mark and on dry sand up to 50km inland.

7 *Haematopota* — Clegs

Medium-sized (wingspan 22mm), all species very similar, grey, with mottled black wings and dull iridescence on eyes. A complex genus, with over 200 African species. **Biology** Closely associated with antelope and cattle, and frequently bite humans. Adults often drink from muddy patches. Larvae live in damp habitats and feed on other insects. **Habitat** Generally in shady scrub and forest. Cosmopolitan.

J. Manning ▶

Family Mydidae
Mydas flies

Large (average body length 20mm), hairless, with long, 4-segmented, clubbed antennae. Tropical species are the largest of all flies. Long proboscis in adults of some species (including those that feed on Aizoaceae); those in very arid parts where there are few flowers do not feed and have reduced mouthparts. Larvae live in loose sand like that found on riverbanks, and prey on other insect larvae. Females have a spine-like ovipositor, an indication that eggs are laid in sand. Most occur in dry sandy parts of the region and are active at the hottest times of day. About 195 species and 29 genera known from the region, the richest concentration in the world.

1 *Cephalocera fascipennis*

Medium-sized (wingspan 17mm). Wings dark brown with darker veins. Abdomen with thin white bands. Like other mydids, mimics wasps, both in appearance and behaviour. Hind femora swollen. **Biology** Settles on the ground and rocks. Active at midday. **Habitat** Fairly common in arid sandy or rocky areas. **Related species** *Cephalocera* includes 27 other very similar regional species.

2 *Afroleptomydas*

Medium-sized to large (wingspan 30mm in species shown), with finely ridged brown wings. Antennae long and club-like. Long abdomen, banded in yellow and dark brown. Mouthparts are long and functional. Tapering white larvae **(2A)** consume other insect larvae (pupa shown top). Genus contains 87 similar regional species. **Biology** Flight is wasp-like. Rest on rocks and feed on flowers of Mesembryanthemaceae. **Habitat** Stony plains and slopes with sparse vegetation, with almost all species concentrated in the Western and Northern Cape.

Family Asilidae
Robber flies

An easily recognizable family. Small to large flies (body length 3–40mm), generally bristly with the top of the hairy head hollowed into a deep groove. Adults prey voraciously on other insects, which are usually caught in flight and subdued by the powerful bristly legs and robust proboscis. Generally take a range of suitably sized prey, including bees, wasps and grasshoppers. Larvae look like typical maggots, with no unusual external features. Most larvae feed on detritus, some feed on locust egg pods, and a few are predaceous. An important family, with at least 500 species known from the region.

3 *Alcimus tristrigatus*
Large Grasshopper Robber Fly

Large (wingspan 25mm), pale brown, elongate, with a dark brown oval mark on the thorax and a dark brown square on each abdominal segment. Females have a long ovipositor. **Biology** Rests on the ground, mimicking short-horned grasshoppers, their major prey, in appearance and flight. Shown here with fresh fly prey. **Habitat** Bare ground in tropical savannas, often on riverbanks or on the tips of bushes. **Related species** There are 11 other regional species in the genus, including *A. porrectus* (body length 26mm) **(3A)**, which is a large species with a reddish-brown thorax; occurs in the coastal parts of the Eastern Cape.

4 *Damalis heterocera*

Small (wingspan 12mm), with smoky black wings, smooth body, brown abdomen (darker on top), and very broad eyes. **Biology** Clings to vegetation with body held out at a right angle to the perch. **Habitat** Common in low herbaceous vegetation. **Related species** Many other similar species occur in the region, including 1 that feeds on winged ants.

5 *Daspletis placodes*

Large (wingspan 40mm), with iridescent eyes, body covered in long yellow-white hairs. **Biology** Ground-resting, bee-like in flight. Feeds largely on bees and wasps (mydid fly prey shown here). **Habitat** Open ground in arid bushveld. **Related species** The genus contains several other very similar bristly species.

6 *Gonioscelis ventralis*

Medium-sized (wingspan 18–28mm), covered in yellow hairs, with smoky black wings and orange legs. Fore legs with very stout femora, each armed with a pad of black spines. **Biology** Feeds on flower-feeding scarabs (usually monkey beetles), which it grasps with the fore legs. **Habitat** Ground, soil, rocks, grass tips and bushes.

1 *Hoplistomerus nobilis*
Golden Robber Fly

Medium-sized (wingspan 19mm), stout, covered in fine shiny hairs, which are golden on cylindrical abdomen. Eyes black, wings with a brown mark, tibiae of hind legs swollen. **Biology** Feeds almost entirely on small scarabs and other dung-visiting beetles. **Habitat** Ground and stones, often perching on cow dung. **Related species** Other bee- and wasp-mimicking species of robber fly, including the red-and-brown *Ancylorhynchus crux* (body length 16mm) **(1A)**.

2 *Hyperechia marshalli*
Carpenter Bee Robber Fly

Large (wingspan 34–44mm), stout carpenter bee mimic, uniformly black with a yellowish or white band of hair across the hind margin of the metathorax. Legs thickly covered with long hair. **Biology** Rests and oviposits on tree trunks. Adults hunt from dead trunks bored by carpenter bees. Feeds on carpenter bees and other bees and wasps. Larvae bore and live in wood tunnels in association with carpenter bee larvae, on which they are reported to feed. **Habitat** Carpenter bee logs. **Related species** *H. imitator* **(2A)**, which is very large (wingspan 50mm) and uniformly black, and occurs at high altitudes in the Western Cape, where it can be seen resting on *Protea* trunks.

3 *Lasiocnemis lugens*
Picture-winged Robber Fly

Medium-sized (wingspan 16mm), slender, usually with dark shading on the wings. Hind legs enlarged and furry. Abdomen slender at the base becoming swollen towards the end. **Biology** A specialist spider hunter, snatching spiders from their webs. **Habitat** Settles on grass tips in open bushveld.

4 *Leptogaster*
Slender robber flies

Small (wingspan 15mm), slender, with a very long, thin abdomen, long legs and brown wings. A very large genus, with many similar-looking regional species. **Biology** Rest on grass stems in a characteristic posture with abdomen held at 90° to the stem. Stationary prey are caught during slow and direct foraging flights. **Habitat** Open woodland and grasslands.

5 *Lamyra gulo*
Wasp Robber Fly

Very large (wingspan 24–60mm), metallic black, a convincing wasp mimic. Wings brown with an orange base and fore margin, abdomen blue with white spots on the sides and top of the first 3 segments. **Biology** Flies with trailing legs, resembling *Belonogaster* (p.476) or *Sphex* (p.490) wasps. Adults live for more than a month, and feed mainly on wasps and flies. **Habitat** Bushes and trees in open patches of subtropical forest and scrub. Very widespread in Africa, extending north to the Sudan. **Related species** There are 3 other similar African species in this genus.

6 *Microstylum*

Large (wingspan 28–40mm), body smooth, grey-brown, with white bristles on the legs, eyes iridescent. Black mark on top of last 2 abdominal segments, wings light brown. A large genus with many similar species. **Biology** Feed predominantly on beetles and grasshoppers. **Habitat** Alight on various substrates in subtropical bush and scrub.

7 *Neolophonotus*

Medium-sized (wingspan 15–17mm), robust, mostly grey and hairy, thorax often very humped and with black median stripe, abdomen short and banded in black or grey. Genus includes a large number of similar species. **Biology** Larger and more common species occur in the fynbos of the southwestern Cape. They prey on honey bees and may thus be of importance to apiarists. Larvae live in rich humus at the base of maize plants, in association with scarab grubs. **Habitat** Widespread, but most abundant in arid areas with sparse vegetation. Alight on various substrates. **Related species** *Promachus* species (body length 20mm) **(7A)**, which are similar but more robust, with patches of white hair on the abdomen; they are common across the region.

1 *Pegesimallus pulchriventris* Green-eyed Robber Fly

Medium-sized (wingspan 16mm), robust, with brown body and wings and green eyes. **Biology** Prefers slow-flying prey (mainly craneflies and bees). Flight brief and, when disturbed, bobbing. Makes characteristic abdominal contractions just after alighting. Eggs are laid superficially in soil. **Habitat** Widespread and locally abundant in margins of shaded subtropical forest and in long grass under *Senegalia* trees. **Related species** Many other species in this large genus; in some, males have feathery hairs on their legs; also, some species have small, pale, fragile bodies.

2 *Philodicus* Grasshopper robber flies

The species shown here is large (wingspan 22–34mm), with a long, tapering abdomen banded in black and white, as in all members of the genus. Thorax has a characteristic dark brown oval marking surrounded by white. Contains 5 regional species. **Biology** Feed mostly on grasshoppers, bees, wasps and flies. Fly rapidly and smoothly from 1 resting place to another, alighting on the tips of shrubs. Adults unusual as they appear to rely on standing water. **Habitat** Primarily wetlands or at the margins of streams.

3 *Proagonistes praeceps* Spider-wasp Robber Fly

Massive (wingspan 60–70mm), convincing spider wasp mimic with black body and wings, orange head and legs, thin, blade-like proboscis and conspicuous palps. **Biology** Flies swiftly from resting places fairly high on tree trunks. Feeds on a wide range of insects, including large wasps. **Habitat** Subtropical coastal bush. **Related species** Genus includes approximately 4 other regional species.

4 *Synolcus dubius* Garden Robber Fly

Large (wingspan 24mm), light orange, with a black stripe along the thorax. Abdomen long and tapering, with a row of brown spots on top. Femora of all legs have black barring. Wings shaded, especially towards the tips. **Biology** Sluggish, feeding mainly on bugs of the family Lygaeidae. Females oviposit on tree bark. **Habitat** Common on vegetation and trees, but may alight on the ground. Abundant in gardens in Cape Town. **Related species** There are 12 other similar species in this regionally endemic genus.

Family Nemestrinidae Tangle-veined flies

Medium-sized to large flies, with large and unusual wings that have a number of veins running parallel to the fore margin. Excellent fliers, producing a loud buzzing while hovering over long-tubed flowers, which they visit for nectar. Virtually nothing is known about the larval stages of species in the region, but in other parts of Africa nemestrinid larvae are known to be quite common internal parasites of locusts. There are 48 species of these rather rare flies known from the region.

5 *Prosoeca peringueyi* Long-tongued Tangle-veined Fly

Large (wingspan 32mm), with grey or brown wings and body (as in all members of the genus), often with black markings. Eyes very large, bronze-coloured. Proboscis usually very long (up to 40mm). **Biology** Produces a deep buzzing when hovering (over red or violet flowers of *Pelargonium*, *Lapeirousia* and *Gladiolus*) to feed. **Habitat** Low vegetation in Namaqualand. **Related species** There are at least 29 other regional species in the genus.

6 *Atriadops vespertilio* Batwing Tangle-veined Fly

Medium-sized (wingspan 28mm), body grey, wings large, black, with a cream-coloured crescent on leading margin, and a few wing veins at wing tip. No proboscis. Females have an unusual, curled ovipositor. **Biology** Adult flies may not feed. **Habitat** Bushveld. **Related species** There is another, still undescribed, regional species with more slender wings.

7 *Moegistorhynchus longirostris* Mega-proboscid Tangle-veined Fly

Large (wingspan 40–45mm), with the longest (70mm) proboscis of any known fly. Body and wings chequered in brown. **Biology** Hovers loudly over long-tubed flowers and pollinates a wide range of white or pink flowers with nectar guides. Larvae unknown. **Habitat** Rocky slopes and flats where suitable host plants occur.

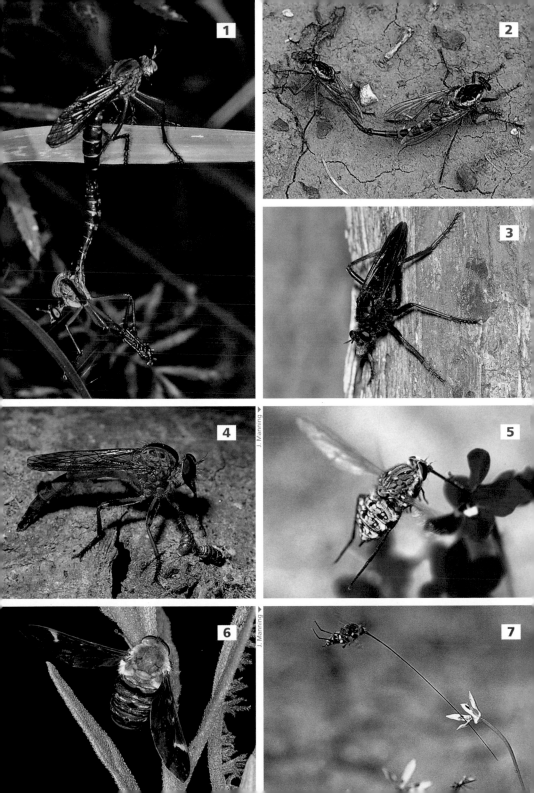

1 *Stenobasipteron wiedemanni* — Bent-wing Tangle-veined Fly

Medium-sized (wingspan 36mm), wings uniform transparent brown, with narrow base. Leading edge of wing has distinctive projecting bulge in male; female (shown here) lacks this bulge. Body cinnamon brown, without markings. Proboscis as long as, or longer than, body, and always tucked between legs when not feeding. **Biology** Pollinates a wide range of forest plants, including orchids, wild impatiens and *Plectranthus*. **Habitat** Closed canopy forest and sunlit forest edges where host plants occur. **Related species** Other – undescribed – *Stenobasipteron* species occur in Limpopo and Mpumalanga.

Family Acroceridae — Small-headed flies

Unusual-looking flies, easily recognized by their very small head, which is attached below an inflated thorax. Proboscis long and wings curiously roughened, with very large white plates (calyptra) that cover the smaller balancing wings (halteres). Adults occasionally seen feeding off flowers, but generally rare, living for 1–6 weeks. Larvae are internal parasites of spiders. Eggs hatch into very active first-instar larvae that find a host (e.g. sac spiders, *Cheiracanthium*) and burrow into it. There are 40 species known from the region.

2 *Psilodera valida* — Yellow Humpbacked Fly

Large (wingspan 22mm), bright yellow, with clear brown wings, large eyes that meet on the head, very humped thorax and long, thin proboscis (see arrow). **Biology** Pollinates *Gladiolus brevifolius*. Adults are short-lived. **Habitat** Coastal dune forest.

3 *Psilodera fasciata* — Banded Humpbacked Fly

Small (wingspan 14mm), easily recognized by the combination of a humped and hairy thorax, long proboscis, black-and-white-banded abdomen and yellow legs. **Biology** Adults feed on *Plectranthus* nectar, although some species in the genus do not feed as adults. **Habitat** Common in fynbos and gardens in the southwestern Cape. **Related species** There are about 9 other regional species in the genus.

Family Bombyliidae — Bee flies

Small to large, aptly named, stout and furry flies. Legs slender, proboscis often long. Usually found basking in the sun. Active at the hottest time of day, and very common in the drier parts of the region. Superb fliers that can hover very accurately while feeding from flowers. Important as pollinators of diverse plant species in succulent karoo vegetation. With a few exceptions, the larvae are parasitic or predaceous on the eggs, larvae or pupae of almost every kind of insect with a pupal stage (Endopterygota). The larvae of a few species attack the egg pods of grasshoppers. One of the largest fly families in the region, with about 940 known species.

4 *Bombomyia discoidea* — White-tailed Bee Fly

Medium-sized (wingspan 20mm), striking, with an orange or black thorax and white-tipped black abdomen. Wings very long, with a black patch at the base. There are a number of colour forms. **Biology** Usually seen hovering over and imbibing nectar from flowers, especially thistles. Parasitizes solitary bees. **Habitat** Grassland and bushveld. Widespread.

5 *Australoechus hirtus* — Golden Bee Fly

Medium-sized (wingspan 22mm), bee-like body, similar to *Systoechus*, but with denser, shaggier, golden or silver-yellow hairs on the body. Thorax not humped. **Biology** Makes a loud humming noise in flight. Usually settles on sand. Feeds on Mesembryanthemaceae flowers. **Habitat** Open sandy areas with low succulent vegetation. **Related species** *Australoechus* includes about 29 other species in the region.

6 *Systoechus* — Woolly bee flies

Small (wingspan 14mm), similar to *Australoechus*, with the broad head and body covered in plush white hairs, giving a frosted or chalky appearance. Many similar species are known from the region. **Biology** Some species of economic importance, as the larvae feed on eggs of the Brown Locust (*Locustana pardalina*) (p.120). **Habitat** Most common in arid, sparsely vegetated areas.

1 *Exoprosopa nemesis*
Phantom Bee Fly

Medium-sized (wingspan 24mm), unmistakable, uniformly black (apart from wing tips). **Biology** Probably parasitizes the larvae of various wasps. (A member of the genus was reared in the nest of a Mud Dauber *Scelifron spirifex*, p.490). Observed hovering over the ground tunnels of digging wasps and the mud nests of small solitary wasps. **Habitat** Open woodland and koppies. Widespread in open woodland in summer-rainfall areas. **Related species** The genus is extremely species-rich and includes the largest bombyliids in the region, with wingspans of up to 40mm.

2 *Notolomatia pictipennis*
Orange Bee Fly

Medium-sized (wingspan 20mm), body typically black with dense tufts of hairs (black, white, orange or yellow) on the sides of the broad, flattened abdomen. Wings long and black, with a dull orange central patch. **Biology** Probably parasitizes the egg pods of locusts. **Habitat** Open grassland and woodland.

3 **Family Mythicomyiidae**
Micro bee flies

Minute (wingspan 2–5mm), smooth flies, often with a humpbacked appearance. *Cephalodromia* **(3)** has 10 regional species. Mythicomyiidae larvae prey on solitary bee larvae, Brown Locust (*Locustana pardalina*) (p.120) egg pods, or live in ant nests. Adults frequent daisies and *Euphorbia* flowers for nectar. Although widespread, most of the diversity is in the semi-arid Western and Northern Cape provinces.

Family Empididae
Dance flies

Smallish flies that resemble small robber flies, but are not covered in hair or bristles. They have a stout thorax and, usually, a long, stout proboscis. Under a microscope, a small dent is visible on the inner edge of the front of the eyes. Adults are found on flowers in spring, feeding on nectar. Their main diet, however, is other insects (especially other flies). In some species swarms are formed over water, where males offer a nuptial gift to females before mating takes place. In others, a small ball of silk is offered instead, and accepted by the female before mating. Empidids are restricted to temperate habitats, and local genera are mostly endemic. Little is known of the biology of local species. About 230 species occur in the region.

4 *Hilarempis*

Small (wingspan 16mm), grey flies, some with black-and-yellow legs. Proboscis fairly long. Genus contains at least 31 regional species. **Biology** Often skate over the surface of water by buzzing their wings. Prey on soft-bodied insects trapped in the surface film. **Habitat** Associated with water and streamside vegetation, and restricted to cool microhabitats. **Related species** *Empis makalaka* **(4A)**, which is small (wingspan 14mm), slender, grey, with black on the upper abdomen, and a long proboscis. Occurs in the Western Cape, and is found on flowers; females also prey on flies. Also, *E. bivittata* (also from the Western Cape), which has 3 thoracic stripes and brownish wings.

Family Dolichopodidae
Long-legged flies

Small, bristly, with long, strong legs. Often metallic green with bronze or blue reflections. Wings may be marked in black. In males the very large genitalia are folded forward under the abdomen. In some species females are attracted with mating dances. Like many groups of inconspicuous African insects, the family is relatively poorly known, the 123 species known from the region certainly being an underestimate.

5 *Condylostylus stenurus*

Small (wingspan 7mm), metallic green, with patterned smoky black wings and a reflective pink abdomen. Small black swelling on underside of each hind tibia. **Biology** Adults perch conspicuously on the upperside of leaves, with the legs outstretched. They feed on other small insects, crushed with the fleshy (sometimes toothed) mouthparts. **Habitat** Vegetation under trees and shrubs or at the edge of water. Fairly common on the leaves of understorey plants in riverine forest. **Related species** There are 4 other regional species in the genus.

The following families are the third major group of flies. All pupate in a barrel-shaped case and have 3 antennal segments, the last segment being large, with a single erect bristle (the arista). Most are compactly built, and their featureless larvae are the well-known maggots.

Family Phoridae
Scuttle flies, coffin flies

Small to minute hunchbacked flies, easily recognized by their rapid, agitated, scuttling movements. Under a microscope, leading edge of wings is seen to be spiny, coloration dull shades of black, brown and yellow. Phoridae includes 58 regional species. Larvae diet varies, often comprising decaying matter or other insects. Some feed on dead animals, including human corpses, hence the alternative common name coffin flies.

1 Megaselia scalaris
Common Coffin Fly

Small (wingspan 4mm), pale yellow with concentric brown rings on abdomen; head with strong backward-projecting spines. Base of legs (the coxa) enlarged and disc-like. **Biology** A cosmopolitan alien species introduced from the Americas. Larvae feed on decaying plant and animal matter, including human and other animal corpses. After feeding on giant forest snails (*Achatina*), they pupate in rows inside the lip of the empty snail shell. Their maggots often infest dairy products and fish. **Habitat** Generally overlooked but ubiquitous in most habitats. Often appear at windowpanes and near urinals.

2 Family Syrphidae
Hoverflies

Conspicuous bee or wasp mimics, with yellow and black stripes. Well known for their precise hovering flight. Distinguishable on close examination by the 'false margin' formed by veins running parallel to the edge of the wing. Body smooth, males with large eyes joined on top of the head. Adults feed on nectar and pollen. Some have distinctive brown or green slug-like larvae **(2)**, many of which feed on aphids. Others have 'rat-tailed' maggots, see **(6A)**, which feed in anaerobic mud in ponds and have a long telescopic 'tail' or breathing tube. Still others occupy an array of dietary niches in between. About 230 species in 38 genera known from the region.

3 Asarkina africana

Medium-sized (wingspan 19mm), with thick, complete, yellow abdominal bands, a brown thoracic stripe and very large red eyes. Abdomen flat and broad. **Biology** Hovers very accurately and conspicuously in patches of sunlight in forest gaps or gardens. Proboscis can be greatly extended to reach into flowers. **Habitat** Forest or scrub; absent from drier parts. **Related species** The genus includes 4 other similar regional species.

4 Allograpta fuscotibialis

Medium-sized (wingspan 18mm). Abdomen parallel-sided with yellow bands, thinner than those of *Asarkina*; first band is divided. Thorax largely shiny black. **Biology** A common visitor to daisies. Larvae have reverted from feeding on decomposing organic matter back to feeding on plants, either as stem borers or leaf miners. **Habitat** Veld and gardens. **Related species** *Allograpta* includes 4 other similar regional species. Also, members of genus *Ischiodon* are similar, but with a more rounded abdomen.

5 Eristalinus taeniops
Bar-eyed Hoverfly

Medium-sized (wingspan 20mm), stocky, honey bee mimic with black-barred eyes, dull orange thorax and yellow-and-black-striped abdomen. **Biology** A common visitor to flat white or yellow flowerheads. Buzzes aggressively when caught, but is harmless. Larvae are rat-tailed maggots, occupying mud or rotten animal corpses. **Habitat** Veld and gardens. A common and widespread native of Africa and Asia; introduced to North America. **Related species** *Eristalinus* includes 14 other regional species.

6 Eristalinus modestus
Marble-eyed Drone Fly

Medium-sized (wingspan 15mm), stocky. Mimics bees in behaviour and appearance. Similar to Bar-eyed Hoverfly (*Eristalinus taeniops*), but eyes marbled (not striped) in black, and pattern of black markings on abdomen differs. Larvae **(6A)** grey, with a long, thin siphon. **Biology** Males perform hovering displays. **Habitat** Larvae are 'rat-tailed maggots' found in putrid water. Adults feed on flowers, especially of carrot and fennel. Common in Africa and in Europe. **Related species** The Drone Fly (*Eristalis tenax*) **(6B)** (wingspan 15mm), which is a honey bee mimic common across South Africa; told from *Eristalinus* species by its black eyes and the hourglass-shaped black pattern on the first section of the abdomen.

1 Simoides crassipes — Bandy-legged Drone Fly

Large (wingspan 20mm), stocky bee mimic with black eyes, grey thorax, orange- and black-banded abdomen (black terminally) and swollen hind femora. **Biology** Slow-moving when feeding (usually on small, yeast-smelling white flowers). **Habitat** Common in veld and gardens in the Eastern Cape.

2 Microdon — Ant nest hoverflies

Medium-sized (wingspan 15mm) bee mimic. Adults **(2)** with dark body and eyes and restricted bands of golden hairs on the abdomen. Antennae unusually long and stout. Genus comprises 13 regional species. **Biology** The remarkable larvae **(2A)**, which resemble little round slugs and were originally classified as such, live and scavenge in ant nests. Oval pupae are glued to the underside of rocks **(2)** (right in image). **Habitat** The adults occur on flowers, the larvae under stones covering ant nests. Common in the eastern parts of the region.

3 Phytomia incisa

Medium-sized (wingspan 15mm) bee mimic with black eyes and a dull brown thorax. Abdomen orange, with indistinct, mostly terminal, black stripes. Hind femora not swollen. Very large eyes in males almost cover the head. **Biology** The very large eyes of males enable them to maintain precise positions in swarms to which females are attracted. **Habitat** Flat flowerheads on trees, shrubs and annuals, in natural and urban habitats. Common in the region and the rest of Africa and Madagascar. **Related species** Phytomia includes 4 other regional species.

Family Pyrgotidae — Fruit beetle-parasite flies

Medium-sized to large, unusual, with a bulge at the front of the head. Wings usually patterned, spotted or pigmented. Females lay an egg under the elytron of a scarab, where the beetle cannot remove it. The larvae enter the beetle's body and feed within, eventually killing the host.

4 Campylocera

Medium-sized (wingspan 14mm), brown, shiny, with small wide-spaced eyes; females with a long ovipositor and the abdomen flattened from side to side. There are 36 species in the region. **Biology** Active at dusk or at night. Eggs laid on the abdomen of a scarabaeid beetle (probable host) while in flight. **Habitat** Subtropical bushveld and forest.

Family Conopidae — Thick-headed flies

Resemble hoverflies, but have a very narrow 'waist', long, forward-projecting proboscis that can be folded away, and disproportionately large head. Includes very convincing wasp mimics. Adults feed on nectar, while larvae are parasitic on various insects, especially wasps. Hooked eggs are attached to the wasp or bee host in flight, and the larva lives inside the host. Adults may or may not resemble hosts. Form mating swarms at hilltops or on vegetation. There are 65 species known from the region.

5 Dacops

Medium-sized (wingspan 16mm), mostly brown wasp mimics, with a narrow waist, yellow halteres and forehead and brown blotches near wing tips. Legs short and thick. **Biology** Mimic paper wasps (Ropalidia, p.474). **Habitat** Bushveld and subtropical forest. Widespread in Africa.

6 Conops zonatus

Medium-sized (wingspan 18mm), light brown wasp mimic, with brown wings, thin white abdominal stripes, and long, stout antennae. Female (shown) has unusual appendages under the abdomen. **Biology** Abdominal appendages may assist in attachment to host for egg-laying. Frequents flowers. Possibly a parasite of bees, as for European species. **Habitat** Bushveld and subtropical forest.

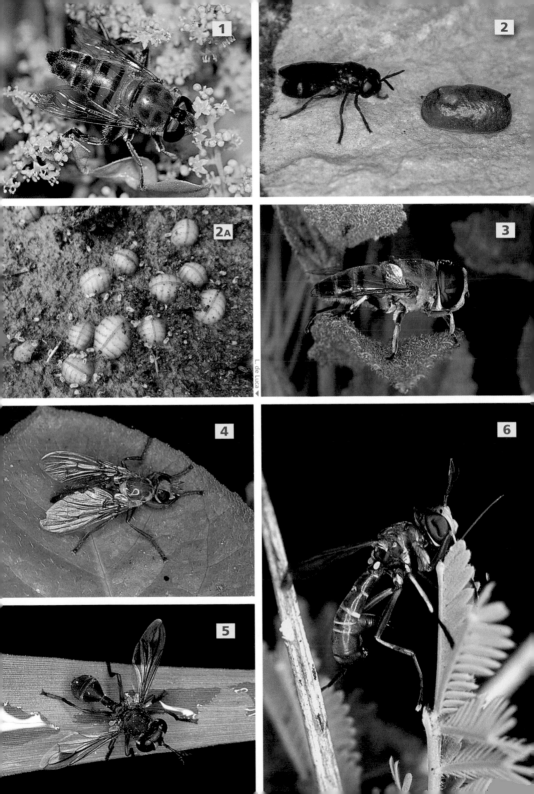

L. de Luca

1 Family Sphaeroceridae
Lesser dung flies

Very small (body length 2mm), and generally associated with dung and other decaying matter. Dull brown or black. First tarsal segment of each hind leg thickened and elongated. Some of the wing veins fade before reaching the wing margin. There are 70 species in the region, including several short-winged and wingless species and a few invasive species with cosmopolitan distributions. **Biology** Attracted to wet food sources, such as decaying vegetable matter, sap flow on trees, and herbivore dung. *Ceroptera* are often carried by dung beetles, as in **(1)**. **Habitat** Occur across the region in a variety of habitats where suitable decaying substrates, including forest litter, occur. A widespread family globally, with many introduced species. More common in the summer-rainfall parts of the region.

Family Tephritidae
Fruit flies

Easily recognized by the pointed, elongate ovipositor of the female and the patterned wings of many species. Adults generally found close to the larval substrate – generally fleshy fruits, but sometimes less familiar fruits, such as the green seeds of daisies. Larvae develop inside the fruit, and many are important agricultural pests. Less conspicuous species live in plant stems or leaves. Most species are fairly, or very, host specific. There are 375 species known from the region.

2 *Ceratitis rosa*
Natal Fruit Fly

Small (wingspan 8mm), stocky, with green iridescence on eyes, broad, patterned wings and yellow stripes at the edge of the thorax. Last part of thorax (scutellum) bears 3 black patches. Tubular ovipositor in females. **Biology** This species and *C. capitata* (below) attack at least 19 varieties of fruit in the southwestern Cape, as well as most tropical fruits. Storing fruit at very low temperatures (cold sterilization) may kill the eggs, allowing fruit to be exported more widely. Chemical control (by spraying) is also employed. **Habitat** Most gardens and a range of veld types where fleshy fruits are present.

3 *Ceratitis capitata*
Mediterranean Fruit Fly

Similar to *C. rosa*, but smaller (wingspan 6mm), with bluish iridescence on the eyes, fainter wing markings and a black scutellum. **Biology** Eggs are laid beneath the skin of fruit, and developing larvae produce soft, pulpy areas in the fruit. A serious pest of deciduous fruit in the southwestern Cape and elsewhere, spoiling many crops, including citrus, mango, litchi, guava, coffee and granadilla. The release of millions of sterile males reduces the population by causing females to produce sterile eggs. Numbers decline naturally in winter. **Habitat** Wherever wild and cultivated fruits are present. Originally from Africa, now a cosmopolitan pest.

4 *Didacus ciliatus*
Lesser Cucurbit Fly, Small Cucurbit Fly

Small (wingspan 12–14mm), wasp-like, brown, with bright yellow markings on the thorax and a thin brown line on the wing tip and margin. Females have a short ovipositor. **Biology** Attacks various species of cucurbits (pumpkins, watermelons, gourds, gem squash), piercing young fruits to lay eggs. Maggots destroy or deform the fruit. **Habitat** Prefers arid areas, natural and cultivated, where cucurbits are growing. Along with *D. vertebratus*, common in the region.

5 *Afrocneros mundus*

Small (wingspan 13mm), thorax brown with a central stripe, abdomen yellow with black bands towards the rear; wings with bold black patterning. **Biology** Males manufacture and use small 'mating balls' of foam to attract females; these are then eaten by the females. Larvae live under the bark of living cabbage trees (*Cussonia*), sometimes killing the tree. **Habitat** Attracted to rotting fruit, especially ripe figs, in the summer-rainfall region.

6 *Dacus apoxanthus*
Milkweed Fruit Fly

Small (wingspan 9mm), body and eyes orange-brown, with a large black dot on each side of the abdomen; wings tinged brown, with a brown spot at the tip. **Biology** Larvae develop in the buds of the milkweed *Ceropegia ampliata* (Apocynaceae), destroying the majority of buds. **Habitat** Arid succulent vegetation where its host plant occurs.

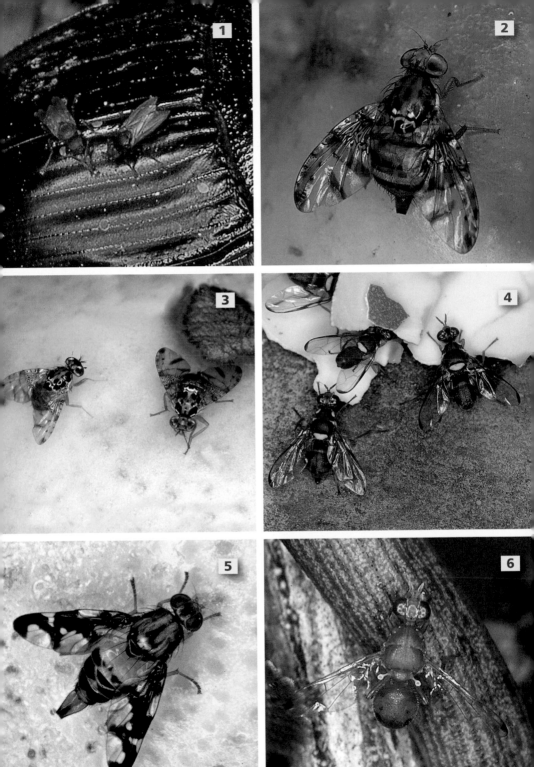

Family Platystomatidae

Signal flies

Unusual, medium-sized to large, very sluggish flies, with some red and orange coloration and generally boldly patterned black wings. Adults and larvae have catholic tastes, frequenting all kinds of decaying organic matter (flowers, fruit, faeces, human wounds and dead snails). About 80 species are known from the region.

1 *Peltacanthina*

Orange signal flies

Medium-sized (wingspan 20mm), stocky, with a distinctive bright orange head and thorax and a black abdomen. Wings blackish with white spots. **Biology** Adults sluggish and bold, sitting in exposed positions on foliage. **Habitat** Subtropical scrub and forest.

2 *Bromophila caffra*

Red-headed Signal Fly

Large, but size highly variable (wingspan typically 30–50mm, but may be smaller). Striking, with body and wings metallic bluish-black, and head sealing wax red. **Biology** Adults sluggish, sitting in exposed positions, often gregariously, where mating occurs. Found on the flowers of various trees and shrubs. **Habitat** Bushveld and subtropical forest.

3 *Lule*

Band-wing signal flies

Small (wingspan 16mm) and stocky, with a humped thorax. Body and wings black, with green and blue irridescence and a white band and spots. Head bright red, with a very large proboscis. **Biology** Adults congregate and feed on faeces. They walk slowly, extending the proboscis at intervals. **Habitat** Adults on water plants adjacent to forest streams, bushes and trees.

4 *Elassogaster*

Scavenger flies

Small (wingspan 6mm), black with a shiny green thorax, large, rounded red eyes, and a dark spot on the wing tips. **Biology** Adults aggregate on dung piles, walking and constantly waving their wings. Typically breed in dung, but recorded elsewhere as attacking the egg pods of migratory locusts. **Habitat** Found on mammalian dung in the warmer eastern parts of the region.

Family Braulidae

Bee lice

Minute (body length under 1.5mm), squat, reddish-brown and wingless. The eyes are reduced and the antennae are hidden in grooves. The legs, which are short and strong, have a comb of microscopic teeth on the claws.

5 *Braula*

Braula flies

Very small (body length 1.5mm), wingless, with short, stout legs ending in well-developed attachment pads **(5A)** (see arrow), with short, comb-like teeth. There are 2 species in the region. **Biology** Cling to bees, especially drones and queens. Agile, rushing to scavenge food from around the queen's mouthparts when she is fed. Larvae feed on pollen and wax, leaving thread-like tunnels under the wax capping. Considered a minor pest of commercial keeping due to damage caused to honeycombs. **Habitat** Found only in beehives.

Family Micropezidae

Stilt-legged flies

Very long-legged flies with slender, elongate bodies and spindly legs, of which the front pair are shorter and placed well forward. Wings often dark.

6 *Mimegralla fuelleborni*

White-footed Stilt-legged Fly

Medium-sized (wingspan 20mm), spindly, with banded and striped grey body, very long, banded legs and smoky grey wings banded in black; holds the white-tipped fore legs straight out, waving them up and down like antennae. **Biology** Probably mimics ichneumonid wasps, often gregarious. Larvae likely to feed on living or decayed plant matter (frequently fruits or ginger rhizomes). **Habitat** Cool, shady and humid forests, on leaves in the undergrowth. May occur near disturbed habitats. **Related species** There are 4 other species in the genus, all occurring in the eastern subtropical parts of the region.

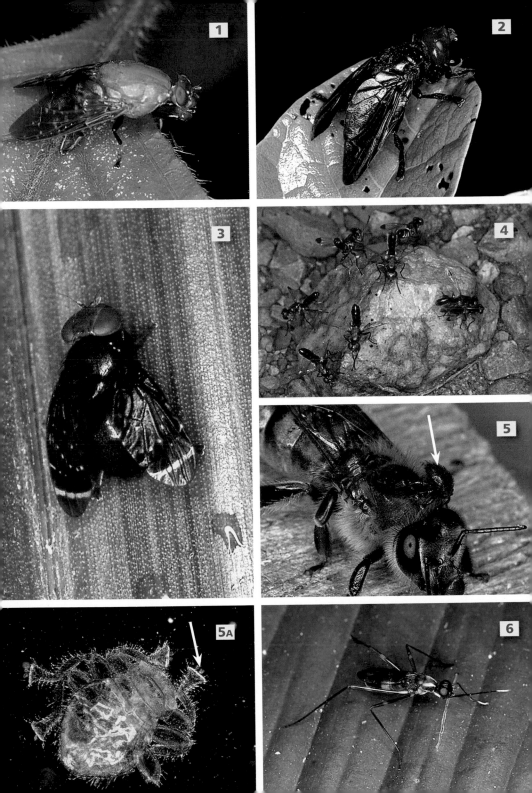

Family Neriidae
Banana flies

Tropical, slender-bodied, with elongate legs and conspicuous, upright antennae with a club-like base and a thin terminal section. Males have longer legs than females, and are known to tussle in competition for mates. Larvae able to leap from their feeding area to a dry tree trunk for pupation.

1 *Chaetonerius*
Striped banana flies

Small (wingspan 8mm), with brown and yellow stripes along head and body. There are 2 regional species in the genus. **Biology** Larvae have been recorded from decaying pumpkin and probably feed on a range of vegetable matter. Adults occur throughout the year, often resting head-downwards on tree trunks or standing in an alert posture on the various kinds of ripe fruit on which they feed. **Habitat** Adults are often associated with human habitation, appearing especially on windows. Fairly common in shade in indigenous forest or in gardens in the moister southern and eastern coastal regions.

Family Diopsidae
Stalk-eyed flies

Easily recognized by the bizarre eyes, which are situated at the end of very long or, in the case of smaller species, shorter stalks. The length of the eye stalks is possibly the result of competition between males (sexual selection), although females have equally long eye stalks. Most species have 2 backward-facing thoracic spines on the scutellum. Larvae are known to bore into young shoots, including those of rice and other cereal crops. Rare in cool or dry areas. There are 33 species known from the region.

2 *Diopsis*
Orange stalk-eyed flies

Small (wingspan 12mm), with an orange head and abdomen. Thorax black with backward-projecting spines, no paired bristles and no apical bristle on the scutellar spines. Stout femora on all legs. Wings usually have dark patches at or near their tips. **Biology** Adults are slow, clumsy fliers and suck up any exudates. Form swarms in shaded areas. **Habitat** Rank vegetation under trees in moist, humid places. **Related species** *Diopsina* species, which are virtually identical except that the scutellar spine has an apical bristle.

3 *Sphyracephala*
Hammerhead flies

Small (wingspan 8–9mm), with brown body, short, thick eye stalks, unmarked wings, and large spines on the thorax. **Biology** Thought to mimic small jumping spiders. Groups gather on moist rocks, or under bridges over streams. **Habitat** Very moist microhabitats, often associated with water. Common in Africa and Madagascar.

4 *Diasemopsis*
Waisted stalk-eyed flies

Small (wingspan 6mm), dark, almost black, body with 'waisted' abdomen characteristically patterned in grey, black and white. Wings clear, eyes red. Have paired rows of bristles on top of the thorax and an apical bristle on the scutellar spines. Genus comprises 2 species. **Biology** Unlike other diopsids, they do not form swarms. **Habitat** Warmer forested areas.

Family Sepsidae
Black scavenger flies

Small, easily recognized by their ant-like appearance, the constricted waist of the abdomen and the almost spherical head. Wing tips have a black spot and males have spines on the front legs. Usually encountered on dung, in which the larvae of most species develop. There are 23 regional species.

5 *Dicranosepsis*

Small (wingspan 6mm), with a black body and red eyes. **Biology** Constantly wave their wings outwards while walking. Usually seen feeding and courting in large groups on fresh dung or on carcasses, although may also be present on low vegetation. Males grasp females with the front legs during mating. **Habitat** Very common in moist forests in the warmer parts of the region. **Related species** *Paratoxapoda* species (wingspan 5mm) **(5A)**, which are black with clear wings and occur in subtropical bushveld.

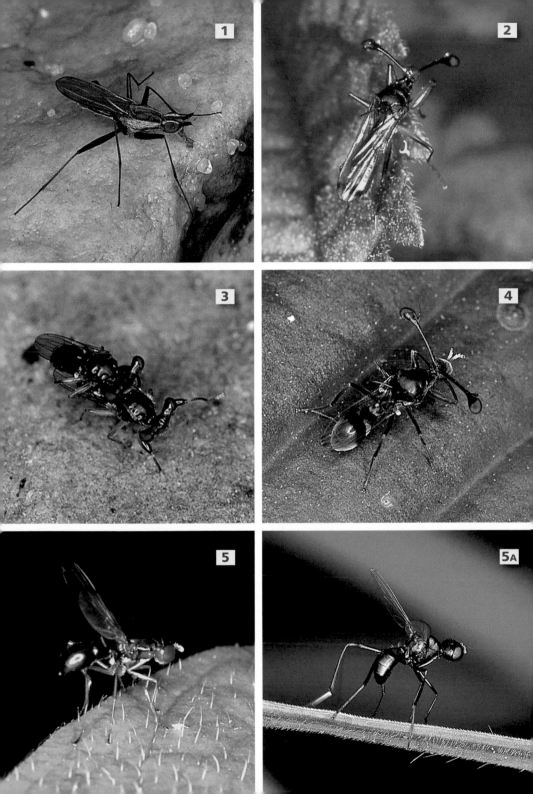

Family Sciomyzidae
Snail-killing flies, marsh flies

Medium-sized to large, yellowish or brownish, often with longitudinal stripes on the thorax and a few spots on the wings. Antennae long and upright or forward-projecting. The larvae are dark brown and warty and feed on aquatic snails. Most species in the region belong to *Sepedon*, a cosmopolitan genus of marsh flies.

1 *Sepedon*
Bilharzia flies

Medium-sized (wingspan 12mm), with long, thickened hind legs. Femora yellow at the base and reddish at the tip. Wings and head dull orange; thorax metallic grey. There are 30 recorded regional species. **Biology** Adults lay eggs on grass growing out of water. Larvae drop into the water, swallow air, and float just beneath the surface, where they attack aquatic snails that cannot close off their shells with a 'lid' (operculum). Prey species include snails (*Biomphalaria* and *Bulinus*) that are intermediate hosts for bilharzia in humans, making them useful agents for biological control. **Habitat** Slow-flowing or standing water with marginal vegetation in subtropical regions of KwaZulu-Natal and Limpopo. **Related species** *Ethiolimnia geniculata* (**1A**), which has mottled dark wings and is possibly the most common snail-killing fly in the region.

2 Family Lauxaniidae
Lauxaniid flies

Smallish, plump flies with large, often brightly coloured eyes and variegated patterns on the wings. Very common and can be distinguished by the lack of facial vibrissae (**2**) – a pair of stout bristles that in other flies project from the bottom part of the face. The region probably has far more than the 42 described species.

3 *Cestrotus*
Rock flies

Small (wingspan 10mm), squat, with wings boldly patterned in smoky black. Head has a black band on top, 2 pairs of white spots on the sides and a white spot at the centre. Base of antennae marked with orange spots. **Biology** Adults aggregate on damp vegetation in shady places, on rocks along rivers, on tree trunks and on the walls of houses. Larvae feed mainly on leaf litter, but have been reared from birds' nests. **Habitat** River margins, shaded or lichen-covered rocks, often along streams.

Family Coelopidae
Kelp flies, seaweed flies

Medium-sized, resembling house flies, but relatively flat and bristly, with stout legs. Wings clear with brownish tinge. Best recognized by their association with wrack (beached seaweed), where they keep company with *Fucellia* (p.350).

4 *Coelopa*
Bristly kelp flies

Medium-sized (wingspan 12mm), dark brown, wings tinged with brown, body and legs covered in bristles. There are 2 species of this cosmopolitan genus in the region. **Biology** Very sluggish and loath to fly. Larvae develop in wet kelp and other seaweeds, where they feed on exudates and mucus released by decay. Adults may occur on flowers and form swarms around suitably aromatic sources. Rising tides may cause masses of the flies to make contact with humans and become an annoyance. **Habitat** Sandy and rocky shores with stranded kelp.

Family Heleomyzidae
Rotting flies

Medium-sized, with mottled or spotted wings that appear to bend in the long axis over the abdomen. Under the microscope, small scattered spines are visible along leading edge of fore wings. There are 31 species in the region.

5 *Suillia picta*
Black-winged Rotting Fly

Medium-sized (wingspan 12mm), with black body, small red head and eyes, and broad, patterned black wings with whitish wing tips, held flat over the body. **Biology** Sluggish. Larvae feed on a wide variety of rotting matter, including fungi, carrion and rotting snails. They have been reared from rabbit burrows, bat caves and birds' nests. **Habitat** Afromontane forest, woodlands and gardens.

Family Ephydridae
Shore flies

Most members of the family are small to minute, blackish, with mottled wings and hairy antennae. Adults are often very common, and feed on aquatic insects. Larvae are generally aquatic, but occur in a variety of habitats. (The larvae of exotic species live in hot springs and petroleum pools.) Larval food sources vary considerably. Most common on the seashore and margins of estuaries. There are 70 species in the region.

1 *Ochthera*
Mantid shore flies

The largest regional shore flies (wingspan 10mm). Grey and stocky; easily recognized by the enormously enlarged and claw-like fore legs, which are used for grasping prey, usually adult midges. **Biology** *Ochthera* flex their fore legs as they walk over wet algal crusts. Larvae are predaceous. **Habitat** Found on the margins of dams, pans and other waterbodies.

Family Piophilidae
Cheese skippers

Small, somewhat metallic-looking, blackish-blue flies, whose larvae feed on decomposing animal matter, either dried-out corpses, or animal products such as cheese. Some breed inside the dried-out marrow of large bones. There are 3 species in the region.

2 *Piophila casei*
Cheese Fly

Small (wingspan 6mm), shiny and black, with yellow legs and large red eyes. **Biology** May cause considerable damage to foodstuffs such as cheese and ham, where adults and larvae typically occur (although they may be found in other protein matter – an impala carcass in 1 instance). Larvae skip by grasping their hind parts with mouthhooks and suddenly releasing their grip. **Habitat** Common throughout Africa, near drying animal protein.

Family Agromyzidae
Leaf-mining flies

Small to minute, blackish or yellowish, elongate and slimly built. Larvae are usually leaf or stem miners of flowering plants, mostly members of the daisy and carrot families. There are about 131 species of the family in the region.

3 *Liriomyza*
Damaging leaf-mining flies

Small (wingspan 4mm), black dorsally and yellow beneath, with a conspicuous yellow spot between the wings. The 2 regional *Liriomyza* species both originated in the Americas and are very similar in appearance. **Biology** Pests of considerable importance; the larvae mine leaves, blemishing foliage and retarding growth. The American Leaf Miner (*L. trifolii*) is a pest of tomatoes, beans and other vegetables, the bright yellow larvae mining the upper leaf epidermis and leaving a trail of faeces in the meandering mine **(3A)**. The Potato Leaf Miner (*L. huidobrensis*) has caused millions of rands in damage to potato crops, and also attacks tomatoes and melons. **Habitat** Fields and gardens, where crop plants are grown. *L. huidobrensis*, a recent invader, has spread from the sandveld region of the Western Cape to other potato-growing areas.

Family Scathophagidae
Dung flies

Medium-sized, hairy, with very small plates protecting the halteres. Males yellow, females greyish and rarer than males.

4 *Scathophaga soror*
African Dung Fly

Large (wingspan 16–20mm), yellow (males brighter), stocky and furry. The only regional species in the genus. **Biology** Larvae feed on dung and pupate in the soil below. Adults mate on dung, where they also prey on flies. Their fleshy proboscis is armed with teeth, allowing them to capture flies as large as bluebottles. **Habitat** Adults found on fresh dung or flowers.

Family Drosophilidae
Vinegar flies

Vinegar flies are occasionally incorrectly called fruit flies, a term that should be reserved for Tephritidae (p.340). Small to minute, brown, yellow or grey flies, with bright red eyes. Under the microscope, a bristle (arista) at the end of each antenna is seen to fork at the tip. Larvae feed not on rotting fruit itself, but on the colonies of yeast that infect it. Exudates from infected grapes can ruin bunches destined for export. Some larvae, however, mine leaves, feed on other insect larvae in aquatic habitats, or live in the spittle of froghopper (Cercopidae) nymphs. There are 61 species in the region.

1 *Drosophila*
Vinegar flies

Small (wingspan 3–4mm), with a tan body and red eyes. Abdomen striped, with a black patch on the underside in males of many species. **Biology** Adults very common around rotting fruit, attracted by the fermentation of colonies of yeast (equally attracted by a glass of wine). They rise in a cloud when approached, but soon resettle. *D. melanogaster* is the most famous species, being the classic model for much of our knowledge of genetics. **Habitat** All habitats, but less numerous in more arid areas. Widespread in Africa. **Related species** Striped fruit flies (*Zaprionus*) **(1A)**, which are small (wingspan 4mm), but a little stockier than *Drosophila* species, and easily recognized by the 4 silvery stripes along the head and thorax. They are common throughout the region and have a similar biology to *Drosophila*.

2 Family Milichiidae
Jackal flies

Minute to small (wingspan 2–4mm), black, best recognized in the field by their unusual habit of loitering around kills made by arthropods. Under the microscope, very long and folded mouthparts are visible. There are at least 13 species in the region. **Biology** Scavengers that hang around crab spiders feeding on bees **(2)** – thought to be attracted to an alarm pheromone given off by dying bees. Also around feeding wasps, robber flies, preying mantids and assassin bugs, sucking up body juices that leak from their prey. Larvae feed on rich plant and animal organic matter, such as carrion and manure. **Habitat** Various habitats, but most abundant in subtropical scrub and bushveld. Widespread, a few species spread around the world by humans.

Family Anthomyiidae
Root-maggot flies

Resemble small, slender grey house flies and, apart from the common kelp flies *Fucellia*, which characteristically occur in large numbers on stranded seaweed (but see also Coelopidae, p.346), are difficult to identify. Larvae may feed on plant or decaying organic matter, or are aquatic. Those that feed on developing seedlings are agricultural pests. There are 27 species known from the region.

3 *Fucellia capensis*
Striped Kelp Fly

Small (wingspan 10–13mm), grey, nondescript, with 4 indistinct stripes on the thorax, and light brown wings. **Biology** The most common fly on kelp, where it breeds in large numbers **(3A)**. Sometimes reach nuisance levels when onshore winds blow them towards coastal dwellings. Larvae feed on stranded decaying kelp, and complete their life cycle before the next spring tide, 28 days later. Adults present throughout the year, peaking in early winter when the largest amounts of kelp are beached. **Habitat** Sandy and rocky seashores with stranded kelp.

4 *Anthomyia*
Striped anthomyiid flies

Small (wingspan 10mm), resembling slender house flies in build, but with contrasting silvery grey and black stripes across the body. There are 3 species in the region. **Biology** Larvae breed in rotting organic matter, or that found in birds' nests. **Habitat** Variety of vegetation types.

Family Muscidae
House flies

Contains the most common and most frequently seen flies in the region, including the ubiquitous House Fly (*Musca domestica*). Not easily recognized in the field, but microscopically identifiable by veins in the anal region of the wing that fail to reach the wing margin, and by the absence of a fan of bristles just below, and in front of, the halteres. Adults of most species mop up surface fluids, although a few are bloodsuckers and have spines on the end of a hardened proboscis or, alternatively, long piercing mouthparts. The latter group is of medical and veterinary importance. Muscids have varied life styles, but the maggot-type larvae all inhabit various kinds of organic matter, a few developing in growing shoots. This diverse family has at least 364 species in the region.

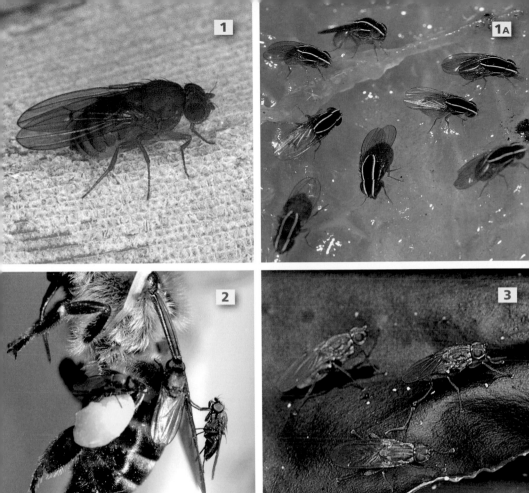

1 *Musca domestica* — House Fly

Small (wingspan 9–14mm), familiar, with 4 dark stripes on the thorax. Wings tinted brown, with 2 veins at the wing tip that meet to enclose a 'cell'. Abdomen is dirty cream with a black central stripe. Light, fleshy proboscis. There are 2 common African subspecies. **Biology** Adults suck up various liquids. Like many in the family, they feed on human food and excrement and thus transmit numerous diseases, including cholera, poliomyelitis, leprosy, typhoid and salmonellosis. As in other species, strains resistant to insecticides have developed. Breeds in fresh manure, garbage or rotting vegetation. **Habitat** Ubiquitous and cosmopolitan, 1 subspecies being nearly always associated with human habitation. The other, however, is associated with natural veld. **Related species** *Helina* species (**1A**) (body length 7mm), which are very similar to the House Fly, but are distinguished by the presence of 3 grey dots on each wing, and by having a plumose terminal bristle on the antennae. The larvae occur in fresh dung, rotten wood and rotting vegetable matter.

2 *Stomoxys calcitrans* — Stable Fly

Small (wingspan 11–13mm), resembling *Musca domestica* (above) but has a hard, black proboscis and a grey abdomen with black dorsal patches. **Biology** An aggressive and persistent bloodsucker, tormenting wild animals (such as buffalo) as well as cattle, horses and dogs. Rests in groups on the shaded bellies of cattle. Ears of bitten dogs eventually become a mass of open sores. Breeds in horse dung mixed with straw. **Habitat** Stables and farms.

3 *Lispe* — Barred mud flies

Small (wingspan 10mm), black, with distinctive white bars on the sides of abdomen and enlarged accessory mouthparts alongside the proboscis. **Biology** Adult males court females by scuttling in front of them in a small arc. Larvae develop in organically rich mud. **Habitat** Common on algal mats at the margins of pans and dams.

4 Family Glossinidae — Tsetse flies

Medium-sized (wingspan approximately 25mm), light brown or grey, easily recognized by the flattened body, short, forward-projecting proboscis, alert posture when resting, and feathery bristle on each antenna. Wings folded scissor-like, with the wing tips crossed over one another. Four species of *Glossina* (**4**) are known from the region. **Biology** Adults active by day, but spend a lot of time resting. Both sexes are host-specific bloodsuckers, humans being attacked only in the absence of game. Grassland and woodland species concentrate on large game. Subtropical forest species feed mostly on reptiles. Females incubate a single maggot in their bodies, which is nourished by a 'milk gland' and pupates soon after it is 'born'. Pupae may be parasitized by mutillid wasps. Of tremendous veterinary and economic significance in Africa, where they transmit species of single-celled organisms (trypanosomes) that cause sleeping sickness in humans and nagana in domestic animals. Their presence historically excluded animal husbandry from large areas. **Habitat** *G. brevipalpis* lives in deep forest; the other species occur in subtropical grassland and forest.

Family Hippoboscidae — Louse flies

Medium-sized, flat, leathery flies, some, like the Sheep Ked (*Melophagus ovinus*) having lost all vestiges of wings. The large head fits into a hollow in the thorax. Proboscis well developed, often forward-projecting. Wing veins well developed on the fore margins, but weaken and are absent towards trailing edge of wing. Legs robust, armed with strong claws. Adults are bloodsucking ectoparasites of birds and mammals, mainly antelope, but also of carnivores, horses and zebras. In very large numbers can cause death through blood loss. All species that parasitize birds retain wings; mammalian parasites show various forms of wing reduction. There are 37 species in the region.

5 *Pseudolynchia canariensis* — Pigeon Louse Fly

Medium-sized (wingspan 16mm), flattened, dark brown, with brown wings and a short, beak-like proboscis. **Biology** Makes crab-like scuttling movements as it hides between feathers to evade its preening host. Flies reluctantly and directly. Extremely robust and difficult to crush and kill. **Habitat** Wild and domesticated pigeons and birds of prey. Has spread to the warmer parts of the world via the Domestic Pigeon.

1 *Hippobosca rufipes* — Cattle Louse Fly

Medium-sized (wingspan 24mm), flattened, robust, dull red, with a characteristic pattern of small ivory triangles on the thorax. Legs crab-like, with well-developed claws. **Biology** A common ectoparasite of horses and cattle, but also recorded from eland, blue wildebeest and steenbok. May alight on, but rarely bites, humans. **Habitat** A range of vegetation types, but absent or rare in forest. **Related species** The Ostrich Fly (*Struthiobosca struthionis*), which is very similar, but smaller and has a different pattern of ivory markings on the thorax.

Family Nycteribiidae — Bat flies, bat louse flies

Bizarre, small, spider-like flies, completely wingless and with eyes reduced or absent. There are 16 species in the region. The related family Streblidae is similar, and its members also parasitize bats. They show varying degrees of eye and wing loss.

2 *Penicillidia fulvida* — Common Bat Fly

Small (body length 3mm) with very robust legs, a tick-like abdomen, and a small turret-like head sunk into the flesh-coloured thorax. **Biology** A bloodsucking parasite of bats, moving agilely through its host's fur and not easily groomed out. Mature larvae are deposited by females on cave walls and pupate immediately. **Habitat** On bats in the southwestern Cape and in KwaZulu-Natal. Pupae occur on the walls of bat roosts.

Family Bengaliidae — Fly pirates

A recently erected family comprising 12 genera and 70 species of large, yellowish-brown flies found in association with ant nests. Active predators of ant cocoons and ant prey, which are snatched from the ants while they are being transported. Mouthparts robust, with the upper lip extending below the other mouthparts. All have a silent flight.

3 *Bengalia* — Fly pirates

Large (wingspan 21–25mm), light brown, with somewhat darker abdomen, thorax streaked in dark brown, and tip of abdomen dark brown. Eyes red. **Biology** Larvae feed in the nests of termites. Adults (shown here taking a termite from an ant) hover over columns of driver ants, snatching their eggs and larvae. **Habitat** Subtropical forest floor.

Family Calliphoridae — Bluebottles, greenbottles, blowflies

Stoutly built, medium-sized to large flies, with brilliant metallic blue or green colouring. Each antenna ends in a feathery bristle. The eyes of males meet in the centre of the head. Adults, especially females, are attracted to decaying flesh and excrement, but males are more likely to be found lapping nectar from flowers. The maggot-like larvae of some species develop inside rotten flesh, carrion and faeces. Others have more unusual larval habits, being associated with termite or ant nests. About 152 species are known from the region.

4 *Chrysomya chloropyga* — Copper-tailed Blowfly

Medium-sized (wingspan 18mm), stout, abdomen with violet and green stripes, last abdominal segment green. Prothorax has a diagnostic black marking that forms an inverted 'U'. **Biology** *Chrysomya* and *Lucilia* feed on dead animals. Most can also develop in moist areas under the wool of sheep, causing large ulcers that may result in the animal's death. Treatment of jagged wounds in humans was originally done by introducing sterile maggots into the wound. This method is being revived, with the maggots debriding dead tissue, resulting in healing of the wound. **Habitat** Ubiquitous. Found near corpses or excrement.

5 *Chrysomya albiceps* — Banded Blowfly

Medium-sized (wingspan 20mm), stout, uniformly green, with black abdominal bands and an extensive silver-white facial pattern. **Biology** The hairy maggots appear late in the succession of corpse colonization, feeding aggressively on other maggot species. Capable of spreading anthrax. **Habitat** Ubiquitous in the region, aggregating on corpses and carcasses in an advanced stage of decomposition.

1 *Chrysomya marginalis* — Regal Blowfly

Medium-sized (wingspan 16mm), with an orange head, red eyes, metallic greenish-black body with violet irridescence, and a black stripe along the fore margin of each wing. **Biology** Like *C. albiceps*, lays egg masses in shaded places on freshly dead animals. Within a day or two the corpse is covered in a mass of larvae, which migrate together into the soil to pupate after 4–6 days. **Habitat** A variety of vegetation types. Aggregates on fresh corpses, but, like others in the genus, also frequents flowers. Very common and widespread in Africa.

2 *Lucilia sericata* — European Green Blowfly

Medium-sized (wingspan 12mm), uniformly green, with bronze cast to thorax. Easily confused with *L. cuprina*, which may display shades of bluish-green. **Biology** Introduced to South Africa, probably over a century ago. Attracted to offal and moist wool. Attacks sheep, the larvae **(2A)** eventually burrowing into the skin, excreting ammonia and attracting more flies. A major pest for sheep farmers around the world. **Habitat** Ubiquitous in domestic and natural areas, summer populations reaching pest levels in the southwestern Cape.

3 *Cordylobia anthropophaga* — Mango Fly, Putzi Fly

Large (wingspan 20mm), squat, yellowish-brown, with honey-coloured eyes. Abdominal segments partially banded in brown. Characteristically has 2 broad charcoal-coloured thoracic stripes. The second abdominal segment is longer than the rest. Larvae oval, yellowish-white and covered in fine spines, usually grouped in 2–3 rows per segment. **Biology** Lays eggs in moist areas smelling of ammonia, urine or faeces (typically on drying clothing and baby nappies). Larvae bore into the skin of mammalian hosts, where they develop and form boils on the skin surface, only emerging when mature to pupate in soil. Newly hatched larvae can survive for days without boring into the host, which can include rodents, dogs, monkeys and humans. Adult flies feed on rotting fruit and faeces, and commonly enter homes. **Habitat** Most subtropical vegetation types, commonly in forest. In very wet years extends its range, even reaching the Namib Desert.

4 Family Sarcophagidae — Flesh flies

Generally robust, medium-sized to large, grey and black, with characteristically striped thorax, and a chequerboard pattern of grey on the abdomen. In males the eyes do not join on top of the head (see Calliphoridae, p.354). Some species are unusual in that the larvae are predators or guests in the nests of wasps and the egg cases of grasshoppers and other insect groups. The more obvious species feed as adults on decaying organic matter, often of animal origin, and on faeces, where females deposit several live larvae. Those in the subfamily Miltogramminae **(4)** are specialized parasitoids of wasps and bees, and can be seen hovering or sitting in front of their burrows. Others feed on the prey of spider wasps, on termites or on locust egg pods. They have very large eyes, and a narrow abdomen with black spots or silvery bands. There are 157 species known from the region.

5 *Wohlfahrtia pachytyli* — Locust Fly

Large (wingspan 28mm), stocky, with thoracic pattern typical of genus, but distinguished by rows of 3 black spots on pinkish-grey abdomen. End of abdomen with orange tip. **Biology** Attracted to odour of moulting Brown Locusts (*Locustana pardalina*) (p.120), depositing small larvae that consume and kill the host. A valuable biological control agent for locust swarms. **Habitat** Attracted to fresh faeces in a variety of habitats. **Related species** Other *Sarcophaga* species, which are very similar, with the typical chequerboard-like, grey-and-black-marked abdomen, tipped with orange.

Family Tachinidae — Tachinid flies

Medium-sized flies, often covered in strong bristles. Recognized under magnification by extra ridge (post-scutellum) under the most backward-projecting part of the thorax (scutellum). Adults feed on nectar, and often have a characteristic buzzing flight. All species are internal parasites of insects or, occasionally, centipedes. Larvae feed first on non-essential organs of the host and attack vital organs only when the hosts are just about to pupate. Butterflies and moths are the most common hosts, but any large larva from a range of groups, including bugs, grasshoppers, beetles and (rarely) wasps and bees, is attacked. Females may lay eggs directly into the host, or on a plant that the host may eat inadvertently. Tachinids are important agriculturally, as they suppress populations of injurious moth caterpillars. There are about 377 species known from the region.

1 *Gonia bimaculata* — Cutworm Fly

Fairly large (wingspan 13mm), stout and bristly, with reddish-brown abdomen (sometimes only the first few segments) and yellowish-white hairs on the cheeks. **Biology** Widespread and useful parasite of the Cutworm (*Agrotis segetum,* Family Agrotidae). **Habitat** Most native and disturbed habitats. **Related species** There is 1 other species in the genus, also with yellowish-white hairs on the cheeks.

2 *Hermya diabolus* — Twig Wilter Fly

Fairly large (wingspan 11mm), black, with shiny white face and side patches. **Biology** Appears to specialize in parasitizing the Twig Wilter (*Anoplocnemis curvipes*) (p.148). **Habitat** Bushveld. **Related species** The genus includes 4 other regional species – all are a uniform shiny black or bright orange and have dark brown or black wings.

3 *Phasia* — Red-eyed tachinids

Small (wingspan 4mm), both sexes with very large eyes that almost meet on top of the head. Shiny, pointed ovipositor in females. Genus comprises 6 regional species. **Biology** Ovipositor used to lay eggs in bug hosts, typically *Dysdercus*, lygaeid bugs and stink bugs. **Habitat** Grassland and bushveld.

4 *Dejeania bombylans* — Bloated Tachinid

Robust and large (wingspan 16–24mm), with yellow-and-brown abdomen. **Biology** Known to parasitize only the African Boll Worm (*Helicoverpa armigera*). **Habitat** Broad-leaved herbs and grasses in the shade of bushveld trees. **Related species** The 3 other *Dejeania* species in the region are similar, with plain wings and an inflated abdomen armed with very prominent spines and bristles.

5 Family Oestridae — Warble flies, bot flies, nasal flies

Adults large (wingspan 18mm), stout, mottled grey, hairy but without bristles. Head large and inflated, with large eyes and reduced mouthparts **(5)** (*Gedoelstia hassleri*). Larvae white or tan, spiny and barrel-shaped, with black mouthparts. There are 17 species in the region. Adult flies do not feed and are short-lived and rarely seen. Larvae are internal parasites of mammals, either in the nasal cavities of horses, antelope, zebras and sheep or under the skin of antelope, hares and rodents. Others live in the gut of herbivores. Cause pus-filled boils (warbles) in cattle that damage the hide. Typically, as in the Sheep Warble Fly (*Oestrus ovis*), living larvae are deposited on the nostrils of the host, where they attach themselves to the sinus membrane and feed on mucus. They are later sneezed out and pupate in the soil. Can also occur in the living pulp cavity of the horns of antelope, sheep, zebras and horses. Larval development is slow (up to 9 months).

6 *Gasterophilus* — Bot flies

Larvae **(6)** are large (body length 8mm), barrel-like, with a narrower head, and armed with rows of spines. They live in and pack the guts of zebras, horses, rhinos and elephants. The adults are dull yellow, bee-like, similar to but stouter than oestrids. Adults lack mouthparts. There are 12 species in the region. **Biology** Eggs laid on the host's fur are licked and swallowed, or bore through the skin, moving to the stomach where they attach by mouth hooks. Larvae have haemoglobin to survive low oxygen levels in the gut. Mature larvae pass out in the host's faeces, to pupate in soil. **Related species** The Rhinoceros Bot Fly (*Gyrostigma rhinocerontis*) **(6A)**, which has very large larvae (body length 2.5cm) that develop in the gut of a rhino. Larvae are passed out in the rhino's dung, where they pupate and emerge as Africa's largest fly, with a wingspan of 70mm. The fly lays eggs on the neck and face of a rhino **(6B)**. These are apparently licked and swallowed by the grooming rhino, where they enter the gut, attach and begin feeding. Apparently the large adults mimic spider wasps, do not feed and are thus short-lived and rarely seen. **Habitat** Widespread wherever there are suitable hosts.

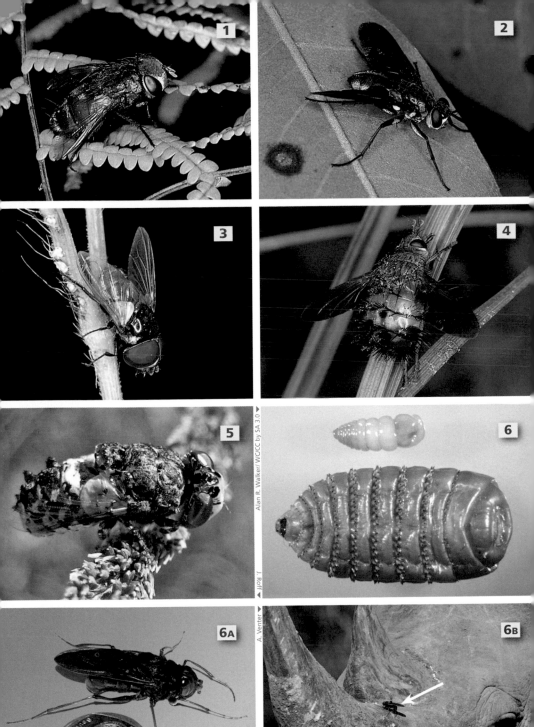

Alan R. Walker/ WCKC by SA 3.0 ▶

◄ J. Roff

A. Venter ▶

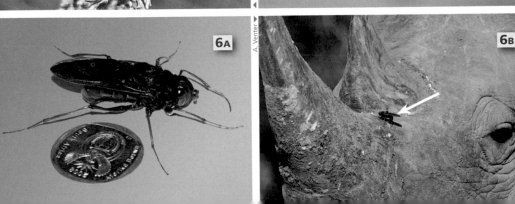

ORDER SIPHONAPTERA • FLEAS

Small, laterally flattened, completely wingless insects, renowned for their remarkable jumping abilities; the adults are all blood-feeding parasites of birds or mammals.

Fleas are unusual, secondarily wingless insects with laterally flattened bodies and sucking-and-piercing mouthparts that they use to feed on the blood of birds and mammals. Their closest living relatives are the hangingflies and scorpion flies (p.310). The hind legs are larger than the first 2 pairs and can generate spectacular jumps, utilizing pads of the elastic protein resilin. Some fleas are restricted to a single host species, but most occur on several hosts. Females require a blood meal before laying eggs. The eggs hatch in 2–12 days and the maggot-like larvae feed on organic debris and undergo 3 moults before pupating. Pupae can remain dormant for months, hatching in response to vibrations caused by the presence of a host. Fleas are of great medical importance, notably as vectors for plague and typhus, and also as intermediate hosts for dog and rodent tapeworms. There are about 2,500 species globally, of which some 100 are known from South Africa.

1 Family Hectopsyllidae
Chigoe fleas

A small family containing only a few genera and characterized by immobile females that remain attached to the host for prolonged periods. The Sand Flea or Chigoe Flea (*Tunga penetrans*) is the only species in the region. It is small (body length 1–3mm) and recognized by its unusual habits. **Biology** Female Sand Fleas burrow into soft areas of skin (especially between toes and under toenails) with only a small opening at the posterior end of the body **(1)**. They cause intense itching and, as the eggs develop, can swell to the size of a small pea, causing inflammation and infection. **Habitat** Males are free-living in sand. Females are permanently embedded in the skin of various mammals, especially that of pigs and humans. Introduced from South America in 17th century and now widespread in tropical Africa.

Family Pulicidae
Common fleas

A large and diverse family with a worldwide distribution, including most of the important pests of domestic animals and humans, among them the vectors of plague.

2 *Echidnophaga*
Sticktight fleas

Small (body length 2–3mm). Recognized by their habits. **Biology** The Hen Flea (*E. gallinacea*) **(2)** occurs widely on birds and small mammals, but is best known as a pest of poultry, clustering densely on the faces of chickens **(2A)** and, in extreme cases, impacting growth and egg production. Female *E. larina* burrow into the skin of warthogs, bushpigs and elephants. Resultant ulcers cause inflammation and discomfort. Eggs fall to the ground and the larvae develop in the soil. **Habitat** Males free-living; females attach themselves to, and partially bury themselves in, the skin of their host.

3 *Ctenocephalides felis*
Cat Flea

Small (body length 2–3mm), distinguished from similar species by microscopic differences in its body spination. **Biology** Adults **(3)** are voracious blood-feeders, injecting anti-coagulant saliva into wounds made by their piercing mouthparts. Bites cause irritation and swelling. Large, white eggs hatch into active, cylindrical larvae **(3A)** that feed on skin scales and flea excreta. They spin a silken cocoon in which to pupate. **Habitat** The most common flea in human dwellings. Despite its name, the main flea pest found on dogs, in human dwellings, and on cats.

4 Family Hystrichopsyllidae
Rodent fleas

A diverse family, found mostly in the nests or on the bodies of rodents. The Rat Flea (*Listropsylla agrippinae*) **(4)** is small (2mm), and distinguishable by its unusually large and hairy fore legs. **Biology** An ectoparasite of the Striped Mouse (*Rhabdomys pumilio*), Bush Karoo Rat (*Otomys unisulcatus*), Namaqua Rock Mouse (*Aethomys namaquensis*) and other rodents. **Habitat** The fur and nests of rodents.

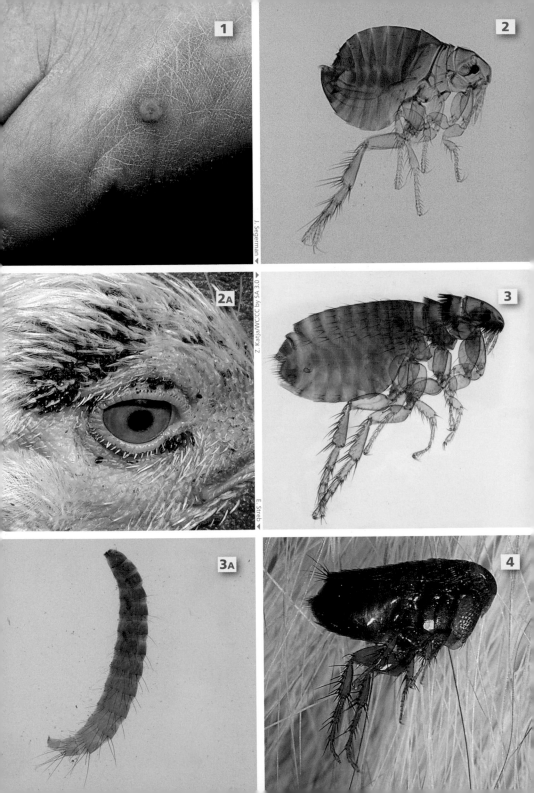

ORDER TRICHOPTERA • CADDISFLIES

Small to medium-sized, moth-like insects, with hairy wings that lie roof-like over the body when at rest. The antennae are very thin, often longer than the body, and held forwards. Adults have reduced mouthparts.

Adults of this small order are generally grey or dull brown, occasionally with markings, the fore wings being hairier and narrower than the hind wings. They are short-lived and not often encountered, hiding in riverside vegetation during the day and flying weakly at dawn or dusk. The larvae are much longer lived and are common in unpolluted streams, rivers and dams. Caddis larvae generally build elaborate cases, some free-living species spinning nets to capture food, or as shelters. Others have portable silk cases, ornamented with sand grains or plant matter arranged to form a miniature mosaic. A third group of fat, grub-like larvae construct food-catching nets, but are not free-living. All larvae have well-developed thoracic legs and a pair of extra prolegs at the end of the abdomen, which hook the larva into its case. The form of the larval case is often a useful guide to identification, although some occasionally colonize the discarded cases of other species. Larvae are often carnivorous. Pupation occurs either in a modified larval case, or a newly constructed case. The very active pupa has free legs, and uses its mandibles to cut its way out of the pupal case. Adults of the 19 families and 230 species that occur in the region are difficult to identify, the mouthparts and leg spurs being used as distinguishing features.

Family Hydropsychidae
Web-spinning caddisflies

Medium-sized to large (wingspan 44–56mm), with clear, pale or hairy, brownish wings marked with brown, yellow or grey, with a spur formula (number of spurs on fore, mid and hind legs) of 2, 4, 4 and a swollen basal antennal joint. There are 13 genera and 23 regional species.

1 *Cheumatopsyche*
Adults medium-sized (wingspan 45mm) **(1)**, often with an orange body and dark grey wings. Larvae are green, with either notched or toothed front margin to the head capsule. Small hairs on the head taper at their tips. The largest genus in the family, with 13 regional species. **Biology** Larvae construct shelters decorated with sand particles and small pebbles, attached to nets that filter food out of the current. **Habitat** Fast-flowing mountain streams and rivers with stony beds.

2 *Hydropsyche*
Adults **(2)** medium-sized to large (wingspan 50mm), with mottled wings. The green larvae **(2A)** are strongly built, with tufted abdominal and thoracic gills and large mandibles. There are 2 regional species. **Biology** Larvae glue stout shelters to rocks and incorporate small stones and a silk net that traps plankton. Adults abundant in summer. **Habitat** Widespread in fast-flowing streams. Probably the most common caddisflies in the region.

Family Dipseudopsidae
Burrowing (large-collared) caddisflies

Larvae very long (up to 40mm), with large heads and maxillae, small, flattened legs armed with brushes, and very large anal prolegs and gills at the end of the body. Under magnification, the enlarged rostral processes of the larval maxillae are visible. There are 3 species in the region.

3 *Dipseudopsis*
Adults medium-sized to large (wingspan 25mm), brown or grey. Pronotum is conspicuous, large and wide (collar-like), without warts. There are 3 regional species in the genus. **Biology** Larvae construct U-shaped tubes of silk with an internal net to filter out suspended food particles. Larvae can undulate the body to generate current through the tube when water flow is minimal. **Habitat** Slow-flowing or still waters.

1 Family Ecnomidae — Tube case net-spinning caddisflies

Adults small (wingspan 5–12mm), inconspicuous, silvery grey to yellowish-brown, with light oval markings on the wings **(1)**. Larvae small to medium-sized, with all 3 thoracic shields hardened. **Biology** Larvae inhabit an ill-defined, fixed gallery of silk covered in debris. **Habitat** Found in slow-flowing or still waters. About 23 species in the region.

2 Family Leptoceridae — Long-horned caddisflies

Adults small to large, slender, with very long (often pale) antennae and narrow fore wings; the hind legs carry 2 spurs, and the thorax of adults has 2 faint longitudinal lines running along it. In larvae the antennae are sufficiently large to be visible under a hand lens. The head is usually a long rectangle and the hind legs are often fringed with swimming hairs. They live in portable cases of sand, cut leaves or twigs, neatly cemented together or made entirely of silk **(2)**. The appearance of the cases is often species-specific. At least 120 species known from the region.

3 *Parasetodes maguira*

Adults small (wingspan 15mm), narrow-winged, cream with white antennae. **Biology** Larvae construct straight cases made up of long strips of vegetation sewn together with silk produced by the labial glands. **Habitat** Slow-flowing rivers and standing bodies of water. **Related species** There are 2 other regional species in the genus.

4 *Athripsodes*

Larvae are long, lacking abdominal gills and live in neat cases of cemented sand granules **(4)** or in very long cases made of fine longitudinal strips of vegetation; the cases usually have a 'hood'. **Biology** Larvae with long cases swim well, rather jerkily. Omnivorous. Very sensitive to pollution. **Habitat** Unpolluted waters, streams and ponds. Most diverse in the Western Cape. **Related species** *Leptocerina* adults **(4A)** are small (wingspan 20mm), dark brown, with lighter markings, and occur in the eastern subtropical parts of the region.

Family Philopotamidae — Finger-net caddisflies

The small, dark adults have flat heads, simple eyes between larger compound eyes, and sub-oval wings. Larvae have narrow heads, with a broad, white (not hardened and dark) upper lip. They also have thin legs and lack tracheal gills on the abdomen. There are 15 known species in the region.

5 *Chimarra ambulans*

Adults small (wingspan 20mm) and dark, with a distinctive white spot on each fore wing. Larvae have a distinctive asymmetrical notch at the front of the head capsule. Large sensory palps are associated with the reduced mouthparts **(5)**, as in all caddisflies. **Biology** The bright yellow larvae **(5A)** glue narrow, fine-meshed, elongate silk nets to rocks, which are used to filter minute diatoms from the water; adults scuttle rapidly across rocks during the day. **Habitat** Fast-flowing mountain streams.

6 Family Glossosomatidae — Saddle (tortoiseshell) case-making caddisflies

Adults small (wingspan 6–8mm), dark, with a spur count of 2, 4, 4 (on fore, mid and hind legs). There are 2 species of *Agapetes* in the region. **Biology** The larvae build characteristic oval cases of sand grains in the shape of a tortoise's shell **(6)**; they move slowly, feeding on algae. Adults emerge simultaneously in large numbers in summer and autumn and are very active, scurrying over rocks and occasionally taking short flights. **Habitat** Occur in fast-flowing mountain streams; larvae are found under rocks.

7 Family Hydrosalpingidae — Golden-tusk case-making caddisflies

Family comprises a single species, *Hydrosalpinx sericea*, adults of which are medium-sized (wingspan 20–30mm), golden brown, with hairy wings and very long maxillary palps. The larvae live in tubes of golden-brown silk, slightly flared at the opening **(7)**. **Biology** Larval cases may be found among underwater plants in rapids. Empty tubes occur in aggregations, persist for a while, and may be adopted further downstream by leptocerid larvae. Appear to have suffered a decline through trout predation. **Habitat** Family is endemic to the southwestern Cape and found only in pristine mountain streams.

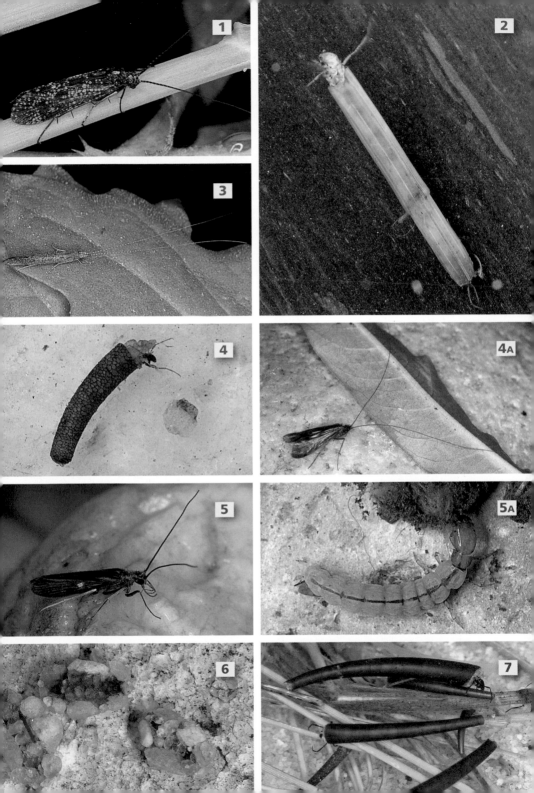

1 Family Barbarochthonidae — Barbarian tusk case-making caddisflies

Adults are small (wingspan 10–15mm) with a dark brown head and pale brown wings. The larvae construct characteristically curved, tapering cases, comprising concentric bands of very fine sand particles embedded in silk. **Biology** The single species in this endemic family, *Barbarochthon brunneum*, is common in streams, where the larvae feed on aquatic mosses and submerged *Scirpus* reeds. The head with its honeycomb markings and ridge on upperside is diagnostic. **Habitat** Reeds and mosses in pristine, fast-flowing streams.

Family Petrothrincidae — Limpet shell case-making caddisflies

Adults are small, with white-flecked wings; the underside of the abdomen is reddish, and the hind wings have a long fringe of hairs. Larvae are flattened, and construct flat, limpet-shaped cases of fine sand granules. Occur in slow-flowing pools in mountain streams, where the larvae browse on diatoms. The single endemic genus contains 3 species.

2 *Petrothrincus circularis*

Adults small (wingspan 11–15mm), grey-brown to silvery, with attractive white patterns on the short, triangular wings and long, hairy palps. The larvae are pale, and show little hardening of the thorax. They live in oval, shield-shaped cases made of fine sand grains incorporating a bottom plate **(2A)**. **Biology** Larvae feed on diatoms and other algal mats on rocks. **Habitat** Common under stones in fast-flowing mountain streams, especially in slow-flowing backwaters. *P. circularis* is the most common species in the Western Cape; its larvae construct almost circular cases.

Family Sericostomatidae — Brushtail case-making caddisflies

Adults generally have an orange abdomen and a fine felt of hairs on the brown wings. The larvae have square to round head capsules, very short antennae, and a projecting spine from either side of the pronotum. Most readily recognized by their larval cases, which are tubular and slightly curved, constructed of sand granules embedded neatly and uniformly in the silk tube. Five genera and 12 species in the region, most occurring in the Western Cape.

3 *Rhoizema saxiferum*

Adults medium-sized (wingspan 35mm), with an orange body and unpatterned, pale brown wings **(3)**. The larvae have a dark brown head covered in fine hairs and may have a few abdominal gills. There are 12 species in the region. **Biology** Larvae make perfect curved, tubular, golden cases of neatly arranged sand grains **(3A)**. The mouth of the case flares slightly. The larvae often burrow in fine sediments and apparently feed on dead plant matter. Males may have scent glands on their antennae, as well as scent pockets on the face. **Habitat** Larvae occur in the slow-flowing sandy stretches of mountain streams or on marginal vegetation.

Family Pisuliidae — Pisuliid triangular case-making caddisflies

Adults have oval, very hairy wings that may be covered in scales. The first antennal segment is longer than the head, very hairy and, occasionally, covered in scales. Larvae have a diagnostic D-shaped hardened plate on top of the third abdominal segment. There are 4 *Silvatares* and 2 *Pisulia* species in the region.

4 *Silvatares*

Adults large (wingspan 20–40mm), light brown, grey or black, with wings attractively marked with light brown patches. The larvae have abdominal and anal gills. **Biology** Larvae feed on fallen leaves. Their cases are large (cylindrical, square or triangular in section) **(4)**, made using cut leaves, bark or mosses. **Habitat** Slow-flowing, shaded streams or pools.

ORDER LEPIDOPTERA •
MOTHS & BUTTERFLIES

Minute to very large (3–180mm) insects with elongate antennae of variable form, and 2 pairs of large membranous wings, the entire body and wings covered in scales (flattened hairs). Mouthparts typically elongated to form a long siphoning proboscis, held coiled under the head when not in use; a few do not feed and have non-functional mouthparts, while small primitive moths have mandibles for feeding on pollen.

Probably the best known and most familiar order of insects. There is no simple distinction between butterflies and moths. However, moths have a distinctive wing-coupling apparatus, are mostly nocturnal, and their pupae are often covered in an elaborate silk cocoon. Only burnet moths (Zygaenidae) have the clubbed antennae characteristic of butterflies. Butterfly pupae are not protected by a cocoon. The larvae (caterpillars) of both groups usually feed on live plant parts, but a few are predaceous. The dramatic colours of butterflies and moths are produced by thin scales that cover the body and wings: pigments within the scales produce the white, red, yellow and orange colours; brighter metallic and iridescent colours are produced by fine grooves and ridges on the scales, which diffract light. These brighter colours do not fade in dead specimens. In males of some species, certain scales have been modified into long scent hairs (androconia) that release airborne chemicals used in courtship. In addition to the normal 3 pairs of thoracic legs, larvae have additional prolegs on their abdominal segments, with a ring of hooks at the base of each proleg to ensure safe attachment to the smooth, waxy surface of leaves.

MOTHS

A few species have clear wings or wingless females. Small moths can be confused with caddisflies (Trichoptera) (p.362), but caddisflies have hairs instead of scales on the wings, and they lack a proboscis. Moth antennae may be thread-like or comb-like. The wings are typically folded roof-like over the abdomen when at rest. The larvae of most species feed on plants. However, many feed on dry organic matter such as hair or wool (keratin), and a few are predaceous. With such a range of feeding habits, it is not surprising that some moths are of economic significance.

Family Hepialidae   Swift moths, ghost moths

Medium-sized to very large moths found mainly in Australia and South Africa. Easily recognized by the combination of very short antennae, wings of equal size, and a longish lobe (jugum) at the base of each fore wing, which overlaps the hind wing and couples the wings in flight. The larvae of a few genera (e.g. *Gorgopis* and *Eudalaca*) feed on grass roots underground and may become pests in lawns. The females scatter their eggs in low flight. There are 66 species known from the region.

1 *Leto venus*   Silver-spotted Ghost Moth, Keurboom Moth

Very large (wingspan 120mm), spectacular, with distinctive, silver-spotted, maroon fore wings, and pink body and hind wings. Females (wingspan 150mm) are among the largest moths in the region. **Biology** Sawdust appears at the mouth of tunnels that the larvae bore into the trunk of a Keurboom (*Virgilia oroboides*) **(1A)** about 1m from the ground. Larval development apparently extends over a few years. Adults emerge in late February or March. **Habitat** Temperate forests in the southern Cape.

2 *Eudalaca exul*   Brown Swift Moth, Burrowing Lawn Caterpillar

Medium-sized (wingspan 50mm), with brown wings. Fore wings have a light brown central triangle. Distinguishable from similar species by the absence of an oblique white band across the fore wings. **Biology** Larvae feed on grass roots. **Habitat** A wide range of vegetation types, including grassland and bushveld.

Family Adelidae
Longhorn moths

Small moths recognized by generally very long antennae in males (some species have short antennae). In both sexes antennae are thicker at the base. Head covered in rough scales. Eyes large. Narrow fore wings generally drab grey or whitish, although some have bright metallic colours. Many diurnal, but usually attracted to lights. Larvae of some species form meandering scribble-like mines in leaves, which increase in diameter as the larva grows. Others make cases like those of small bagworms. At least 68 species known from the region.

1 *Ceromitia turpisella*
Dusted Longhorn

Small (wingspan 15mm), with exceedingly long antennae, white body and wings, and 2 distinct oblique bands on the fore wings. A third band (usually broken up) may be present in subtropical specimens. **Biology** Adults feed on flowers and are attracted to lights. Larvae recorded from Black Wattle (*Acacia mearnsii*). **Habitat** Occurs in a range of vegetation types, but is absent from arid parts of the region.

2 Family Cossidae
Goat moths, carpenter moths

Medium-sized to large moths, often very bulky. Grey, brown or white wings speckled with black; fore wings long and narrow. Their short, curled antennae are often branched (comb-like) for about half their length, and they lack a proboscis. Larvae are cream-coloured grubs, elongate, cylindrical and stout, with a large sclerotized shield over the prothorax. They bore into the heartwood or roots of trees of various types, attacking growing wood. Some species bore into reeds, and a few construct cases like bagworms do. Larval development extends over years, the nutritive content of the host tree determining the size of the adults. The long, cylindrical pupae have rows of spines on the abdomen and protrude from the larval tunnel **(2)**. The name 'goat moth' is derived from the strong smell of the larvae of certain species. About 100 species are known from the region.

3 *Coryphodema tristis*
Sad Goat, Apple-trunk Borer

Medium-sized (wingspan 38mm), with a pale brown thorax and fore wings, the latter with grey and white markings and thin black lines over the wing veins. Hind wings translucent brown. Easily confused with other species in the genus. **Biology** Larvae bore into the trunks of a very wide range of unrelated shrubs and trees, including bush willow (*Combretum*), apple and quince. **Habitat** Bushveld and forest.

4 *Macrocossus toluminus*
Giant Goat

Large (wingspan 110mm), very bulky, with black markings on the thorax. Fore wings mottled in grey and brown, with a chestnut-coloured inner patch. Hind wings dark brown. **Biology** Larval foodplant is Sweet Thorn (*Vachellia karroo*). **Habitat** Forest and open woodland; its range includes much of sub-Saharan Africa.

5 *Azygophleps inclusa*
Leopard Goat

Medium-sized to large (wingspan 26–80mm); highly variable in size. Fore wings grey, with a cream or white longitudinal stripe and a very narrow grey area near margin. Hind wings white. May be confused with similar species in the genus. **Biology** Typical borer larvae. Foodplant unknown; Asian species in the genus bore into the stems of Corkwood (*Sesbania grandiflora*) and *Indigophora* trees. **Habitat** Common in bushveld and subtropical forest.

6 *Phragmataecia irrorata*
Brindled Goat

Medium-sized (wingspan 25–50mm) with an unusually long body (especially in females). The fairly long antennae have branches on either side and narrow sharply towards the tips. The fore wings are very pale brown, with faint, barely discernible markings; hind wings white or grey. **Biology** The genus has a cosmopolitan distribution. Foodplant in South Africa unknown; in Europe, the larvae feed destructively in the leaf tips and, later, stems of the Common Reed (*Phragmites australis*). They also attack Sugar Cane (*Saccharum officinarum*) and sorghum. In these species larval development is complete in approximately 2 years. **Habitat** Open woodland and savanna.

Family Metarbelidae
Tropical carpenterworm moths

Small to medium-sized moths, considered by some to be part of the goat moth family (Cossidae). Head small, legs and comb-like antennae short. Body squat with rounded white, pale grey or brown wings. Adults probably fly at dusk. Approximately 40 regional species.

1 *Salagena obsolescens*
Obsolescent Kid, Bark Borer

Medium-sized (wingspan 30–45mm), body and broad fore wings pale white, the latter covered in numerous brown ocelli circled with slate blue scales (some specimens are darker brown). Thorax with unusual tufts of large, slate-blue-and-orange scales, which are also present at the end of the abdomen. **Biology** The larvae are commercially important, boring into and damaging a range of fruit trees such as guava, litchi, mango, and avocado, as well as *Eugenia*, bush willow (*Combretum*), jacket plum (*Pappea*) and water berry (*Syzygium*). Eggs are laid on tree bark. The somewhat hairy brown caterpillars construct superficial feeding tunnels on the bark, consisting of webbing bound to chewed bark. Larger (30mm long) caterpillars then bore beneath the bark, constructing large tunnels, ring-barking trees and causing die-back of branches. **Habitat** Forest, woodland and orchards.

2 *Salagena tessellata*
Tesselate Goat

Small (wingspan 20mm), stocky and bizarre, with long, white hairs on the legs and body and at the end of the abdomen. Wings white, boldly patterned with small orange and black stripes. **Biology** Adults very active and attracted to lights. In Kenya, a species of *Salagena* has been known to destroy mangrove trees (*Sonneratia*). **Habitat** Bushveld and subtropical forest.

Family Tortricidae
Leaf rollers

Small and distinctive moths. Wings very broad, almost rectangular, typically dull brown, ending in short fringes of hair. Resting posture is highly characteristic, see *Lozotaenia* below **(3)**. Larvae are cryptic and generally feed in the seclusion of a tube, which they construct by rolling a leaf and binding it with silk. The larvae of other species are more diverse in their feeding habits, ranging from leaf miners to gall formers. The pupae project from the larval shelter. There are 40 species known from the region.

3 *Lozotaenia*
Leaf rollers

Small (wingspan 22mm), with fine, net-like, brown markings on the sandy-coloured wings **(3)** (moth to left of pupa), which overlap when at rest to form a squarish edge. The larvae **(3A)** are small, worm-like and black, with rows of white dots. **Biology** Adults fly at dusk with rapid, erratic flight. The larvae of this species of *Lozotaenia* feed in leaf tubes constructed from buds of the Tickberry (*Chrysanthemoides monilifera*), but as they mature they begin feeding on older leaves. Introduced (unsuccessfully) for biological control in Australia, where the Tickberry is an invasive alien. **Habitat** Fynbos and strandveld.

4 *Cydia pomonella*
Apple-Codling Moth, Codling Moth

Small (wingspan 16mm), brownish or grey, with square-tipped fore wings. Body grey or pinkish-brown, with fine white speckling and a dark brown spot, half encircled by a metallic coppery ring, on the wing tip; hind wings brownish-white, with a cream marginal line. Larvae pink and worm-like. **Biology** A serious pest of deciduous fruit. *C. pomonella* females lay 40–60 eggs at sundown on apple leaves or young fruit. After 10 days, the larvae enter the fruit, and leave 4–6 weeks later to pupate in bark or earth. Adults are active from September to December. **Habitat** Cultivated land.

5 *Cryptophlebia peltastica*
Litchi Moth

Small (wingspan 16mm). Fore wings grey, each with a brown leading edge and tip, a single dark triangular patch on the inner margin, and folds towards the tip. Hind wings dirty grey. **Biology** Feeds on a very wide range of plant families; larvae bore into the fruit, feeding on the seed. Also found in fungal galls on Sweet Thorn (*Vachellia karroo*) and Port Jackson Willow (*Acacia saligna*). Damages litchi fruit, macadamia nuts, *Senegalia* and *Bauhinia* seed. Of commercial importance on litchi. **Habitat** Common across tropical Africa and on the Indian Ocean islands.

Family Somabrachyidae
Flannel moths

A small family of unusual-looking African moths, with very hairy bodies and broad, round wings, which may be translucent or lightly covered in black scales. Males fly by day; females are wingless. Somabrachyidae includes 2 genera and approximately 6 species in the region, which occur coastally from the Western Cape to KwaZulu-Natal.

1 *Psycharium pellucens*
Clear-winged Flannel Moth

Medium-sized (wingspan 30–40mm), with the entire body clothed in long, dense white hairs. Wings clear, with yellow veins and small, sparse black hairs instead of scales. **Biology** Larvae feed on a wide range of plants, including various restios (*Restio triticeus* and *Wildenowia*), *Erica* and the invasive Monterey Pine (*Pinus radiata*) and Silky Hakea (*Hakea sericea*). **Habitat** Adults rarely seen. The slow-moving larvae **(1A)** are found on the ground in pine plantations after winter storms. Occurs infrequently in the Western Cape, on the Cape Peninsula and in the Stellenbosch area.

Family Brachodidae
Little bear moths

Small to medium-sized, with the hind wings shorter and broader than the fore wings, which have pointed or squared-off tips. Antennae thickened and may have comb-like bristles. Adults active by day. There are approximately 8, similar-looking species in the region.

2 *Phycodes punctata*
Leaden Grey

Small (wingspan 22mm) and sturdily built. Body lead grey, with white stripes on the abdomen. Fore wings lead grey, with black dots and squared-off tips; hind wings dark brown with a white fringe. Large plate-like scales on the mouthparts and thorax. The larvae have abdominal spines that protrude from the cocoon. **Biology** Adults feed on pollen and are active and alert during the day. Larvae feed on grass or grass seeds, covering themselves in a silken web. **Habitat** Grassland.

Family Scythrididae
Flower moths

Small, colourful moths with long fringes at the wing tips. Larvae conceal themselves under webbing and skeletonize the leaves, buds and flowers of a range of host plants. Only a few species are known from the region, although around 50 are known from across Africa.

3 *Eretmocera*
Small (wingspan 12mm), with thin black fore wings and very narrow orange or red hind wings with long fringes. The antennae have whip-like tips, and the basal two-thirds are thickened. Larvae long and slender. **Biology** Diurnal. The spurred hind legs are raised in characteristic position when at rest. Larval foodplant unknown; larvae of foreign species of *Scythrididae* consume a variety of plants, but chiefly grasses, Asteraceae, Amaranthaceae and legumes. **Habitat** Common in grassland, weedy patches and subtropical bush, but absent from arid parts.

4 Family Psychidae
Bagworms

Males very stout, dull-coloured, with transparent (often short, strong) wings that lack scales. They have comb-like antennae and no mouthparts **(4)**. Females are wingless, frequently eyeless, legless, lack antennae, and remain in their larval cases. Newly hatched larvae use vegetable debris, leaves or thorns to construct portable silk-lined cases, which are often very diagnostic. Both males and females pupate within the case. A few species are parthenogenetic, males not being required for fertilization of eggs. In most species males are nocturnal, but the fragile males of genus *Monda* fly diurnally. About 134 species known from the region.

5 *Chaliopsis junodi*
Wattle Bagworm

Larvae (shown) make characteristic pear-shaped bags (43mm long) covered with thin twigs and leaves. Adult females yellow, grub-like, with black eyes and no appendages, and never leave the larval bag. Adult males have pectinate antennae, hairy black bodies and clear grey wings with hairy veins. **Biology** Adults rarely encountered, but larvae are common. Lacking mouthparts, adult males are short-lived, flying rapidly at night from August to October, searching for females; are also attracted to lights. Mated females turn around in the case, lay eggs within it and die. It is a pest of Australian Black Wattle (*Acacia mearnsii*); also feeds on various species of *Senegalia*. Parasitized by the Inchneumonid wasp (*Sericopimpla sericata*). **Habitat** Bushveld and wattle plantations.

1 *Eumeta cervina* — Common Bagworm

Larvae make characteristic large cases (up to 80mm long) of thin twigs or thorns bound together with very strong silk. Males stocky, with large, feathery antennae, and short, translucent grey-and-brown wings. **Biology** Males nocturnal; as they grow, the case is enlarged. Feeds on a very wide range of trees and shrubs, including Pomegranate (*Punica granatum*), Australian Black Wattle (*Acacia mearnsii*) and Sugar Bush (*Protea caffra*) in summer-rainfall regions. **Habitat** Various natural vegetation types.

Family Tineidae — Clothes moths

Small, often yellow or white. Wings long and narrow, with a golden or silvery sheen and long fringes of hair. Proboscis reduced or absent. In nature, they breed in the nests of birds or small mammals, and even in owl pellets. Larvae are unusual in that they can digest the protein in hair and wool (keratin). Many species have become pests, feeding on clothing or carpets containing wool. At least 150 regional species.

2 *Ceratophaga vastella* — Horn Moth

Small (wingspan 25mm), elongate, with golden-yellow fore wings and grey hind wings, both fringed with yellow hairs, and a conspicuous tuft of yellow hairs on the head. Larvae fairly squat, cream-coloured, with the head and tip of the abdomen brown. **Biology** Larvae tunnel into and feed on the keratin in dead antelope, rhino and cattle horns, or on fruit and fungi. Pupal tunnels form characteristic branched masses **(2A)**, covered by faecal pellets (frass). **Habitat** On the horns of dead antelope, rhinos and cattle in warmer summer-rainfall parts, agricultural land and most veld types. **Related species** There are 11 other similar-looking regional species of *Ceratophaga*.

3 *Tinea pellionella* — Case-making Clothes Moth

Small (wingspan 10mm), with buff-coloured fore wings marked with 3 black spots. **Biology** *T. pellionella* larvae avoid light and live in a flattened silken case **(3A)**, with fore and rear openings for the head and feet. They take between 5 months and 2 years to pupate, and damage a range of fabrics, clothes and upholstery. **Habitat** Ubiquitous in houses. An alien species from Europe, now widespread across the world.

Family Sesiidae — Clear-wing moths

Easily distinguished, as their wings lack scales and are transparent, with a black, brown, yellow or orange border. Body often banded, and legs may be heavily cloaked in hairs. Abdomen is often slender, with a fan of long hairs at its tip. Excellent mimics of wasps and bees, day-flying but rarely encountered as they fly very rapidly and are never common. Larvae are borers, feeding on various trees and shrubs, and, in temperate regions, may attain pest proportions in orchards. Sesiidae probably includes far more than the 90 species known from the region.

4 *Chamanthedon* — White-banded clearwings

Small (wingspan 14mm), wasp-like, with a blackish-blue body and wings, and white bands behind the head and on the abdomen. The abdomen ends in a brush of white hairs. Wings clear, except for black veins. **Biology** Very active and alert day-flying mimics of small wasps. **Habitat** Rocky hilltops in bushveld.

5 *Synanthedon platyuriformis* — Dark Clearwing

Small (wingspan 20mm), wasp-like, with short wings. Body brown, with 4 white bands on the abdomen. Fore wings partially clear, with brown veins, a characteristic orange inner border and a circular area of orange veins. **Biology** Larvae bore into the flowerheads of *Protea* and *Leucadendron* and possibly also Common Mimetes (*Mimetes cucullatus*), causing damage to the seeds. Adults occur on ball-like, white flowers of *Berzelia* and may display aggressively when disturbed. **Habitat** Fynbos.

Family Yponomeutidae — Ermine moths

Medium-sized, usually lead grey or white, with black dots on the wings and long scales on the head. Wings long and often fringed with hairs. The larvae of most species skeletonize leaves or enclose them in webbing; others are borers or leaf miners. In Europe, the speckled larvae live gregariously in thin webs, feeding largely on Celastraceae and Boraginaceae. There are over 40 regional species.

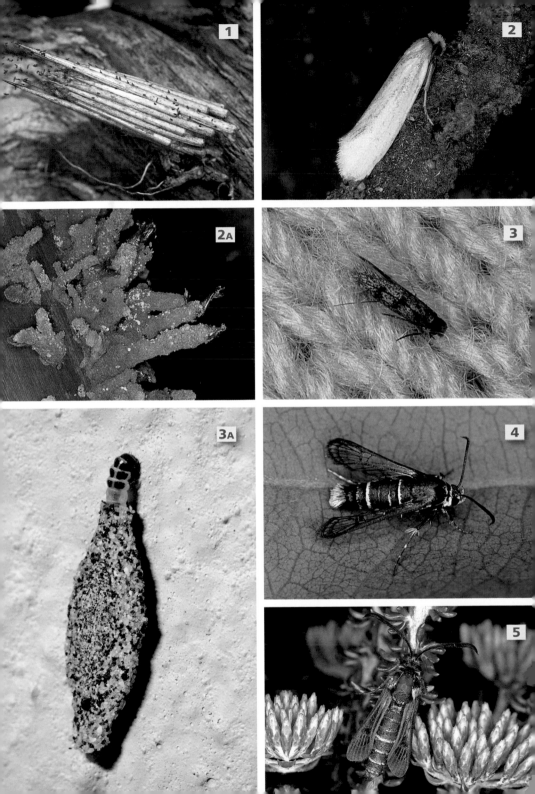

1 *Yponomeuta fumigatus* — Speckled Ermine

Small (wingspan 20mm), elongate, silver, with a sprinkling of black spots on body and fore wings. Hind wings dark grey. **Biology** Larvae feed in groups under silk tents that may cover entire trees **(1A)**. The larvae of all South African *Yponomeuta* species feed on Common Spike Thorn (*Gymnosporia buxifolia*) and *Cassine*. **Habitat** Fynbos to bushveld and subtropical forest. **Related species** About 10 other similar species in the genus with which it may be confused.

2 *Yponomeuta strigillatus* — Lesser Speckled Ermine

Small (wingspan 20–25mm) with a white body. Fore wings white, with rows of black spots running along their length; hind wings dark grey. **Biology** Unknown, but the majority of ermine moth larvae feed on 1 or more species of Celastraceae, and other South African *Yponomeuta* feed on Spike Thorn (*Gymnosporia buxifolia*). **Habitat** Open woodland and savanna.

Family Choreutidae — Metalmark moths

Larvae feed on daisy species (Asteraceae), including the Cape Daisy (*Arctotheca calendula*). The larvae are leaf miners, although older larvae may live freely on the underside of leaves. Family contains at least 12 regional species.

3 *Tebenna* — Metalmarks

Small, with narrow fore wings, often with metallic spots. Antennae thin and unbranched. **Biology** Day-flying moths, which hold the fore wings in an unusual upright position and move around in a jerky fashion – possibly to mimic jumping spiders, with the eyespots on the wings simulating the spiders' eyes. South African *Tebenna* species feed on fig trees, including Cape Fig (*Ficus sur*). **Habitat** Subtropical forest and bushveld.

Family Oecophoridae — Concealer moths

A large family of small moths. Although some are bright, they're more usually dull coloured, with rounded wing tips and a flattened appearance. Some fold their wings roof-like over the body. Many have a tuft of small hairs at the base of the antennae (as in some other families), which at rest are often held along the sides of the wings. Conspicuous curved labial palps near the proboscis are held upright. Females of some species are short-winged and flightless. Larvae construct and hide within various cases or tubes, hence the family's common name. They often bind leaves together with silk, but many construct portable cases and can tunnel into wood or flowers or live in galls. May enter homes and feed on fabric and stored foods. Over 150 regional species in the family.

4 *Schiffermuelleria pedicata*

Small (wingspan 28mm), with triangular orange wings marked in brown, white and yellow. **Biology** *S. pedicata* larvae feed on rust fungus galls on Sweet Thorn (*Vachellia karroo*). **Habitat** A variety of urban and natural habitats.

Family Xylorctidae — Grassland moths

A small family with only a few species in the region – considered by some to be part of Oecophoridae. Small to medium-sized, with hind wings broader than the fore wings. Palps are held upright. Antennae of males are simple or branched.

5 *Eupetochira xystopala* — Grassland Moth

Small (wingspan 26mm), with a yellow longitudinal stripe on the white fore wings, and conspicuous upright palps. **Biology** Usually flushed in the daytime and is highly camouflaged when it resettles on grass stems. **Habitat** Grassland.

Family Zygaenidae
Burnets, foresters

Small, easily recognized, usually with dark metallic blue or green fore wings (often banded in red) and clubbed antennae. Some have a brightly banded abdomen. A few species lack the metallic coloration and are yellow or orange. Larvae are short and plump. Adults generally found feeding on flowers. Larvae feed in an exposed position on *Gymnosporia*, *Leucadendron*, *Putterlickia* and *Ptelocelastrus*. Pupation takes place in a tough, yellowish cocoon attached to the host plant or to grass stems. Pupae are active and break through the cocoon to emerge as adults. About 100 species known from the region.

1 *Neurosymploca affinis*
Belted Burnet

Small (wingspan 26mm), with a grey thorax. Fore wings grey, with 4–5 red or whitish spots. Hind wings crimson, with a broad black border. The lime green larva are short and oval, covered in fine bristles. **Biology** Day-flying. Known to harbour cyanide, which is sequestered from alkaloids in larval foodplants. Fly slowly, but with rapid wingbeats, displaying their warning (aposomatic) colours as a sign to would-be predators that they are poisonous. *N. affinis* larvae feed on *Leucodendron*; others in the genus feed on Spike Thorn (*Gymnosporia buxifolia*) and Cape Saffron (*Cassine peragua*). **Habitat** *Neurosymploca* species occur in open areas with low vegetation.

Family Himantopteridae
Long-tailed burnet moths

A small family, formerly included in the Zygaenidae (above). All are small, slow-flying, diurnal moths with triangular fore wings and characteristically long hind wing tails. Most are brightly coloured in orange and black, with short, comb-like antennae. The slug-like larvae feed on the leaves of Dipterocarpaceae trees. They skeletonize the leaves and often mass in large numbers, migrating at the same time down the trunk of the host tree to pupate at its base. Occur in the moist subtropical parts of the region. There are 13 regional species.

2 *Semioptila fulveolans*
Streamer Tail

Small (wingspan 20–25mm), bizarre-looking moths. Fore and hind wings are bright orange at the base, becoming black, each bearing a large orange dot. Hind wings are elongate, broad at the base, narrowing about halfway along their length into twisted tails. **Biology** Slow, fluttering flight, especially on overcast days. Larval foodplant unknown, but a related species feeds on the tropical miombo woodland tree *Julbernardia globiflora*. **Habitat** Moist forest and woodland.

3 Family Limacodidae
Slug moths

The larvae and very squat, short-winged, hairy-bodied adults are easily recognized. Adults of many species are emerald green; others are red, brown or white. The wings are rounded. In males the antennae are usually comb-like. The larvae are often bright green, decorated with other colours. The body is short, plump and slug-like, with the head hidden **(3)**, and long antennae. The thoracic legs are reduced and prolegs are either absent or, sometimes, replaced by a broad, adhesive sole with ventral suckers. Larvae are often covered in stout stinging hairs grouped in clusters, which can cause severe pain. The hard, oval pupae, with a circular lid-like opening at 1 end, are also fairly diagnostic of the family. About 120 species are known from the region.

4 *Afrobirthama flaccidia*
Flaccid Slug

Small (wingspan 22mm), squat, with reddish-brown thorax and broad wings. Fore wings are divided into dark brown and grey halves. Hind wings cream, with a dark border. In males the antennae are comb-like only at the base. Female has unbranched antennae **(4)**. **Biology** Active and a rapid flier. **Habitat** Bushveld and subtropical forest.

5 *Parapluda invitabilis*
Netted Slug

Small (wingspan 20mm), with a white thorax and wings and an orange abdomen. Wings marked with a pattern of brown lines. Larvae are pale greenish-yellow, with blue stripes. **Biology** Possibly mimics the Reticulate Bagnet (*Anaphe reticulata*) (p.412). Larvae feed on Sweet Thorn (*Vachellia karroo*) and Hook Thorn (*Senegalia caffra*). **Habitat** Widepread in a range of habitats.

1 *Coenobasis amoena* — Rayed Slug

Small (wingspan 26 mm), squat, with emerald green wings and thorax and an orange abdomen. Fore wings have a straw-coloured branching stripe and spot. Hind wings whitish. Larvae short and rectangular, green and yellow, with small blue rings dorsally and along the sides, and short branched spines. **Biology** Larvae feed on wattle, *Senegalia* species and cotton; can sting severely. **Habitat** Bushveld and subtropical forest.

2 *Latoia latistriga* — Broad-banded Latoia, Plum Slug

Small (wingspan 30mm), short, squat, with an emerald green thorax. Fore wings emerald green, with a brown band at their tips; hind wings orange-yellow. Larvae green or yellowish-green, with spines. **Biology** Larvae feed on a wide variety of trees and shrubs including apricot, peach, plum, castor oil, False Spike Thorn (*Putterlickia*), *Celastrus* and macadamia. Related species feed on rose and privet bushes. **Habitat** A variety of urban and undisturbed habitats. **Related species** *Latoia* includes 11 other similar species.

3 *Ectropa ancilis* — Angular Slug

Small (wingspan 26mm) and squat. Wings short, very broad, with scalloped margins fringed with golden hairs. Body and wings goldish-cream with black flecks. Larvae are slug-like, green, with rows of long, white-tipped black hairs running along the body. **Biology** Have a strong, buzzing flight. **Habitat** Bushveld and open woodland.

4 *Chrysopoloma varia* — Spotted African Slug-caterpillar Moth

Small to medium-sized (wingspan 30–50mm). Wings and body pale orange. Fore wings with large, bright yellow scales, a broad fringe of large scales and a white central spot; hind wings greyish-white. Body short and stocky. Antennae black and comb-like. **Biology** Unknown, but related species, such as *C. bicolor*, feed on Common Spike Thorn (*Gymnosporia buxifolia*). **Habitat** A range of vegetation types. **Related species** There are approximately 9 other species in the genus, including *C. similis*, which is known to have expanded its range into the southern Cape recently, and which has a flattened larva with a short brush of hairs along the margins of the body.

Family Thyrididae — Lattice moths, picture-winged leaf moths

Most have broad wings with clear windows. Body and wings usually yellow, orange-yellow or brown. Wings, especially the clear areas, are covered in a network or lattice of lines and stripes. Larvae short, cylindrical and sparsely haired. The larvae of some species are stem borers, while others live in galls. A small family of about 60 regional species.

5 *Cecidothyris pexa* — Clouded Lattice

Small (wingspan 28mm), with a pale apricot body and wings; fore wings indented along the front margin. Clear windows on the hind wings are covered in a brown latticework. **Biology** *C. pexa* larvae feed on the Silver Cluster-leaf (*Terminalia sericea*). **Habitat** Sandy bushveld and subtropical forest, wherever host plant occurs.

6 *Arniocera auriguttata* — Gold-spotted Burnet

Small (wingspan 25–30mm), brilliantly coloured moths, with a red or orange head, metallic blue-green body and red legs. The orange spots on the velvety, royal blue wings may be circled with green scales. **Biology** Adults fly slowly by day and aggregate on moist mud, presumably to imbibe salts. As with other thyridids, adults probably toxic and may store hydrogen cyanide. Larvae feed on crossberry (*Grewia*). **Habitat** Dry open woodland and grassland.

Family Plutellidae — Diamond-back moths

Small, with long, narrow wings with rounded tips; light markings along the fore margins form diamond-shaped spots when the wings are folded at rest. Wing tips are upturned at rest, and their long fringes emphasize this effect. Larvae are mostly leaf miners. A small family with about 10 species in the region.

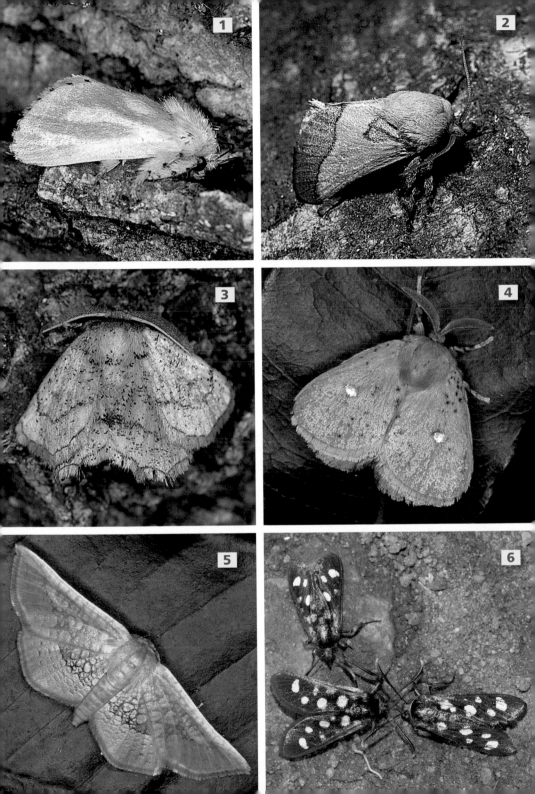

1 *Plutella xylostella* — Diamond-backed Moth

Small (wingspan 12mm), with brown wings and a broad, white-edged buff stripe along the centre of the body and wings. Larvae are green. Believed to have originated in the Mediterranean region and subsequently to have spread worldwide. **Biology** *P. xylostella* is a cosmopolitan pest of cabbage. Populations can be very large, but a variety of parasitoids control numbers to some extent. Adults emerge from winter hibernation to lay eggs beneath leaves, on the midrib. Larvae feed on or under the leaves, leaving characteristic transparent patches in the epidermis. The life cycle is completed in 14 days. **Habitat** Ubiquitous, wherever cabbages and related vegetables (Brassicaceae) are grown.

Family Pyralidae — Snout moths

Small to large moths, not all of which are easily recognizable in the field. The large, straight labial palps form a snout, a useful field character for the family. Generally delicately built, with narrow fore wings and broad hind wings, both with a short fringe of hairs. In many species the wings are satiny and translucent. The antennae are generally thread-like. Larvae are recognizable by their vigorous wriggling when disturbed. They feed on a wide variety of foodstuffs, including all plant parts, and often live in silken tubes and tunnels. A number are commercial and domestic pests. Pupation takes place in the larval shelter.

Subfamily Galleriinae

Adults may lack ocelli (simple eyes) and proboscis. Males produce high-piched chirping sounds, possibly used in courtship. The larvae and adults of many species have parasitic associations with bees and wasps.

2 *Eldana saccharina* — Eldana Borer, Sugar-cane Borer

Small (wingspan 35mm), with uniformly brown wings. Larvae brown, tough and leathery. **Biology** Adults live for a week, but only to mate, and do not feed. Eggs are laid on dry material and hatch into very active larvae that bore into the stems of cultivated sugar cane or, in the wild, on large sedges, such as *Cyperus*. Faecal pellets (frass) accumulate at bored holes. Cannot be controlled by insecticides and can cause severe crop losses. **Habitat** Cane fields and wetlands. Indigenous.

3 *Galleria mellonella* — Greater Wax Moth

Medium-sized (wingpan 30–45mm), wings dull grey and brown, appearing greasy. Larvae (**3A**) white and grub-like, with a brown head and brown sclerite (plate) on the thorax, and a row of small black dots along each side. **Biology** Female moths enter beehives and lay eggs on the comb, where the larvae hatch and feed destructively; of commercial concern. **Habitat** Originally from Eurasia, now cosmopolitan. In urban areas, in association with beehives.

Subfamily Pyralinae

Broad-winged and fairly nondescript moths, whose larvae feed on a wide range of foodstuffs.

4 *Zitha laminalis* — Laminated Pyrale

Small (wingspan 25mm), with a brown thorax. Fore wings a lustrous pinkish-buff, with brown at the base, brown zigzag lines and a brown spot. Hind wings cream. **Biology** Active larvae aggregate under rotten logs and construct cocoons covered in faecal pellets (frass). **Habitat** Forest.

Subfamily Phycitinae

The largest group of snout moths. Usually small and slender, with drab coloration. Males have a swollen basal antennal segment and a row or tuft of hairs or scales on the fore wings. Larvae of phycitids feed on a range of substrates, emerging from silken tunnels to feed at night. Some pest species feed on stored products (chocolate, tobacco, dried fruits, biscuits); others feed on buds. The famous jumping-bean moths have larvae that bore into hard seeds and wriggle if the seed is dropped, causing it to 'jump'. Most of the larvae are leaf rollers, but a few are predatory. Over 80 species are known from the region.

1 *Ephestia* — Dried fruit moths, chocolate moths

Small (wingspan 14mm), drab. Fore wings grey, with dark half bands halfway along their length. Hind wings clear with a grey tinge at their borders. Larvae are small cream maggots. There are 3 species in the region. **Biology** May be seen flying slowly indoors or in cupboards. Larvae found crawling in containers of dried foods. Larvae feed on a wide range of stored food products, including chocolate, dried fruit, raisins, nuts, cornmeal and even seeds. **Habitat** A very common and widespread pest in areas of human habitation.

2 *Plodia interpunctella* — Indian Meal Moth

Small (wingspan 16mm). Creamish-white basal area of fore wings is separated from the reddish-brown tip by a black stripe. Hind wings are dirty white, with a thin brown marginal line. Larvae whitish, with a yellow-brown head. **Biology** Larvae are cosmopolitan pests of a wide variety of stored products, spinning webs over dried fruit, cereals, nuts, flour, chillies and even fur. **Habitat** Indoors, in domestic or agricultural products.

3 *Cactoblastis cactorum* — Prickly Pear Moth

Small (wingspan 30mm), elongate, silvery grey, with black flecks on the wings, a shadowy band halfway along each wing, and a row of small black spots near the wing tips. Narrow zigzag band near wing tip in most specimens **(3)** (see arrow). Palps long and projecting. Larvae bright orange with black spots. **Biology** Introduced from South America to control prickly pear (*Opuntia*). Females lay a characteristic curved 'egg stalk' made of a number of eggs glued end-to-end to a cactus spine **(3)**. Larvae **(3A)** bore into cactus pads, leaving a mushy, dying pad. **Habitat** Valley bushveld and most other arid vegetation types where prickly pears grow.

Family Crambidae — Crambid snout moths

Small to medium-sized and slender. Wings often translucent, satiny and finely patterned. Many are long-legged and delicate, and inhabit wet or flooded areas, their larvae being semi-aquatic borers of stems or leaves. One group spins a loose web around leaves, fruit or stems, sometimes damaging vegetable crops. Over 300 species are known from the region.

Subfamily Spilomelinae

A very large group with few diagnostic features. Wings broad, often patterned. Legs long. Larvae roll leaves or cover them in a silk sheet.

4 *Haritalodes polycymalis*

Small (wingspan 20–30mm). Wings satiny, pale yellow, and intricately marked with brown scribble-like markings; fore wings marked with 2 brown circles. **Biology** In tropical Africa it is an agricultural pest of kola trees (*Cola*). **Related species** The Cotton Leaf Roller (*H. derogata*), which feeds on a range of *Hibiscus* species and is a pest of cotton leaves. The thin green larvae roll leaves and live inside these tubes. Their feeding on the leaf margins causes the leaves to wilt, and the plant to form cotton bolls prematurely. Also feeds on tomato and aubergine leaves and occurs very widely across Africa and Asia.

5 *Aethaloessa* — Florid pearls

Small (wingspan 32mm) but spectacular. Body and wings orange, with purplish-brown bands. Hind wings deeper orange than fore wings. **Biology** Tropical species. **Habitat** Subtropical bushveld and forest. Also occurs in West Africa.

6 *Obtusipalpis pardalis* — Snow Leopard

Small (wingspan 18mm). Body and fore wings greenish or yellowish-brown, patterned with evenly spaced white spots, and with a fine dotted margin on the wing tips. Hind wings white with brown border. **Biology** Larvae feed on wild figs (*Ficus*). **Habitat** Subtropical forest and bushveld.

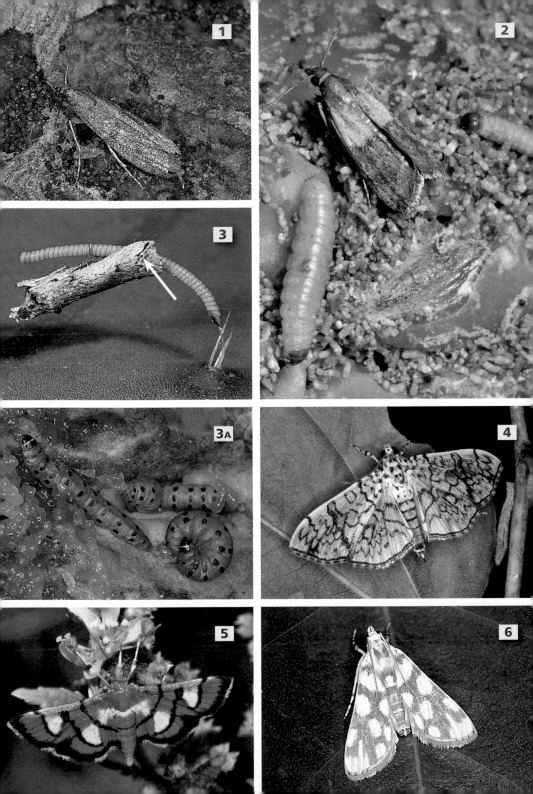

1 *Notarcha quaternalis* — Fourth Pearl

Small (wingspan 32mm). Body white, with yellow bands. Wings opaque white, with wavy orange lines. Each fore wing has a row of 3 black spots, with a fourth spot below it. **Biology** Adults frequent herbaceous vegetation beneath trees. Larvae feed on the herb *Sida* (Malvaceae) and on Rough-leaved Raisin (*Grewia flavescens*). **Habitat** Subtropical forest. **Related species** *N. obrinusalis* (wingspan 30mm), which is similar, but has a uniformly orange body. Its larvae have been recorded from maize.

2 *Palpita unionalis* — White Pearl

Small (wingspan 28mm). Body white. Wings clear, white, with a brown or gold line on the fore margins. (Most species in the genus have similar translucent, opalescent wings.) **Biology** Larvae feed on jasmine and Olive (*Olea europaea*). **Habitat** Widespread and common in a range of vegetation types and in gardens.

3 *Glyphodes bicoloralis* — Bi-coloured Pearl

Small (wingspan 20 mm), wings fringed with long white hairs. Fore wings marked with translucent white triangles; hind wings have a white basal band. **Biology** Commonly flushed from herbage beneath trees. Larvae feed on a wide range of foodplants, including wild figs (*Ficus*), oleander, *Carissa*, and tick clover (*Desmodium*). **Habitat** Natural and disturbed vegetation. **Related species** The female Indian Pearl or Cucumber Moth (*Diaphania indica*) (wingspan 30mm), which has dark-edged, pearly white wings that form a large white triangle when at rest. Females release pheromones by expanding a gland comprising flattened scales at the end of the abdomen (**3A**). This species is of Asian origin and now occurs widely across the Old World. It is an agricultural pest that feeds on members of the cucumber family and on cotton.

4 *Agathodes musivalis* — Painted Pearl

Medium-sized (wingspan 38mm). Fore wings pinkish, with a brown basal stripe, a purplish central stripe and, in the male (shown here), a hooked tip. Hind wings cream. **Biology** Larvae feed on the Coral Tree (*Erythrina caffra*) and on Wild Plum (*Harpephyllum caffrum*). **Habitat** Forest.

Subfamily Pyraustinae

A very large group of long-legged, broad-winged and brightly coloured moths. The larvae of many species are stem and fruit borers.

5 *Loxostege frustalis* — Karoo Moth

Small (wingspan 20mm), uniformly brown, with white markings, 1 marking edged with a brown zigzag. Hind wings brown, with thin brown line along the margins. Familiar, blackish-green larvae. **Biology** Very abundant, seen in large numbers on vegetation or feeding on pollen, often during the day. Numbers have increased with the change in vegetation from sweet grasses to perennial woody daisies, as a result of human activity. Larvae feed on stinkbush (*Pentzia*), reducing browse for cattle. **Habitat** Drier habitats, particularly succulent karoo and grassland.

Subfamily Crambinae — Grass moths

Very common in grassy areas, where they wrap their straw-coloured wings tightly around their bodies when at rest, head-upwards on a grass stalk. The wings often have longitudinal streaks, providing perfect camouflage in grass. Two horn-like palps (part of the mouthparts) protrude from the head. Larvae feed in silken galleries on grasses and reeds. A few (e.g. *Chilo*) are pests of cereal crops and sugar cane. There are 30 species known from the region.

6 *Ancylolomia perfasciata* — Grass Moth

Medium-sized (wingspan 38mm). Fore wings pale tan. The blunt wing tips have dark lines on the veins and a whitish band. **Biology** When flushed, adults settle on restio stems. **Habitat** Restio-dominated veld.

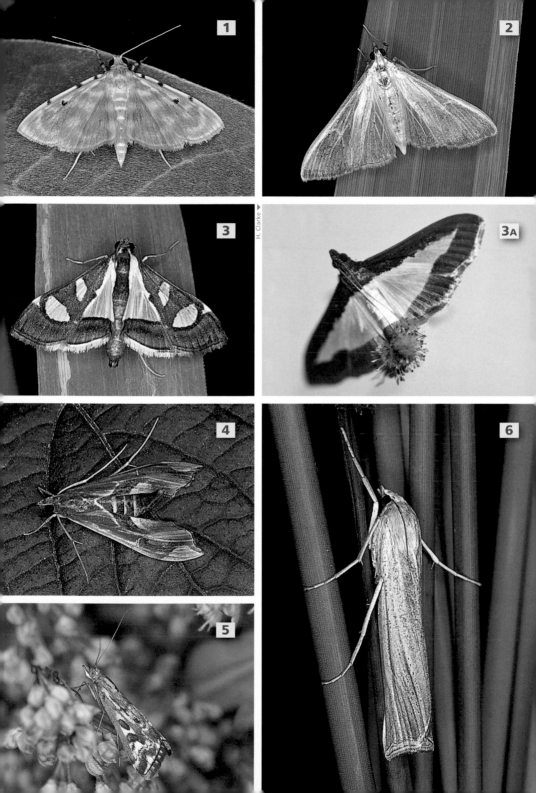

1 *Crambus sparsellus*   Dotted White Veneer

Small (wingspan 20mm), thin, chalk white, with black, grey and brown speckling on the fore wings, and creamish-white hind wings. The prominent palps are useful field characters. **Biology** Flushed adults settle on grass with tightly furled wings, mimicking grass stalks. Larvae make silk webbing on their foodplants (grass, moss and reeds). **Habitat** Grassland.

Family Pterophoridae   Plume moths

Small and delicate moths, very easily recognized by the division of the wings into plumes, with 2–4 lobes in the fore wing, and 3 lobes in the hind wing (if divided). They fly slowly and rest with the wings and prominently spined legs outspread. In some species the fore wings are rolled around the hind wings and held at 90° at rest, the insect resembling the letter 'T'. Larvae are long and covered in hairs, with long abdominal legs (prolegs). They feed on flowers and leaves, and occasionally on stems and seeds, usually those of daisies (Compositae). Over 70 species are known from the region.

2 *Agdistis*   Pustule plume moths

Small (wingspan 20mm), appearing delicate and mosquito-like at rest. The narrow grey wings are not divided and are marked with a few dark dots. Larvae green, with silver-white speckling and a spiky, plain green shield over the head and thorax. End of abdomen also with spines. **Biology** Larvae feed on *Exomis*. **Habitat** Open areas in a wide range of vegetation types.

3 *Stenodacma wahlbergi*   Orange Plume Moth

Small (wingspan 14mm), with orange body and wings. The wings are divided into plumes with black fringes. **Biology** Adults live in fields of weedy annuals and are common in rank vegetation under bushveld trees. Feed on a wide range of unrelated herbs, including *Oxalis* and *Ipomoea*. **Habitat** Grassland and open fields.

Family Alucitidae   Many-plumed moths

A small family of moths, easily recognized by their feather-like wings, which split into about 12 separate veins. Wings are generally held open like a fan. The larvae bore into buds, flowers and young shoots, occasionally causing galls.

4 *Alucita spicifera*   Many-plumed Moth

Small (wingspan 12mm), with brown-and-white-barred wings that are characteristically divided into feather-like plumes. Black stripe on thorax. Larvae grub-like, living in galls or unripe seed pods. **Biology** Day-flying, often seen feeding on nectar. **Habitat** Common on annuals in weedy fields. **Related species** Other species of Aluctidae, with which it may be confused; also, members of the related family Pterophoridae (above) look very similar, but have just 2–4 wing divisions.

Family Geometridae   Geometrid moths

Adults of this very large family are fairly easily recognized by the combination of a thin body, small thorax and very broad wings that are spread flat when at rest, although a few have narrow wings and thick bodies. They may be confused with some pyralids. Mostly small to medium-sized, weak fliers, and usually cryptically coloured, blending very well with tree bark. Many are shades of green or have wavy lines and patterns that enhance their camouflage. Larvae are twig-like and equally characteristic, with long, slender bodies, usually well camouflaged, and the first few pairs of extra abdominal legs (prolegs) reduced or absent. Larvae move in a very deliberate fashion, bringing the end of the body up against the front, so that it forms an upright loop. The front part of the body extends and a loop is formed again (hence their common name 'looper' or 'inchworm'). Over 1,000 species are known from the region.

5 *Piercia*   Carpet wings

Small moths (wingspan 16–20mm) with cryptically coloured fore wings, often in shades of pale to olive green, with grey bars. Some species have darker coloration, others are reddish-brown. Fore wings oval and intricately marked. **Biology** The many similar and variable species in the genus are all camouflaged to resemble lichen or moss on tree trunks and branches. **Habitat** Most vegetation types in undisturbed habitats.

1 *Callioratus abraxas*
Dimorphic Tiger

Large (wingspan 50–70mm), with two colour morphs, some of which resemble pierid butterflies (p.448). In females, wings orange, with broad black lines enclosing a slate grey central stripe; males largely white. Abdomen black. Larvae sparsely haired, fairly short and squat, with a white stripe running the length of the body, bordered on each side by a black and then broader orange region. **Biology** All species of *Callioratus* are obligate feeders on the leaves of *Encephalartos* cycads, and probably sequester the cycad toxins (macrozamins) to protect themselves from predation. Their bright coloration and slow daytime flight probably advertise their toxicity to potential bird predators. Males of *C. abraxas* are unusual among moths, as they aggregate in small groups (leks) in the treetops and release pheromones that attract females. **Related species** There are 6 other African species in the genus, with *C. millari* considered 1 of the region's most endangered moths.

2 *Chiasmia simplicilinea*
Oblique Peacock

Medium-sized (wingspan 32mm), with pale yellowish-brown wings held outspread at rest. Speckles on wings and stripe across both pairs, forming disruptive lines. Hind wings triangular. Larvae are green loopers with fine white lines and a cream lateral stripe with black central dots on each segment. **Biology** Larvae feed on wattle. Also occurs in the Karoo, where adults are known to suck lachrymal fluid from the eyes of sheep and cattle. **Habitat** Adults very common in most habitats. **Related species** The Variable Peacock (*C. brongusaria*) **(2A)**, which is medium-sized (wingspan 30mm) and similar in shape, but with white, cream, yellow or brownish-yellow wings, the straight line across the fore wings curving upwards at each end. Its larvae feed on *Vachellia* and it is common in most habitats in the region.

3 *Chiasmia rectistriaria*
Oblique Barred Semiothisa

Medium-sized (wingspan 35–40mm) with white wings covered in brown flecks. When wings are outspread at rest, fore and hind wings each show a continuous double stripe. Each wing bears a single small black dot. Short, oblique bars on leading edge of fore wings. **Biology** Adults commonly flushed from grass and rest on the ground; larvae feed on flat-crowns (*Albizia*). **Habitat** Open woodland and savanna.

4 *Pingasa distensaria*
Duster

Medium-sized (wingspan 40mm), pale grey or greenish-grey, speckled, with a characteristic resting posture. Wings marked with thin, black scalloped lines – double on the fore wings, single on the hind wings. Abdomen quite stout. Larvae are thick, pale grey-green loopers, with a lemon-coloured lateral stripe. **Biology** Larvae feed on cabbage trees (*Cussonia*), pepper trees (*Schinus*) and Natal plum (*Carissa*). **Habitat** Bushveld and subtropical forest. **Related species** *Mimandria cataractae* (wingspan 38mm), which is similar, but with orange spots on the hind wings.

5 *Zamarada plana*

Small (wingspan 26mm), with clear yellow or yellowish-green wings with brown borders (as in all members of the genus). **Biology** Larvae feed on Weeping Boer-bean (*Schotia brachypetala*). Larvae of other species in the genus feed on *Senegalia* and sickle bush (*Dichrostachys*). **Habitat** Common in bushveld.

6 *Erastria madecassaria*
Variable Scallop

Medium-sized (wingspan 38mm), with brown wings patterned in green, pale yellow or reddish brown. Hind wings with black dots and a notched margin. **Biology** Larvae feed on Buffalo Thorn (*Zizyphus*). **Habitat** Bushveld and subtropical forest.

7 *Rhadinomphax divincta*
Two-phase Emerald

Small (wingspan 30mm), with uniformly pale green fore wings. Hind wings white with a green marginal tinge. Wings folded in a steep roof-like position at rest, unusual for a member of family Geometridae. **Biology** Not known. **Habitat** Fynbos and subtropical forest.

1 *Comostolopis germana* — Tickberry Emerald

Small (wingspan 20mm), with bright green wings and a cream line along margins of fore wings. Pale whitish zigzag line across the wings in some species; in others, a row of white dots. **Biology** Has been deliberately introduced to Australia as a control agent for its primary foodplant Tickberry (*Chrysanthemoides monilifera*), which is a common weed accidentally introduced there from South Africa. **Habitat** Succulent karoo, bushveld and subtropical forest vegetation. **Related species** There are 2 other species in the genus.

2 *Heterorachis devocata* — Red-lined Emerald

Small (wingspan 22mm), with plain green wings. Rusty red line on margins of fore and hind wings, abdomen and head. **Biology** Larvae feed on wild medlar (*Canthium*) and rock alder (*Vangueria*). **Habitat** Common in subtropical parts, in bushveld and forest. **Related species** There are many other very similar species in the genus, some with wingspans of up to 48mm.

3 *Victoria fuscithorax* — Victoria Emerald

Medium-sized (wingspan 50mm), with stout brown body and legs. Thorax edged with green hairs. Wings white, each with an orange or red central dot, scalloped margins and olive green blotches and speckling. **Biology** Adults lichen mimics. Larvae feed on *Erianthium*, *Plicosepalus*, mistletoe (*Tapinanthus*) and *Tieghemia*. **Habitat** Temperate and subtropical forest.

4 *Drepanogynis cambogiaria* — Gamboge Thorn

Small (wingspan 32mm). Female (shown) has sulphur yellow fore wings, each with a brown dot and a thick brown marginal band; hind wings whitish. Male plainer, yellow or yellowish-green. **Biology** Larvae feed on Spike Thorn (*Gymnosporia*) and False Spike Thorn (*Putterlickia*). **Habitat** A range of vegetation types, often forest.

5 *Rhodometra sacraria* — Vestal

Small (wingspan 25mm), with triangular bleached yellow fore wings and white hind wings. Fore wings marked with a faded rust-red oblique line. Wings held at rest in a steep roof-like position. Larvae thin and pale greenish-white. **Biology** Larvae recorded in South Africa on sorrel (*Rumex*); in Europe larvae feed on *Polygonum*, *Searsia* and other plants. Adults often seen settling on Kikuyu Grass. **Habitat** Common in grassland.

6 *Drepanogynis villaria* — Pink Thorn

Medium-sized (wingspan 40mm), with rich pinkish-brown thorax. Fore wings divided by a sinuous cream line into a basal and a darker marginal zone. Hind wings pale pink. Black spot on each wing. **Biology** Larvae feed on 'gonna' (*Passerina*) and 'rooiels' (*Cunonia*). **Habitat** Fynbos, bushveld and subtropical forest. **Related species** *Palaeonyssia trisecta* (wingspan 31mm) **(6A)**, which has mottled brown wings divided into 3 colour bands. Feeds on Cork Thorn (*Vachellia davyii*).

7 *Rhodophthitus commaculata* — Pied Magpie

Medium-sized (wingspan 40mm), with white thorax. Fore wings with large grey blotches and small black dots on wing bases. Hind wings white, with a few black dots. Abdomen yellow with black markings, resembling that of a tiger moth. **Biology** Possibly a mimic of (distasteful) tiger moths. **Habitat** Subtropical forest and bushveld.

8 *Argyrophora trofonia* — Gold Lattice

Small (wingspan 24mm). Fore wings brown with bold silver markings. Hind wings cream. Wings held in steep roof-like posture at rest. **Biology** Larvae of other species of Argyrophora recorded feeding on Drakensberg Heath (*Erica drakensbergensis*). **Habitat** Unknown for this species, but others in this genus occur in a range of vegetation types, from fynbos to forest. **Related species** Silver Lines (*Pseudomaenas alcidata*) (wingspan 30mm), which has 2 long, unbroken silver stripes along the length of the fore wing.

1 *Xanthorhoe poseata* — Barred Carpet

Small (wingspan 22mm), variable, with broad wings. Fore wings intricately marbled in shades of brown on a background of green, yellow, cream or grey. Hind wings white with faint grey markings. Marbling of fore wings characteristic of the genus. **Biology** Larvae feed on lettuce. **Habitat** Diverse vegetation types. May also be found in disturbed habitats.

2 *Coenina poecilaria* — Wisp Wing

Medium-sized (wingspan 40mm), stout. Resting posture unmistakable, with fore wings furled and hind wings spread flat. Fore wings orange and buff with black-and-white markings; hind wings grey with bright orange border and whitish windows. **Biology** Larvae recorded on potato plants. **Habitat** Bushveld and forest.

3 *Xenimpia maculosata* — Ragged Thorn

Medium-sized (wingspan 36mm), with folded fore wings and ragged, incised hind wings in female (shown). Wings reddish-brown with black speckling. Like other males in this dimorphic genus, male has less dramatically notched wings. **Biology** Larvae may feed on Forest Elder (*Nuxia floribunda*). **Habitat** Widespread in forest and riverine vegetation. **Related species** *X. erosa* (wingspan 35mm), which has lines and pale markings on fore wings.

4 *Zerenopsis lepida* — Leopard Magpie

Medium-sized (wingspan 40mm); body orange, with black stripes on thorax, and rounded orange wings evenly covered in round black spots. Red larvae (**4A**), called Cycad Loopers, heavily marked with black longitudinal and transverse lines. **Biology** Larvae feed gregariously on cycads, *Apodytes*, *Carissa* and *Maesa*; can cause extensive damage to garden cycads. **Habitat** Forest, and Gauteng gardens.

5 *Eulycia grisea* — Grey Pennant

Medium-sized (wingspan 46mm), uniformly grey-brown. Very elongate fore wings have a thin black line demarcating a reddish outer band. Hind wings with wavy margins. **Biology** Resting posture of adults mimics a dead twig. Larvae are loopers and feed on bush willow (*Combretum*). **Habitat** Subtropical bushveld and forest. **Related species** *Omphalucha albosignata* (wingspan 38mm), which has shorter fore wings and is browner.

Family Bombycidae — Silk moths

Medium-sized moths similar to the domesticated Silk Moth *(Bombyx mori)*. Antennae comb-like in both sexes, body squat, and wings triangular, with a hook at the end of each fore wing. There are 4 similar African silk moth species.

6 *Racinoa ficicola* — Small Silk Moth

Medium-sized (wingspan 30mm), wings light yellow or reddish-brown, with fine wavy lines. Anal region of hind wings has a concave crescentic margin. Well-camouflaged larvae resemble domesticated Silkworms (*Bombyx mori*), and are brown or pale green, with a swollen thoracic region; the last segment bears tubercles and a retractable horn. **Biology** Eggs are laid in tiers in groups. Emerging larvae feed on both wild and cultivated fig trees. Larvae spin a yellow silk cocoon, very similar to that of *B. mori*. Domesticated Silkworms feed on Mulberry (also a species of fig). **Habitat** Subtropical bush, forest and gardens.

Family Saturniidae — Emperor moths

Giants among moths, including the largest moths in the world. South African species can attain wingspans of 180mm. Many have elaborate ring-like eyespots on the hind wings, reduced in some to thin, clear, crescentic windows. Antennae (especially of males) are obviously comb-like. The huge wings are cryptically coloured and often have a hooked tip. Most species are nocturnal and readily attracted to lights. Adults of both sexes do not feed and are short-lived, mating, laying eggs and dying in 3–5 days. Larvae often change appearance between instars, and are as diagnostic as the adults, being brightly coloured and covered in spines or tubercles (a few are hairy or entirely smooth). They are large and consume vast quantities of green material. Pupae usually rest in an earthen cell or (rarely) construct a cocoon. Some larvae (like Mopane worms) are edible; others are pests on pine or blue-gum plantations (range of foodplants very large). Most of the 74 species in the region are abundant and widespread in subtropical parts of the country, although endemics also occur in limited areas.

1 *Eochroa trimenii* — Roseate Emperor

Small (wingspan 65mm) and bright pink. Wings with yellow borders and large eyespots. Males have disproportionately large antennae. Larvae **(1A)** are cream, with irregular black stripes and short black spines. **Biology** Larvae feed on various species of highly toxic 'truitjie' (*Melianthus*). Adults appear in autumn. **Habitat** Arid rocky 'klipkoppe' and mountain passes where host plants grow.

2 *Tagoropsis flavinata* — Gold-marbled Emperor

Medium-sized (wingspan 93mm). Male bright yellow; female (shown) richer yellow-orange with a larger number of wavy brown lines and blotches on the wings. Both sexes have a single eyespot on each fore wing. **Biology** Larvae feed on Dune False Currant (*Allophylus natalensis*). **Habitat** Subtropical bushveld.

3 *Pselaphelia flavivitta* — Leaf Emperor

Medium-sized (wingspan 90mm), varying from bright to rusty yellow. Fore wings have pointed and slightly hooked tips. Each wing has an eyespot. Long spike at end of the pupa. **Biology** Larvae feed on Natal Mahogany (*Trichilia emetica*) and wild ginger (*Aframomum*). **Habitat** Subtropical bushveld.

4 *Gonimbrasia tyrrhea* — Zigzag Emperor, Willow Emperor

Large (wingspan 120mm). Wings fawn, speckled in black, eyespots isolated by 2 zigzag bands, inner band on fore wings very jagged, hind wings pinkish at base. Black larvae have short black spines on black tubercles and are covered in bright scales that form whitish-yellow patches. **Biology** Larvae feed on a very wide range of trees, including Australian Black Wattle (*Acacia mearnsii*), Sweet Thorn (*Vachellia karroo*), willow and apple. **Habitat** A wide range of vegetation types. **Related species** The genus includes 14 other species.

5 *Argema mimosa* — Luna Moth

Large (wingspan 120mm), emerald green, with yellow-and-red eyespots on the wings, and long tails on the hind wings. Larvae green, with thin white bands and rows of long projections on the back. Cocoon silvery and pitted with small holes. **Biology** Larvae **(5A)** feed on corkwood (*Commiphora*), Marula (*Sclerocarya birrea*), Tamboti (*Spirostachys africana*) and other trees. **Habitat** Subtropical bushveld. Breeds twice a year in the southern parts of its range.

6 *Cirina forda* — Pallid Emperor

Large (wingspan 100mm), pale fawn. Males have short-tailed hind wings and hooked fore wings. Small eyespots on hind wings only. Larvae creamish-white, with a black blotch on top of each segment and short white hairs (similar to *Gonimbrasia cytherea*). **Biology** Larvae feed on many woody plant species, including domestic plum and apple. **Habitat** Bushveld, forest and farms.

7 *Bunaea alcinoe* — Common Emperor, Cabbage Tree Emperor

Large (wingspan 160mm), with brown body, centre of fore wings and base of hind wings. Wing borders brown and pink; sharp brown stripe cuts through wing. Fore wings have clear eyespots; those on hind wings bright orange circled with black. Larvae striking jet black, with thick white spines and orange spots along the body. **Biology** Larval host plants include *Bauhinia*, *Croton*, Common Cabbage Tree (*Cussonia spicata*), *Ekebergia*, *Harpephyllum* and other trees. **Habitat** Bushveld and forest.

1 *Gynanisa maja* — Speckled Emperor, Wattle Emperor

Large (wingspan 130mm). Fore wings have a dark brown marginal band, are not hooked, and have a straight, not wavy, margin. Each hind wing has a large, complex eyespot comprising concentric black, orange and pink rings. Male has large, feathery antennae. Larvae pale green with ridged segments, covered in silver spines tipped with yellow, and have purple spiracles. **Biology** Larval foodplants include a very wide range of trees, among them Sweet Thorn (*Vachellia karroo*), alien Black Wattle (*Acacia mearnsii*), *Elephantorrhiza* and peach. **Habitat** Subtropical bushveld. **Related species** *Gynanisa* includes a few other similar species, some occupying the same range, as well as more arid parts of the Cape.

2 *Gonimbrasia belina* — Mopane Moth

Large (wingspan 120mm); highly variable. Wings may be fawn or shades of green, brown or red, with 2 black and white bands isolating the eyespots. Each wing has an orange eyespot. Larvae **(2A)** black, peppered with round scales in indistinct, alternating bands of whitish-green and yellow; they are armed with short black or reddish spines covered in fine white hairs. **Biology** Larvae eat a wide range of plants including Mopane (*Colophospermum mopane*), Natal Plum (*Carissa macrocarpa*), *Diospyros*, *Ficus*, *Searsia*, Marula (*Sclerocarya birrea*), *Terminalia* and *Trema*. They form an important constituent (after evisceration and drying) of the traditional diet in some parts of the region. Outbreaks defoliate shrubs, depriving game of browse. **Habitat** Widespread and common in semi-desert, bushveld and grassland. **Related species** Carnegie's Emperor (*Nudaurelia carnegiei*) (wingspan 150mm), which occurs in light and dark forms, feeds on *Protea* and miombo (*Brachstegia*) species and occurs in the warmer eastern parts.

3 *Gonimbrasia cytherea* — Pine Emperor

Large (wingspan 120mm), variable, typically with red thorax and abdomen, and yellow-orange wings with conspicuous yellow eyespots isolated by white and brown bands. In the Cape form (shown here), the wings are mainly grey, with a little pink and yellow. Larvae **(3A)** red-brown, covered in bands of light blue and yellow scales (resembling *Gonimbrasia belina* larvae). **Biology** Larvae very common and well known, feeding on a very wide range of plants including *Euclea*, *Protea*, *Searsia*, *Watsonia*, alien Black Wattle (*Acacia mearnsii*), gum trees, apple and quince. Often a pest of pines. **Habitat** Ubiquitous, but absent from arid parts. **Related species** *Gonimbrasia* includes about 13 other species.

4 *Epiphora mythimnia* — White-ringed Atlas Moth

Large (wingspan 120mm), with huge, clear eyespots, each entirely and clearly circled in yellow; those on fore wings elongate. Wings brown or reddish-brown speckled with white; wing margins with complex light brown markings. Abdomen banded in white. Legs and antennae orange. Larvae milky white with hairy blue spots. **Biology** Larvae eat Forest Fever-berry (*Croton sylvaticus*), Ubhubhubhu (*Helinus integrifolius*) and Buffalo Thorn (*Zizyphus mucronata*). **Habitat** Subtropical bushveld and forest. **Related species** There are 3 other species in the genus, in all of which the yellow circling of the eyespots is incomplete. Relatives of the world's largest (Asian) moths.

5 *Pseudaphelia apollinaris* — Apollo Moth

Medium-sized (wingspan 80mm), variable, with broad, clear or whitish wings with distinct veins, yellow eyespots, and marginal dots. Larvae smooth, white, with a black head, black-and-brown or black-and-orange side stripes, and a horn at the end of the abdomen. **Biology** Adults fly slowly by day or at dusk, and are attracted to lights. Larvae feed on lowveld honeysuckle (*Turraea*) and, possibly, bush willow (*Combretum*). **Habitat** Subtropical bushveld and forest.

Family Eupterotidae — Monkey moths, Giant lappet moths

Medium-sized to very large moths, fairly easily recognized by their very broad, often oval, wings fringed with long hairs. The wings have long, dense scales, often intricately marked and with a silky sheen. Adults are hairy and plump, ranging in colour from off-white, to dull yellow, to various shades of brown. They have a slow, flapping flight, and become immobile if molested. The antennae are branched and comb-like, but the branches do not protrude much to the sides. Larvae are dark, densely covered in long, fine hairs and short bristles. Little is known about larval habits, but they feed on the leaves of a number of unrelated plant families, including the grasses. The flimsy cocoon has long larval hairs woven into it. There are 75 species known from the region, mostly from less arid parts.

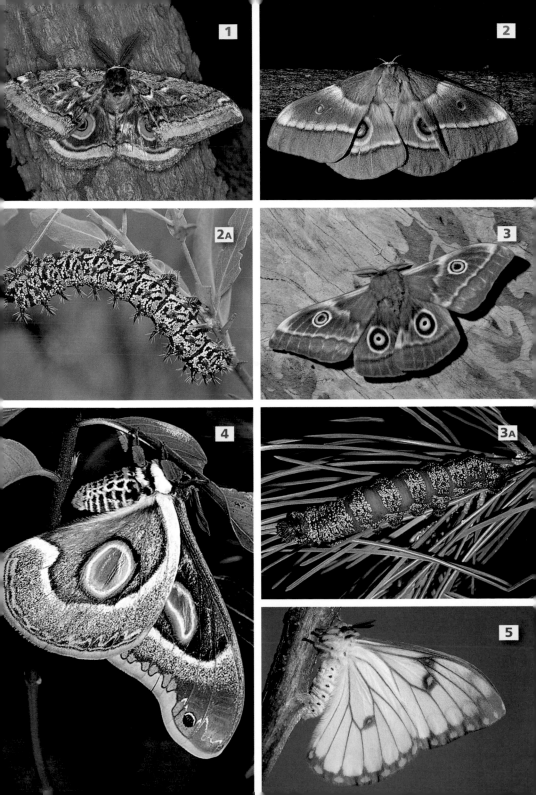

1 *Tantaliana tantalus* — King Monkey

Large (wingspan 100mm), with a black thorax. Fore wings greyish-black, with 1 black and 2 brown bars. Hind wings fawn, banded in black. Wings and body clothed in long hairs. The female of this geographically variable species is paler than the male. **Biology** The larvae feed on Septee Tree (*Cordia caffra*) and wild jasmine (*Jasminum*). **Habitat** Bushveld and subtropical forest.

2 *Phyllalia patens* — Clay Monkey

Medium-sized (wingspan 30mm), wings pale to pinkish-brown, without markings. **Biology** The larvae feed on grass (*Cynodon dactylon*, *Ehrharta calycina* and), like some others in the genus, have irritant hairs. **Habitat** Strandveld to bushveld.

3 *Phyllalia alboradiata* — Fynbos Monkey

Large (wingspan 54mm), with a brown body, heavily clothed in hairs. Wings brown, with prominent white veins. **Biology** The larvae feed on grass and, like some others in the genus, have irritant hairs. **Habitat** Fynbos.

4 *Poloma angulata* — Angular Monkey

Medium-sized (wingspan 50mm). Fore wings pinkish, with a brown central patch and 2 whitish stripes. Hind wings pale pinkish-brown but darker at the margins. **Biology** The larvae feed on Rock Alder (*Canthium mundianum*) and Unarmed Turkey-berry (*Canthium inerme*). **Habitat** Bushveld.

5 *Phiala incana* — White Monkey

Medium-sized (wingspan 45mm), with a white thorax and wings and yellow abdomen and antennae. **Biology** The larvae of related species feed on button stinkweed (*Pentzia*), sedges (*Cyperus*) and crop asparagus. **Habitat** Bushveld. **Related species** *Phiala* includes a number of other similar-looking species.

6 Family Lasiocampidae — Eggar moths, lappet moths

Like their relatives – monkey moths, silk moths and emperor moths – eggar moths lack a proboscis. Most are fairly easily identified, being very hairy, medium-sized to large, and with a very stout body. Fore wings elongate and narrow, frequently pale brown with wavy lines, bands and spots. Hind wings are rounded and reduced in size, paler than fore wings and unmarked, and may have an enlarged anterior margin which is wavy, cryptically coloured and exposed at rest. Antennae very comb-like, often only in the bottom half, the terminal half being toothed and bent backwards. In some females, the antennae are merely toothed. Some species have a snouted appearance (resulting from prominent labial palps). Males of some species smaller than females and narrow-winged; females are much bulkier and thus poorer fliers. Eggs laid in a spiral or ring on a twig, protected in some species by detached abdominal hairs. Other species can apparently lay unfertilized eggs that develop into males. Larvae (6) are distinctive, covered in hairs, and often have shorter, brightly coloured bristles. These detach very easily and can cause severe irritation to human skin. Larvae often live gregariously, resting by day, feeding at night. They form very tough cocoons on their foodplants. In *Gonometa* the short stinging hairs are woven into the cocoon. Of the 150 known species, some are very widely distributed, but most occur in the west of the region.

7 *Bombycopsis bipars* — Divided Eggar

Medium-sized (wingspan 30–45mm). Body and wings light brown, with 2 dark brown stripes and 2 lighter parallel stripes on each fore wing. Snout prominent and hairy. Hind wings brown or yellow-brown. **Biology** Recorded larval foodplants include Blackjack (*Bidens pilosa*), Canary Creeper (*Senecio deltoideus*) and a *Geranium* species. **Habitat** Widespread across the region, in a range of vegetation types, as well as in disturbed fields.

1 *Bombycomorpha bifasciata*　　Barred Eggarlet, Pepper Tree Caterpillar

Medium-sized (wingspan 30–40mm), body white, cream or light brown. Fore wings have 2 faint brown stripes (sometimes only 1 in female), with a small dark brown dot between them. Caterpillars gregarious, black, with thin white longitudinal stripes and a covering of long white hairs with mustard-coloured bases **(1A)**. **Biology** Larvae feed under silken tents on Nana-berry (*Searsia dentata*) and alien Brazilian Pepper Tree (*Schinus molle*). May become cannibalistic after they have defoliated the trees. Larvae take 2 months to develop and pupate, the pupae hatching after 14 days or only in the following spring. **Habitat** Widespread in the Western Cape and wetter summer-rainfall parts of the region.

2 *Eutricha capensis*　　Cape Lappet

Large (wingspan 70mm), bulky, fore wings dull grey, with broad, diagonal tan band; hind wings uniform reddish-brown. Female larger and paler than male. Larvae **(2A)**, as is typical for the genus, have a central black stripe with spines and white side stripes, and a row of long coppery tufts of hairs along the side of the body, with 3 ginger and 2 mauve tufts near the head. **Biology** Larvae aggregrate conspicuously on tree trunks, feeding on (Australian) *Acacia* and *Senegalia*, white stinkwood (*Celtis*), bush willow (*Combretum*), *Bauhinia* and many other trees in nature, and on trees such as mango, peach and the Brazilian Pepper in gardens. Adults emerge in early summer and are very passive when disturbed. **Habitat** A range of natural and garden habitats.

3 *Braura truncatum*　　Truncate Lappet

Large (wingspan 88mm) with narrow wings. Fore wings red-brown, paler near margins, with wavy black and cream lines along margins. Hind wings reddish cream-brown. A tuft of reddish hairs at the end of the abdomen. Larvae grey, with a pair of reddish lines and tufts of hairs on the sides of the body. **Biology** Larvae feed on Sweet Thorn (*Vachellia karroo*), Paperbark Thorn (*Vachellia sieberiana*) and Black Wattle (*Acacia mearnsii*). **Habitat** Bushveld, fynbos and subtropical forest.

4 *Eucraera salammbo*　　Salammbo

Medium-sized (wingspan 50mm), with yellow wings shading to greyish-olive at the edges. Fore wings crossed by 3 white lines. Leading margin of hind wing is curled and bears black bristles (see arrows). Resting posture typical of family, with scalloped and decorated fore margin of hind wing protruding beyond fore wing. **Biology** Larvae feed on *Eucalyptus* and Marula (*Sclerocarya birrea*). **Habitat** Bushveld and subtropical forest.

5 *Lebeda mustelina*　　Incised Lappet

Medium-sized (wingspan 30mm), with hooked fore wings in both sexes. Male light brick red; female yellowish-brown. **Biology** Related species feed on bush willow (*Combretum*), Cape ash (*Ekebergia*), *Macaranga* and *Pinus*, as well as *Plectranthus* and spinach. **Habitat** Bushveld and subtropical forest.

6 *Grammodora nigrolineata*　　Black-lined Eggar

Medium-sized (wingspan 50mm), entirely cream, with fine black parallel lines and a few red lines on the fore wings. Hind wings cream (in male) or grey (in female). Larvae are pale yellowish-brown 'woolly bears', with brown and orange speckling and blue streaks on the body, and long tufts of hair behind the head. **Biology** Larvae feed on *Cassia*, false thorn (*Albizia*) and *Brachystegia*. The tough, light brown cocoon is attached to dry twigs. **Habitat** Bushveld and subtropical forest.

7 *Odontocheilopteryx myxa*　　Hairy Eggarlet

Small (wingspan 26mm). Fore wings variegated in cream, grey, brown and yellow; trailing edges have extensions that stick up when at rest. Hind wings cream. As in many lasiocampids, the palps form a prominent 'snout'. Abdomen ends in brown hairs. **Biology** Larvae feed on Sweet Thorn (*Vachellia karroo*), Fever Tree (*Vachellia xanthophloea*) and *Eriosema*. **Habitat** Bushveld.

Family Sphingidae

Hawkmoths, sphinx moths

Large and streamlined, triangular in resting posture, with narrow wings, a spindle-shaped body and large thorax. The wing tips are pointed and the eyes very large. Wing coupling differs in the sexes, but allows in both for very powerful, accurate (hovering) flight. Antennae are long and toothed, with a curved tip. Proboscis is generally far longer than the body (although very short or absent in some species). Many are active at dusk, taking nectar with great accuracy from flowers, such as those of papaya and other plants with fragrant white flowers, including *Turrea* and wild jasmine. Truly nocturnal species are clumsier than day-flying species, which are accurate and fast fliers. Females lay eggs singly on the host plant (occasionally causing minor damage to garden plants). The characteristic larvae are smooth and bear a short, upright horn at the end of the abdomen. Although cryptically coloured, they withdraw the head into the fleshy, inflated thorax when disturbed and display thoracic eyespots that may startle predators. Pupation normally occurs in an earthen cell or flimsy cocoon. Sphingidae includes 100 species regionally, 4 of them cosmopolitan. Many are widespread in the subcontinent and occur in subtropical bushveld.

1 *Agrius convolvuli*

Convolvulus Hawk

Large (wingspan 80–90mm), body and wings grey, with subtle mottling. Abdomen with characteristic black and pink bars. Larva green to brown, with a row of black spots along the bottom of the body, usually with an angled light or dark line running diagonally above each black spot. **Biology** Larvae feed on a wide range of foodplants, including beans, convolvulis, potato, sweet potato, *Rumex*, bamboo and soya beans. Adults are strong fliers and migratory. **Habitat** Widespread across Africa, Europe and Asia, and probably among the most common hawkmoths in the summer-rainfall region, occurring in a range of veld types as well as in gardens.

2 *Batocnema africanus*

Harlequin Hawk

Medium-sized (wingspan 70–85mm), body and wings flesh-coloured, with olive green markings on thorax and fore wings. Hind wings pale yellow, with a grey border. Larvae green, with oblique blue stripes and a thin horn at the end of the abdomen. **Biology** Larval foodplants include mango and Marula (*Sclerocarya birrea*). **Habitat** Warmer parts, in open woodland, grassland and agricultural land.

3 *Macropoliana natalensis*

Natal Sphinx Hawkmoth

Large and bulky (wingspan 110–120mm), body and fore wings peppered in fine grey and black markings. Hind wings dark brown. Thorax encircled by a thin black-and-yellow ring. Wings broader and darker in female. Larvae pale olive green, with oblique brown or mauve stripes. **Biology** Larvae feed on *Brachystegia*, *Markhamia lutea* and African Tulip Tree (*Spathodea campanulata*). Like most hawkmoths, pupates in the ground. **Habitat** Forest and moist woodlands.

4 *Neoclanis basalis*

Medium-sized (wingspan 30–40mm); male smaller than female. Fore wings and body straw-coloured, without markings or else with a thin diagonal stripe. Hind wings straw-coloured, peppered with small brown spots and with a large blood red spot at each wing base. **Biology** Larvae feed on Apple Leaf Tree (*Philenoptera violacea*). **Habitat** Dry parts of the warmer, summer-rainfall region.

5 *Rufoclanis jansei*

Jansen's Hawk

Medium-sized (wingspan 60mm). Fore wings and body grey, with a single dark scalloped band dividing the fore wing in half, and smaller, fainter bands on either side of it. Hind wings peach, with 2 small dark markings near lower edge. Abdomen striped grey and brown. A dark stripe runs along centre of thorax. Some specimens brighter and more boldly marked. **Biology** Larvae of the related *R. numosae* feed on species of crossberry (*Grewia*). **Habitat** Dry bushveld.

6 *Lophostethus dumolinii*

Arrow Sphinx

Very large (wingspan 140mm). Fore wings with scalloped margins and patterned with boldly contrasting brown patches and white arrow-shaped markings. Larvae huge, pale green and yellow, covered in black spines. **Biology** Larvae (**6A**) feed on Natal plum (*Carissa*), wild pear (*Dombeya*), raisin berry (*Grewia*) and *Hibiscus*. **Habitat** Thick bush in various vegetation types.

1 *Acherontia atropos*   Death's Head Hawkmoth

Large (wingspan 110mm), bulky. Fore wings cryptic, brown and dark grey. Abdomen and hind wings with bright orange bars. Skull-like marking on thorax. Larvae **(1A)** variable, often greenish-yellow, with yellow and blue (or slate) chevron markings on each segment, and a rough, curved horn on the tail. **Biology** Adults raid beehives for honey. They can squeak through the proboscis, producing a sound that may serve as protection by mimicking the sound of the queen bee, thus immobilizing worker bees. Can also avoid getting stung when raiding hives by mimicking colony odour. Larvae feed on potato, tomato and other nightshades such as Cape gooseberry, lantana, cotton, cannabis and at least 10 other species. **Habitat** Most natural and urban habitats.

2 *Platysphinx piabilis*   Measly Hawkmoth

Large (wingspan 118mm). Fore wing coloration highly variable. Hind wings orange with a sprinkling of red dots and a larger black spot at wing base. Larvae with alternating white and yellow stripes. **Biology** Larvae feed on *Craibia*, umzimbeet (*Millettia*), cork bush (*Mundulea*), *Ostryoderris* and wild teak (*Pterocarpus*). **Habitat** Open bushveld.

3 *Daphnis nerii*   Oleander Hawkmoth

Large (wingspan 110mm) and intricately marked in olive green and pink. Larvae green with 2 circular bluish-black-and-pink eyespots near head. **Biology** Larvae feed on jasmine, oleander, mango, *Gardenia*, *Carissa*, *Acokanthera*, *Adenium*, *Vinca*, *Rauwolfia* and *Voacanga*. **Habitat** Riverine forest and bushveld. Extends to Europe.

4 *Coelonia fulvinotata*   Fulvous Hawkmoth

Large (wingspan 110mm), similar to Death's Head Hawkmoth, but with less yellow on abdomen and hind wings, and without skull-like thoracic marking. Proboscis very long. Larvae variable in colour, with first 3 segments crinkled. **Biology** Larvae **(4A)** feed on *Convolvulus*, tobacco, tomato, *Bignonia*, *Clerodendron* and at least 8 other plant species, including *Salvia*, *Lantana* and Cape honeysuckle (*Tecomaria*). **Habitat** Shrubby and bushy areas.

5 *Pseudoclanis postica*   Mulberry Hawkmoth

Medium-sized (wingspan 72mm), with variably coloured wings. Fore wings yellowish or greenish-brown. Hind wings yellow, with a black basal mark. Brown central stripe on thorax. Larvae green, but white on top and covered in white dots. **Biology** Larvae feed on mulberry, fig, mistletoe, White Stinkwood (*Celtis africana*), pigeonberry (*Trema*) and *Chaetachme*. **Habitat** Very common in warmer bushveld and forest areas.

6 *Nephele comma*   Comma Nephele Hawkmoth

Medium-sized (wingspan 74mm), uniformly drab olive green or brown. Fore wings with fawn tips and, usually, a diagnostic comma-shaped white mark. Larvae green or brown. The most common species in the genus. **Biology** Larvae feed on *Carissa* and Hornpod (*Diplorrhynchus condylocarpon*). **Habitat** Widespread in Africa and the Indian Ocean islands.

7 *Hippotion celerio*   Silver-striped Hawkmoth, Vine Hawkmoth

Medium-sized (wingspan 76mm), with long, light brown and olive brown stripes on the body and wings and silver flecks on the abdomen. Hind wings with a black stripe and pink bases. Larvae **(7A)** smooth, brown or green, with prominent eyespots on the head. **Biology** Larvae feed on grapevines, *Arum*, *Impatiens*, carrot tops and, in the wild, on *Senegalia*, *Rumex*, *Galium* and *Epilobium*. Adults are migratory. **Habitat** Abundant in a wide range of natural and garden habitats. **Related species** There are a number of similar-looking *Hippotion* species. Also similar to *Hyles virescens*, which has more pink on the hind wings, and black bars on the abdomen.

1 *Theretra capensis* Cape Hawkmoth, Grapevine Hawkmoth

Large (wingspan 92mm), variable, body and fore wings greenish to reddish-brown. Hind wings pink with green margins. Larvae variable, greenish-brown or pinkish-brown, with pale stripes and white speckles, and with eyespots on the neck. **Biology** Larvae feed on cultivated and wild grapevines (*Rhoicissus*, *Cissus* and *Ampelopsis*). Although only certain owlet moths are known to feed on the lachrymal secretions of cattle, this unusual type of feeding also occurs in hawkmoths, which occasionally drink lachrymal secretions of antelope such as steenbok **(1A)**. **Habitat** Forest.

2 *Temnora fumosa* Smoky Temnora

Medium-sized (wingspan 58mm), plain brown body and wings, the latter with 2 darker brown bands. **Biology** Feeds on wandering Jew (*Tradescantia*) and Yellow Commelina (*Commelina africana*). **Habitat** Coastal areas, most common in warmer, eastern part of its range.

3 *Cephonodes hylas* Oriental Bee Hawkmoth, Coffee Clearwing

Medium-sized (wingspan 70mm), unmistakable. Wings clear with black veins. Thorax and head olive green, separated from the yellow abdomen by a chestnut band. Bluish cast on head. Larvae **(3A)** variably coloured (green, black or brown), with yellow and white lines and orange spiracles. **Biology** Larvae feed on wild pomegranate (*Burchellia*), *Gardenia*, *Pavetta*, rhino-coffee (*Kraussia*), *Vangueria* and coffee. Adults feed during the day, while hovering with a spread fan of black abdominal hairs. **Habitat** A range of vegetation types.

4 *Macroglossum trochilus* African Hummingbird Hawkmoth

Small (wingspan 38mm). Fore wings banded, thorax reddish-brown, abdomen orange shading to brown. Hind wings yellow at the base, shading to orange and then brown. Black-and-white abdominal fan is opened in flight. **Biology** Adults feed in full sunshine during the day, hovering like hummingbirds and flitting between various tubular flowers, especially perennial basil and lavender growing in gardens. Larvae feed on *Galium*, *Rubia* and the exotic mirror bush (*Coprosma*). **Habitat** A range of veld types. Common in gardens.

5 *Odontosida magnificum* Magnificent Scalloped Hawkmoth

Small (wingspan 50mm). Thorax and fore wings marked in contrasting shades of brown. Hind wings with yellow base and an orange-and-brown patch towards the edge. **Biology** Larvae feed on Kei Apple (*Dovyalis caffra*). **Habitat** Semi-arid succulent karoo and bushveld.

6 Family Notodontidae Prominents, puss moths

Medium-sized moths named after a tuft of hair at the end of the fore wing that forms a ridge when the moth is at rest. Adults are usually well camouflaged and often mimic twig stumps. They are nocturnal and attracted to lights. The larvae are smooth or lightly haired and are probably the most varied and spectacular of any moth species. Their shapes are bizarre, the posterior part of the body often greatly inflated, as in *Amyops ingens* **(6)**, with horns and other weird projections. European species have menacing eyespots and whip-like tails, display threateningly and are reputed to eject formic acid. Over 200 species are known from the region.

7 *Phalera imitata* Imitating Bufftip

Medium-sized (wingspan 56mm). Fore wings grey, shading to brown, with buff tips. Hind wings white and brown, becoming yellow at their base. Abdomen yellow. **Biology** At rest, mimics a broken twig. Larvae feed on grasses and the twining legume *Rhynchosia viscosa*. **Habitat** Bushveld and subtropical forest.

8 *Phalera lydenburgi* Lydenburg Buff

Medium-sized (wingspan 42mm). Fore wings mostly grey, with a toothed margin, a buff tip, and a small brown crescent. Hind wings grey-brown. Larvae whitish with a yellow side stripe and 2 long tufts of black hair behind the head. **Biology** Larvae recorded on maize leaves and Guinea Grass (*Panicum maximum*). **Habitat** Subtropical bushveld and forest.

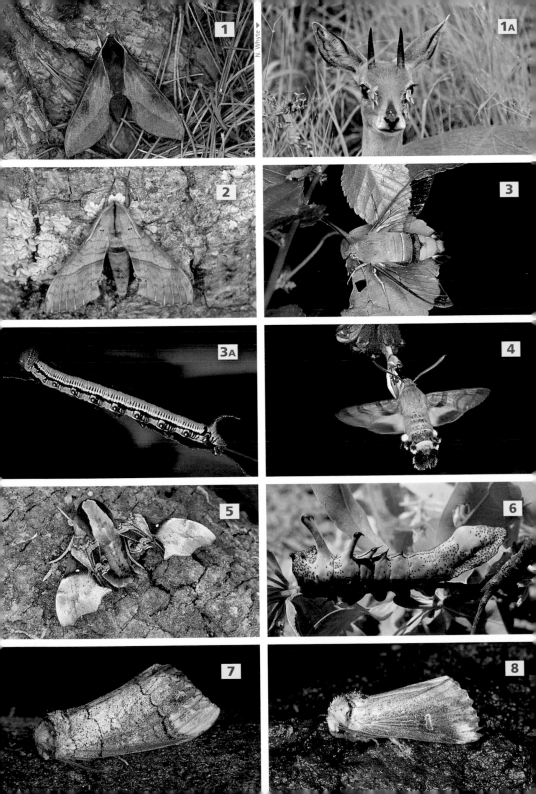

N. Whyte ▶

1 *Desmeocraera basalis* — Basal Prominent

Medium-sized (wingspan 45mm), mainly plain grey or greenish-grey, often with a short, dark curved line at the base of the fore wings. Hind wings white and grey. Has a tuft of hair on the head and at end of the abdomen, and a brown spot on the thorax. **Biology** Larvae feed on *Julbernardia*, *Mimusops*, *Chrysophyllum*, *Ficus*, *Eugenia* and many others. **Habitat** Subtropical bushveld and forest. **Related species** There are a number of other species in the genus, which have patterned green wings, resembling lichen.

2 *Anaphe reticulata* — Reticulate Bagnet

Medium-sized (wingspan 42mm), with silver-white fore wings marked with a bold network of brown lines, cream hind wings and 2 white patches on the thorax. Larvae are hairy. **Biology** Larvae are far more conspicuous than adults, owing to their gregarious habits, and are usually seen crowded together on the trunk of a foodplant – a wild pear (Dombeya) or horn-pod (Diplorrhynchus) tree – occasionally moving in trails from tree to tree (**2A**). One leads and the rest maintain bodily contact, hence they are known as 'processionary caterpillars'. They also pupate together, forming up to 600 cocoons in a mass purse-like structure called a 'bagnet'. The silk in these bagnets is contaminated with larval body hairs, making them unsuitable for silk harvesting. **Habitat** Bushveld and subtropical forest. **Related species** Anaphe includes 5 other species in the region, including A. panda, which is similar, but larger (wingspan 50mm), with brown wing stripes that do not extend close to the wing bases.

Family Erebidae — Erebid moths

One of the largest families of Lepidoptera, containing species that show a huge range of coloration, behaviour and ecology, making field recognition of family members challenging. Many of the species included in the family Noctuidae, while others were in families that have been downgraded to subfamily level and whose members are more easily recognized than members of Erebidae itself.

Subfamily Aganainae — Snouted tigers

Medium-sized colourful moths, with whip-like antennae. Larvae are cylindrical and sparsely haired. Aganainae includes 18 species from the region.

3 *Asota speciosa* — Specious Tiger Moth

Large (wingspan 60mm) and spectacular. Fore wings mostly brown, with white veins, and an orange basal area with black dots. Hind wings usually orange, occasionally white or with a black border. **Biology** Larvae feed on *Ficus* (including domestic figs) and poison bush (*Acokanthera*). Since figs produce a sticky defensive latex when damaged, the larvae first cut the main veins on the underside of the leaf to interrupt the flow of latex, before consuming the soft tissue of the leaf. **Habitat** Forest and gardens.

Subfamily Arctiinae — Tiger moths, footmen, wasp moths

Day-flying moths, slow in flight, with large, narrow fore wings and reduced hind wings. Wings have clear spots and patches. Body often stout, with colourful bands. May be confused with the more primitive Zygaenidae. Larvae short and cylindrical, with tufts of long hairs. They are commonly known as tiger moths, footmen, lichen moths or wasp moths. The larvae of tiger moths are covered in long hairs and are known as 'woolly bears'. The larvae of each species feed on a wide range of plants, and some species are commonly encountered in gardens. Footmen are smaller species, with narrower wings and a thinner abdomen, and fly by day and at dusk. Their weakly haired larvae feed on lichens. Over 300 species of this very large group are known from the region.

4 *Amata atricornis* — Devil's Maiden

Small (wingspan 24mm), slender. Wings black with clear spots. Body black. Hind wings with orange patches. **Biology** Flies slowly by day. Larvae reared on Dandelion (*Taraxacum officionale*). **Habitat** Moist grassland, where they are found by day, clinging to grass stalks. Also recorded from Cape Town.

1 *Amata cerbera* — Heady Maiden

Small (30mm). Wings blackish-blue with clear spots. Body bluish-green with 4 red or orange bands. **Biology** Sluggish, day-flying. In Angola, larvae have been recorded on clover, in South Africa on grass, and, in the Western Cape, on clover (*Trifolium*). Also found on *Rumex*, *Corylus*, *Plantago* and *Rubus*. **Habitat** Succulent karoo to subtropical forest. **Related species** *A. alicia*, which has a white spot on the head and on each leg.

2 *Amata simplex* — Simple Maiden

Small (wingspan 25–30mm) and dark. Body and wings with a metallic blue sheen; fore wings each with 2 large white dots and a small white dot at the tip. Number of white spots of fore wings varies geographically. First abdominal segment has a single orange dot; sides of abdomen bear additional orange dots. **Biology** Larval foodplant unknown. **Habitat** A range of vegetation types and open habitats.

3 *Ceryx anthraciformis* — Yellow-sleeved Maiden

Small (wingspan 24mm). Wings bluish-black with clear spots. Body brown, with 3 rows of yellow patches along the abdomen. **Biology** Probably stores plant toxins to use for chemical defence. **Habitat** Occurs in open bushveld and coastal grasslands.

4 *Euchromia amoena* — Pleasant Hornet

Medium-sized (wingspan 42mm), brilliantly coloured. Fore wings black with bright orange windows. Abdomen metallic black with 1 whitish-yellow and 1 orange band. **Biology** Flies slowly by day, congregating on pollen-rich whitish flowers to feed and mate. Bright coloration suggests it is poisonous. Larvae feed on *Carissa*, *Ipomoea*, *Secamone* and *Sarcostemma*. **Habitat** Subtropical bushveld and forest; most common in coastal areas.

5 *Apisa canescens* — Greyling

Medium-sized (wingspan 50mm), with pale grey fore wings and white hind wings. Creamish-white larvae have tufts of long hair in rows. **Biology** Adults nocturnal. Larvae recorded feeding on alien Sulphur Cosmos (*Cosmos sulphureus*), but also found in the depths of various caves. **Habitat** A range of vegetation types.

6 *Metarctia rufescens* — Matron

Small (wingspan 38mm), fluffy, with broad wings. **Biology** Nocturnal. Larvae feed on wandering Jew (*Tradescantia*) as well as grass roots and litter. **Habitat** Bushveld to forest. **Related species** There are many other very similar species of *Metarctia* with a fluffy thorax and abdomen, small hind wings and, often, contrastingly coloured wing veins.

7 *Thyretes caffra* — Bar Maiden

Small (wingspan 28mm). Wings long and grey, with a continuous row of white spots. Small hind wings. Body slender and grey, with white spots at base of the abdomen, followed by a row of orange dots. **Biology** Larvae feed on Sweet Thorn (*Vachellia karroo*). Adults fly sluggishly by day and rest on grass stems. **Habitat** Grassland.

8 *Thyretes hippotes* — Equine Maiden

Small (wingspan 34mm). Wings grey with white patches. Very broad thorax covered in thick white and grey hairs. Abdomen with a central row of orange dots and side rows of white dots. Larvae hairy. **Biology** A sluggish day-flier, found on grass and shrubs. Larvae feed on Button Stinkweed (*Pentzia incana*). **Habitat** Fynbos and succulent karoo.

1 *Rhodogastria amasis* — Tri-coloured Tiger

Medium-sized (wingspan 60mm), robust, appearing mainly silvery white when at rest, with white and yellow thorax. Fore wings streaked with thin black lines. Hind wings white on top, yellowing towards base, with a black crescent. Underside of wings yellow with black crescents. Abdomen, sides of thorax and legs are red. Larvae are large, long-haired, black-and-ginger 'woolly bears'. **Biology** Adults curl up when molested, exposing bright undersides. Larvae feed on a vast range of (usually soft-leaved) plants, including *Chrysanthemoides* and *Tecomaria*. Replaced by *R. similis* in the summer-rainfall region. **Habitat** Various vegetation types in habitats that include gardens, where they may become pests.

2 *Cyana pretoriae* — Pretoria Red Lines

Small (wingspan 30mm), distinctive, with pure white wings and body, red spots on head and thorax, red-banded legs, and 2 complete red bands crossing fore wings. **Biology** Larvae feed on lichens and *Ipomoea*. **Habitat** Bushveld and forest.

3 *Amerila bauri* — Baur's Frother

Medium-sized (wingspan 60mm), pale tan, with clear windows in fore wings, black dots on thorax, and red abdomen and legs. **Biology** Larvae feed on Water Berry (*Syzygium cordatum*). If handled roughly, adults produce a large ball of presumably toxic foam. **Habitat** Succulent karoo to subtropical forest. **Related species** There are 9 other similar species in the genus, all smaller than *A. bauri*.

4 *Popoudina linea* — Streaked Ermine

Medium-sized (wingspan 40mm), stout, with yellow body and wings, yellow thorax with a central streak, and a row of dots along each side of the abdomen. Black streaks on fore wings. **Biology** Larvae feed on *Tradescantia*, bitter teabush (*Vernonia*) and grasses. Adults become passive when molested, curling the abdomen to display warning colours. **Habitat** Most vegetation types. **Related species** Other members of the genus, which are all white, with streaked fore wings and thorax; 1 species has streaked hind wings.

5 *Afromurzinia lutescens* — Translucent Ermine

Medium-sized (wingspan 40mm) and stout, with clear, smoke grey fore wings with yellow veins. Hind wings white in male, yellow in female. Usually has black bands on abdomen. **Biology** The typical 'woolly bear' larvae feed on a huge range of unrelated plants, including lucerne, maize, *Lantana*, privet, *Tecomaria* and *Mesembryanthemum*. **Habitat** Bushveld, subtropical forest and gardens.

6 *Paralacydes arborifera* — Branched Ermine

Small (wingspan 30mm), stout. Fore wings white with thick, branched grey-brown markings. Hind wings white, with a few dark dots. Abdomen yellow with central and lateral rows of dots. Larvae are 'woolly bears', with yellow side markings and black hair tufts on top. **Biology** Larvae feed on the bulb *Ledebouria*. **Habitat** Bushveld.

7 *Sozusa scutellata* — Plated Footman

Small (wingspan 30mm), slender, with an orange head. Thorax lead grey, with an orange spot. Fore wings in male are fawn, with metallic grey at the base. Hind wings cream. Female generally pale. **Biology** Larval foodplant appears to be lichen. **Habitat** Fynbos to subtropical forest.

8 *Utetheisa pulchella* — Crimson-speckled Footman

Medium-sized (wingspan 35mm), slender, unmistakable, with cream fore wings speckled in red and black. Larvae dark grey, with a white band along the back, a yellow head, and red dots along the sides of the body. **Biology** Day-flying. Larvae feed on *Heliotropium*, grasses, plantain, *Myosotis* and cotton, and pupate in a folded leaf. Both caterpillars and moths accumulate alkaloids and are toxic and unpalatable to birds, hence their bright warning coloration. **Habitat** Disturbed weedy fields, gardens and agricultural land. Cosmopolitan; very common throughout Africa.

1 *Chiromachla leuconoe* — White Bear

Medium-sized (wingspan 52mm), boldly coloured, fore wings with a single oblique cream stripe, hind wings white with a black border. Head and thorax black, with white dots. Antennae whip-like. **Biology** Adults fly slowly by day and settle on leaves. Larvae feed on *Senecio* and coffee. **Habitat** Subtropical forest. **Related species** There are many other similar species in the genus.

2 *Alytarchia amanda* — Cheetah

Medium-sized (wingspan 44mm). Fore wings orange, with bars and black dots (surrounded by white rings in males); hind wings orange with an irregular black border. Female shown. **Biology** Adults fly by day. Larvae feed on ratttlepod (*Cotolaria*). **Habitat** Subtropical forest and bushveld.

3 Subfamily Lymantriinae — Tussock moths, gypsy moths

Medium-sized moths, with a reduced proboscis, feathery antennae, and broad white, cream or yellow wings, both pairs similarly coloured and often translucent. Some, e.g. *Aroa* and *Orgia*, have wingless, very grub-like females (3). Abdomen and thorax are densely covered in barbed hairs. Eggs are laid in clusters, often with hairs from a tuft at the end of the female's abdomen attached to them. Larvae (3A) have 4 neat hair tufts ('hair pencils') on the abdomen, hence the common name 'tussock moth'. Pupal cocoons incorporate larval hairs. Contact with larvae or the barbed hairs of the adults causes skin irritation. Some adults are capable of producing a toxic white foam. In contrast with North America's destructive Gypsy Moth, few South African species are pests.

4 *Laelia fusca* — Milkwood Tussock Moth

Small (wingspan 25–35mm), drab grey moth. Wings marked with faint, wavy grey stripes and a dotted grey border. Head and legs grey and hairy. Larvae (4A) black, with orange sides and a white central stripe, densely covered in stiff white hairs, running the length of the body. Longer hair tufts protrude from head and rear of body. **Biology** Larvae feed on both White Milkwood (*Sideroxylon inerme*) and Red Milkwood (*Mimusops obovata*). In outbreak years in the Western Cape they strip these trees of all leaves. The larvae are eaten by Diederick Cuckoos. **Habitat** Coastal forest where host trees occur.

5 *Dasychira octophora* — Eight-spot Tussock

Medium-sized (wingspan 32mm). Fore wings silvery white with lines made of sharp black crescents. Hind wings white with some grey markings. **Biology** Larvae feed on lichens. **Habitat** Forest. **Related species** Other tussock moths, which have similar but fainter markings. Also, may be confused with *Sommeria strabonis* (p.430).

6 *Bracharoa dregei* — Brown Vapourer

Small (wingspan 24mm). Fore wings in male yellow-brown or orange, each with 2 black dots; hind wings orange at base, darker towards edges. Female apparently grey and wingless. **Biology** Larvae feed on a wide range of plants including *Osteospermum, Dimorphotheca, Aspalathus, Pelargonium* and even pine trees. **Habitat** Low open vegetation, especially coastal scrub.

7 *Aroa discalis* — Banded Vapourer

Medium-sized (wingspan 36mm). Fore wings in male (shown here) reddish-brown, with an oblique orange band. Hind wings orange, shaded with dark brown. Female has orange fore wings, with brown lines, and orange hind wings. Larva is blackish-brown and hairy. **Biology** Male flies by day, female by night. Larvae feed on grasses, and often bear white cocoons of parasitic wasps that have emerged from them. **Habitat** Grassland and bushveld.

1 *Knappetra fasciata* Banded Euproctis

Small (wingspan 34mm); fore wings yellow or orange with a central band of black speckling. Female has a thick tuft of black hair at the end of the abdomen. Larvae black on top, with white hairs between the body segments, a white lateral line with hairs on red warts, and orange lateral tufts with red markings between them. **Biology** Larvae feed on a very wide range of woody plants, including apple, bush willow (*Combretum*), *Senegalia*, *Protea* and *Cassia*. **Habitat** Very common in a range of vegetation types. **Related species** The genus includes a number of similar moths that are white or yellow with markings on the fore wings.

2 *Lacipa pulverea*

Very small (wingspan 11mm). Fore wings white with 2 yellow bands and a row of small black spots at the tip. Hind wings white. **Biology** Some *Lacipa* species feed on Eland's Wattle (*Elephantorrhiza elephantina*). **Habitat** Occurs in many vegetation types, from fynbos to bushveld. **Related species** There are a number of other species in the genus, including several that look very similar.

3 *Naroma varipes* Fig-tree Moth

Small (wingspan 28mm), stout, silvery white. Margin of fore wing bends abruptly near tip. Larvae white and thinly haired. **Biology** Larvae are easily located, feeding in groups on various species of fig tree. The slightly hairy cocoons can be found at the base of the host tree. **Habitat** Subtropical forest and bushveld.

4 *Palasea albimacula* White-barred Gypsy

Medium-sized (wingspan 40mm). Fore wings grey, with a white bar consisting of large white clustered spots. Hind wings yellowish, with brown veins. Abdomen yellow. Larvae orange-brown, covered in orange hairs, with black side stripes. Larvae have 2 long, forward-projecting brown tufts near the head, and long, thick lateral tufts near the end of the body. **Biology** Larvae feed gregariously on corkwood (*Commiphora*). **Habitat** Bushveld and forest.

5 *Pseudonaclia puella* Girl Maiden

Small (wingspan 18–22mm), with brown body. Fore wings dull brown, with 2 pale yellow bands of irregular spots. Hind wings orange, with a large, dark brown marking on each border. **Biology** Flies by day, but short-winged (brachyterous) forms are flightless. **Habitat** Forest and woodland.

6 *Asura sagenaria* Crossed Footman

Small (wingspan 20–30mm) with variable coloration. Fore wings typically orange, with complex dark brown lines that cross and form a rough rectangle, with 3 shorter, separated bands at wing tip, which has a thin, dark brown border. Hind wings yellow, with thin, or in some very extensive, grey border. Thorax with 3 evenly spaced black dots. **Biology** Larvae feed on lichen and Cinnamon Tree (*Cinnamomum verum*). **Habitat** Forest and woodland.

Subfamily Erebinae

Medium-sized to large moths, often with contrastingly coloured fore and hind wings. Have well-developed ears (tympanal organs) for detecting bats in flight.

7 *Dysgonia properans*

Medium-sized (wingspan 30–40mm); fore wings boldly marked with 4 broad alternating bands in brown and tan. At rest, the front dark brown bands form a triangle. Second brown band has characteristic arrow-shaped marking on trailing edge. Hind wings buff-coloured, unmarked. **Biology** Larvae feed on Forest False Nettle (*Acalypha glabrata*). **Habitat** Moist forest and woodland. **Related species** The Jigsaw (*D. torrida*) (wingspan 35mm), which is similar in appearance and occurs across Africa.

1 *Trigonodes hyppasia* — Triangles

Medium-sized (wingspan 35–45mm). Body and hind wings grey, fore wings grey, each with a pair of large, yellow-bordered brown triangles. Larvae cream, with a number of stripes running the length of the body. **Biology** Larvae parasitized by ichneumonid wasp *Enicospilus* (p.462). They are occasional pests of Soya Bean leaves (*Glycine max*); also feed on *Indigofera* species and Lucerne (*Medicago sativa*). **Habitat** Common and widespread; also occurs across Africa, Madagascar, Asia and Australia, in both undisturbed veld and crop lands.

2 *Achaea violaceofascia* — Fruit-sucking Moth

Medium-sized to large (wingspan 40–60mm). Fore wings grey-brown with a silvery sheen; hind wings black with a white spot on the trailing edge. **Biology** Larvae are semi-loopers and feed on a range of foodplants, including many unrelated crops. **Habitat** Widespread across Africa and adjacent islands. **Related species** Larvae of the Asian Croton caterpillar (*A. janata*) feed on Castor Oil Plant (*Ricinus communis*), and in South Africa the common *A. ienardi* is a fruit-sucking moth that feeds on juices exuding from fruits damaged by fruit-piercing moths (*Serrodes*).

3 *Sphingomorpha chlorea* — Sundowner Moth

Large (wingspan 60mm), resembling a drab hawkmoth. Fore wings coloured in shades of brown, with fine black stripes and a weak eyespot. Thorax dark, with a cream central stripe and, in the male, a thick tuft of yellowish hairs. Larvae are twig-like semi-loopers, with black-edged red and yellow spots dorsally, near the head. **Biology** Larvae are often gregarious and feed on *Senegalia*, *Azanza*, *Burkea* and Marula (*Sclerocarya birrea*). Adults are attracted to the smell of overripe fruit, beer and sherry. **Habitat** Various; most common in bushveld and forest.

4 *Calliodes pretiosissima* — Wavy Owl

Medium-sized (wingspan 44mm). Wings brown above, with bright orange to red undersides; fore wings each with an eyespot and intricate wavy brown lines. **Biology** A subtropical species. **Habitat** Common in bushveld and forest. **Related species** *Erebus macrops* (below), which is similar, but larger.

5 *Erebus macrops* — Walker's Owl

The largest South African erebid (wingspan 120mm). Wings brown with wavy margins (hind wings darker) and an eyespot on each fore wing. Wings of live moths have a purplish gloss. Front of hind wings twisted over in male to cover a hairy glandular region. **Biology** Larvae are large semi-loopers and feed on Giant Sea Bean (*Entada abyssinica*). Adults generally flushed from between fallen logs, where they roost. May also rest in sheds. **Habitat** A common subtropical species that also occurs in equatorial Africa, Asia and Madagascar.

6 *Cyligramma latona* — Cream-striped Owl

Large (wingspan 75mm). Both pairs of wings transected by a cream stripe. Fore wings each with a prominent eyespot. **Biology** Larvae are semi-loopers and feed on *Vachellia*, Giant Sea Bean (*Entada abyssinica*) and apple. Adults rest with wings open, or closed upright, and are attracted to lights and the smell of alcoholic drinks or overripe fruit. **Habitat** Very common and familiar, especially in bushveld and subtropical forest. Range extends to equatorial Africa.

7 *Grammodes exclusiva* — Black-and-white Lines

Medium-sized (wingspan 30mm). Fore wings marked with a dark chocolate brown patch divided by a straight cream stripe. Hind wings grey, divided by a white line. **Biology** Larvae feed on citrus and *Polygonum*. **Habitat** Bushveld and forest. **Related species** *G. congenita*, which is very similar, but larger, with a weaker white line on the hind wings.

8 *Grammodes stolida* — Stolid Lines, Dubbeltjie Caterpillar

Medium-sized (wingspan 34mm). Sharp inward kink in second white stripe on fore wing. Prominent spines on hind tibiae. Larvae yellow-grey, with thin brown lines. **Biology** In Europe larvae feed on oak, *Tribulus* and *Rubus*. **Habitat** Common in most natural and disturbed areas in Africa, Europe and Asia. **Related species** *G. exclusiva*, which is similar, but lacks a kink in the fore wing stripe.

1 *Hypopyra capensis*   Red Tail

Large (wingspan 72mm), cryptically coloured. Wings brown, yellowish or reddish-brown, with a light yellow marginal line. Red abdomen, base of hind wings and wing undersides unmistakable. **Biology** Reveals orange-red body and underwings in flight, in contrast with cryptic upperside visible at rest. Larvae feed on false thorn (*Albizia*) and Small Knobwood (*Zanthoxylum capense*). **Habitat** Most common in bushveld and subtropical forest. Range extends into equatorial Africa.

2 *Egybolis vaillantina*   Peach Moth

Medium-sized (wingspan 50mm), bright orange and metallic blue-green. Hind wings black with a metallic blue margin. Antennae comb-like in males. **Biology** Larvae (**2A**) feed on peach and soapberry (*Sapindus*) and, occasionally, on the leaves of cotton. Adults fly quite slowly in the treetops during the day. **Habitat** Subtropical bush and forest, especially near the coast. Range extends into equatorial Africa.

Subfamily Pangraptinae

Comprises a small group of drab moths with large, scalloped wings.

3 *Gracilodes caffra*   Orange Drab

Medium-sized (wingspan 30–40mm). Wings and body dirty brown to orange, with 3 thin black lines running across both pairs of wings; border of hind wing with thin white sickle-shaped area and a small patch of peacock iridescence. Some more brightly coloured forms have yellow-and-brown wings. **Biology** Resting posture characteristic, with long, spidery fore legs stretched out, and wings held flat and open. Larvae feed on alien Blackjack (*Bidens pilosa*). **Habitat** Forest and open woodland.

Subfamily Calpinae

Includes piercing moths, which have a barbed proboscis capable of piercing fruits or, in the case of the vampire moths, mammalian skin.

4 *Eudocima materna*   Dot-underwing Fruit-piercer

Large (wingspan 90mm), striking erebid moth. Fore wings brownish with wavy margins and, in female (shown), a white stripe over a black triangle. Continuous black border runs across the bright orange hind wings, which each have a central black spot. **Biology** Larvae are semi-loopers. Adults readily pierce ripe fruit. **Habitat** Common in subtropical bush and forest. Wide tropical range includes Asia and Australia. **Related species** There are 2 other similar species in the genus, with smooth margins to the fore wings and different patterns on the hind wings.

5 *Rhanidophora cinctigutta*   Dice Moth

Smallish (wingspan 35mm), yellow, with 3 immaculate, round white spots on each fore wing. Larvae are semi-loopers, boldly striped in white and dark brown, with bizarre long clubbed hairs (**5A**). **Biology** Larvae feed on various *Thunbergia* species, and wave the club-like hairs on their heads while feeding. **Habitat** Subtropical bushveld and forest. **Related species** There are a few similar, smallish, regional species in the genus with identical fore wing spots and similar larvae.

Family Nolidae

Nolid moths

Moths with mostly subdued colouring (many are green), characteristically emerging from a vertical slit in the pupa. Many have tufts of hairs on the triangular fore wings, which often end in a point and may be scalloped.

6 *Earias cupreoviridis*   Cupreous Spiny Bollworm

Small (wingspan 18mm). Fore wings bright apple green with orange spots halfway along length. Wing tips yellow shading to brown. Hind wings translucent white. All *Earias* larvae are humpbacked, spiny and yellow. **Biology** The larvae feed on the buds of a wide range of plants, as well as cotton bolls. Like its close relatives of similar appearance, it is a pest of cotton and related *Hibiscus*. **Habitat** Any natural or disturbed area with appropriate foodplants.

Family Euteliidae

Recognized by their characteristic resting position, in which they resemble dead leaves. The wings are folded and pressed downwards, while the abdomen is held up in the air. Abdomen has small lateral extensions or a terminal prong.

1 *Colpocheilopteryx operatrix* Worker Turntail

Medium-sized (wingspan 34mm), with body and wings beautifully marked in pastel shades of yellow, purple and apricot. Tip of abdomen ends in a small fork. **Biology** Rests with wings partially folded and spread flat and outwards, with the abdomen curled over the body. **Habitat** Grassland and bushveld.

Family Noctuidae   Owlet moths

The largest and most successful family of moths in the world, containing medium-sized to large night-flying moths, usually with dull grey or brown, often triangular, fore wings, and plain white or grey hind wings. A few species are brightly coloured. Most have whip-like antennae. Adults of a number of species are attracted to overripe fruit and can pierce the skin of fruit, even of oranges, with a saw-like proboscis. In the subtropics, some species suck lachrymal fluid from the eyes of livestock and antelope. Larvae are generally smooth, but may have hairs on raised bumps. Often destructive, and include the well-known cutworms, army worms, bollworms and stalk borers. Some are semi-loopers, resembling geometrid larvae. Most feed on leaves, although some species bore into stems and a few eat mealy bugs. Pupation occurs in the soil or in a silken cocoon without larval hairs. Over 1,700 species are known from the region.

2 *Acontia* Droplet moths

Includes many similar small moths (wingspan 20–30mm), most of which are remarkable mimics of bird droppings (although a few have bright warning coloration). Thorax and wings often pearly white, with olive and brown markings, so that in the resting position the moth resembles a small, raised, oval bird dropping. **Biology** Larvae feed on *Hibiscus*, *Abutilon*, cotton (*Gossypium*) and various other mallows (Malvaceae). **Habitat** A range of vegetation types, as well as crop lands.

3 *Brithys crini* Lily Borer

Medium-sized (wingspan 30–40mm), satiny black moth. Tips of fore wings rich brown. Hind wings of males white, of female white with smoky wing edges. Legs striped in black and white. The distinctive larvae **(3A)** are banded in black and yellow, with each band comprising a number of smaller patches of colour. **Biology** Feeds on parts of various Amaryllidaceae lilies, including *Amaryllis* and *Eucharis*, and even *Clivia*. Can be very destructive when feeding in groups, as the smaller larvae burrow beneath the leaf epidermis, killing the leaves. Ultimately they migrate downwards and feed on the bulb/rhizome. Will also feed on flower buds and flower stems. Eggs laid in spring, with most of the damage caused by mature larvae in summer. Many generations per year in warmer parts, with pupation at the base of the bulb. **Habitat** In all parts where host bulbs grow, and especially abundant in gardens where large concentrations of bulbs occur. Widely distributed across Africa, southern Europe and Asia.

4 *Neuranethes spodopterodes* Agapanthus Borer

Medium-sized (wingspan 30mm). Wings short, broad, intricately marked in grey and black, with a broad, light grey tip and 2 small grey circles on each fore wing. Larvae **(4A)** rotund and grub-like, grey and covered with rows of conspicuous black tubercles. **Biology** Feeds on all species and hybrids of *Agapanthus*, especially the dark blue *A. inapertus* and its hybrids. Adults present in summer, laying egg clusters on the stems of agapanthus inflorescences or on the leaves. First-instar larvae typically feed on flower buds at first, then move down the flower stem and feed on the leaf bases and, ultimately, rhizome, which may be destroyed. A major horticultural pest. **Habitat** Originated in high-altitude grassland in Mpumalanga and parts of KwaZulu-Natal, but since 2010 has become established in the Western Cape and Gauteng. Distribution apparently limited by its intolerance for frost.

5 *Thysanoplusia orichalcea* Burnished Brass, Golden Plusia

Medium-sized (wingspan 40mm). Fore wings brown, with a gold triangle. Hind wings straw-coloured. A tuft of plumes extends from the thorax. Larvae **(5A)** are green or blue-green semi-loopers, with fine white longitudinal lines. **Biology** Larval foodplants include bean plants, chicory, sunflower, Blackjack (*Bidens pilosa*), most legumes, maize, radish, lettuce and potato. Adults emerge from silken cocoons after 10 days. **Habitat** Most natural and disturbed areas, including gardens. Common and widespread.

1 *Mentaxya albifrons* — White Quaker

Small (wingspan 25–32mm,) with black-and-white fore wings, the base of each bearing a distinct white cross. Hind wings white. **Biology** Light brown cutworm-type larvae feed on potato, but probably attack a wide range of vegetable crops. Also recorded from arum lilies (*Zantedeschia*) and *Pelargonium*. **Habitat** Occasional in most veld types, as well as agricultural and garden settings.

2 *Chrysodeixis acuta* — Silver U, Tomato Semi-looper

Medium-sized (wingspan 34mm). Fore wings mottled brown and marked with a small silver 'U' and a silver circle. Hind wings grey-brown. Larvae are yellow-green semi-loopers, with white and yellow longitudinal lines. **Biology** Larvae feed on leaves or bore into a wide range of plants, including canna lily, banana, tobacco and potato. **Habitat** Most natural and disturbed areas, including gardens. **Related species** *C. chalcites*, which has lighter fore wings and a more tropical distribution.

3 *Spodoptera littoralis* — Tomato Moth, Cotton-leaf Worm

Medium-sized (wingspan 37mm). Fore wings with intricate geometric pattern of brown, grey and tan. Hind wings plain white. Larvae are greyish-brown cutworms, with lines of crescentic spots and a yellow stripe on the black back. **Biology** Like other *Spodoptera* species, a major pest, feeding on a wide range of vegetable crops, and also on sunflower, cotton leaves and tobacco, as well as on diverse plants such as *Hibiscus*, *Lantana* and even *Eucalyptus*. Thousands of eggs are laid on the underside of leaves and covered by brown or yellow down from the moth. Larvae are small and gregarious. **Habitat** Abundant and widespread across most habitats.

4 *Spodoptera exempta* — African Army Worm

Small (wingspan 24–32mm), pale to dark brown or nearly black; hind wings white and translucent (moth not shown). Larvae green, with bold, velvety black stripes (but gregarious-phase with aggregating larvae more black than green). **Biology** Widespread and destructive. Larvae usually feed on grasses and pupate in an earthen cell in the soil. Gregarious-phase larvae often move en masse. **Habitat** Occurs over most of Africa. **Related species** The Fall Army Worm (*S. frugiperda*) **(4A)**, which is a recent invader from the Americas, and a very damaging crop pest. Its larvae are green, turning dark olive green, with a characteristic Y-shaped marking on the head **(4B)**.

5 *Spodoptera cilium* — Cape Lawn Moth

Smallish (wingspan 30mm), stout, difficult to identify. Fore wings plain grey-brown. Hind wings whitish, with a thin brown border. Easily confused with similar lawn moths. **Biology** Larvae live in soil, feeding on grass roots and rice plants. Eggs are laid in bunches on grass and hatch into small green larvae that feed in the thatch of the lawn at ground level. Pupation occurs in soil. Often a pest of lawns. **Habitat** Ubiquitous. Occurs over most of Africa.

6 *Helicoverpa armigera* — African Boll Worm, Corn Ear Worm

Medium-sized (wingspan 38mm), variable in colour. Fore wings generally pale or reddish-brown, occasionally tinged with green. Hind wings whitish, with grey-brown marginal band and brown veins. Larvae **(6A)** brown to greenish-brown, slender, with a fine, black-edged, white central stripe and cream side stripes. **Biology** A major pest, feeding on pods, stems and fruits of a wide range of vegetables and trees, and especially on green tomatoes, causing rot. **Habitat** Ubiquitous in most habitats. Cosmopolitan.

7 *Diaphone eumela* — Cherry Spot

Medium-sized (wingspan 36mm) and striking, fore wings grey and white, banded in black, with a small red C-shaped mark. Hind wings white in male, grey in female. Thorax fluffy and grey, with 6 orange spots. Larvae yellow, with black bands, broken into dots in the Cape form. **Biology** Larvae feed on lilies such as *Ornithogalum*, *Ornithoglossum*, *Albuca* and *Scilla*. **Habitat** Open areas in a wide range of vegetation types. Range extends into Malawi and Zimbabwe.

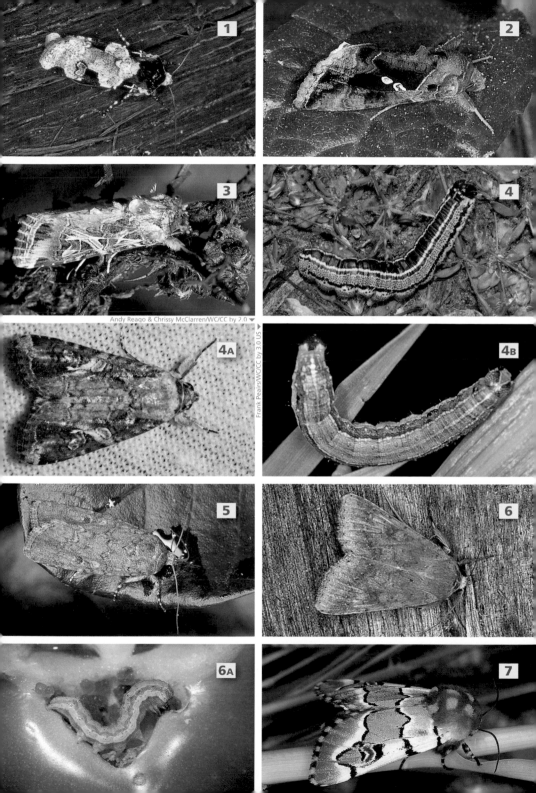

1 *Leucania* — Maize borers

Small (wingspan 28mm). Fore wings straw-coloured, with a brown longitudinal streak in some species. Hind wings white. Larvae yellowish-brown with darker brown stripes and distinctive yellowish-brown 'eyes' marked with a network of fine brown lines. **Biology** Larvae often occur in huge numbers feeding on young leaf bases. In maize fields they feed on the succulent leaves of young maize cobs, fouling the cobs with their droppings. Also attack cereal crops and wheat seedlings. Pupation occurs in the soil. Produce several broods per year. **Habitat** Ubiquitous. Range extends into Europe, Asia and Australia.

Subfamily Agaristinae — Forester moths

Medium-sized, brightly coloured, day-flying moths, with wings boldly marked in combinations of red, orange, yellow and black. Whip-like antennae may have a terminal hook. Males of certain species are capable of producing sound (stridulating) by rubbing the spines of the legs against the wings. Larvae have a hump at the end of the body. Like adults, they have warning colours, being banded in black and white, with red or orange markings. Larvae are very host specific, and pupate at the base of the foodplant.

2 *Agoma trimenii* — Trimen's False Tiger

Medium-sized (wingspan 54mm), with bold markings. Fore wings brownish-black with 1 small cream spot and 2 large cream patches (1 of which is triangular). Hind wings orange, yellow or white, with a black border. Abdomen orange with a black central stripe. **Biology** Larvae feed on wild grape (*Cissus* and *Rhoicissus*) and cultivated grapevine (*Vitis vinifera*). Adults fly at night and are attracted to lights. **Habitat** Subtropical forest. **Related species** There are many other very similar species in the genus.

3 *Heraclia* — Superb false tigers

Medium-sized (wingspan 60mm). Fore wings black with large yellow spots. Hind wings red or yellow with black border. Larvae yellow, with dense black hairs. Genus includes a number of very similar species. **Biology** Sluggish day-flier. Larvae feed on wild pea, *Impatiens hochstetteri*, *Pentas*, wild grape (*Cissus, Cyphostemma*) and cultivated grapevine (*Vitis vinifera*). **Habitat** Mostly subtropical forest.

4 *Sommeria strabonis* — Squinting Digama

Medium-sized (wingspan 30mm), unmistakable; fore wings white, crisply marbled in black. Tips of wings (also cream hind wings) have a grey border. Abdomen yellow with a central row of black dots. **Biology** Larvae of related species feed on poison bush (*Acokanthera*). **Habitat** Forest.

BUTTERFLIES

Among the best known and most familiar of insects. Adults are often brightly coloured and usually active during daylight hours. Most have clubbed antennae and hold their wings together vertically when at rest. There are about 20,000 species globally and over 660 known from South Africa.

Family Hesperiidae — Skippers

Small to medium-sized, hairy and heavy-bodied. Most are relatively drab, brownish or black with yellow, orange or white markings. The antennae are set wide apart and end in long, curved or hooked clubs. At rest, wings may be held open or half open, sometimes upright. The rapid, darting flight gives the family its common name. Larvae have a prominent head, with a constricted neck region, and feed at night, hiding by day in rolled or folded leaves. Pupation often occurs in these shelters.

5 *Sarangesa motozi* — Elfin Skipper

Medium-sized (wingspan 36–40mm). Wings mottled brown. Fore wings with several squarish clear spots. Best identified by the single round, clear spot at the centre of each hind wing. **Biology** Settles on the ground or on foliage with wings open. Flies all year round. Larval host plant is *Peristrophe hensii*. **Habitat** Forest and savanna woodland, especially near the coast. **Related species** There are 3 other similar species in the genus, but they lack the clear hind wing spot.

1 *Gomalia elma* — Green-marbled Sandman

Small (wingspan 30–36mm), with 3 radiating white lines on the back; wings brown with narrow white bands on fore wings and broad ones on hind wings. Larvae feed on *Abutilon grandiflorum* and related species. **Biology** Males patrol their territory, pausing to perch with wings open. **Habitat** Forest edges, woodland clearings, parks and gardens. Widespread, but seldom common.

2 *Spialia* — Sandmen

Small (wingspan 20–30mm), dark brown to blackish butterflies, with numerous creamish-white spots on the wings; wing margins have brown and white fringes. Undersides of wings pale brown, also white-spotted or barred. Genus includes 13 very similar South African species, some localized, others widespread. **Biology** Fast fliers with buzzing wingbeats, settling periodically on foliage, flowers or the ground. Larvae feed on various plants, including doll's rose (*Hermannia*) and *Hibiscus*. **Habitat** Variable; frequently on mountains and hillsides.

3 *Metisella metis* — Gold-spotted Sylph

Small (wingspan 30mm); body golden-yellow. Wings dark brown, with a few bright golden-yellow spots and golden-yellow leading edge. **Biology** Slow, hopping flight, settling frequently with wings open. **Habitat** Found widely in gardens and moist grassland. Larvae feed on various grass species. **Related species** There are 5 other regional species in the genus, which are differentiated by the number and pattern of spots and streaks on their wings.

4 *Kedestes callicles* — Pale Ranger

Small (wingspan 27–33mm). Upperside dark brown with yellow spots; underside striking golden-yellow, with white spots outlined in black. **Biology** Skipping flight, usually in the shade of trees. Flies in summer; males territorial. **Habitat** Thick bush and forest in tropical parts. **Related species** There are 9 other regional species of *Kedestes*, each best identified by the unique patterns on the undersides of the wings.

5 *Gegenes niso* — Common Hottentot Skipper

Small (wingspan 29–35mm). Male (shown) dark brown above, dusted with yellow, more so towards wing bases. Female with yellowish patches on fore wings, and a diffuse main patch on hind wings. Undersides of wings yellowish. **Biology** Rapid fliers, often returning briefly to a favourite perch. **Habitat** Common and widespread in forest, fynbos, savanna and grassland. Frequently visits flower gardens. **Related species** There are 2 other very similar but more range-restricted species in this genus; also easily confused with several other related genera of mostly brown skippers, which have pale spots on their wings, including *Platylesches, Pelopidas* and *Borbo*.

6 *Moltena fiara* — Banana-tree Night-fighter

Medium-sized (wingspan 45mm) and unusually large and robust for a skipper: moth-like, with disproportionately large body and head. Uniformly brown, more chestnut towards wing bases. Conspicuous white antennae in male (shown). **Biology** Active at twilight. Has noisy buzzing flight and may be attracted to lights. Larvae **(6A)** feed on Natal Banana (*Strelitzia nicolai*) and other *Strelitzia* species, and make shelters by folding over the edge of a leaf and fastening it with silk. **Habitat** Coastal bush and forest, also gardens and parks.

7 *Netrobalane canopus* — Buff-tipped Skipper

Medium-sized (wingspan 40–45mm). Wings creamish-white with clear patches and an unmistakable marbled pattern, probably mimicking bird droppings. Tips of fore wings reddish-brown with black markings; hind wings reddish-brown at base. Long hairs on body and inner area of hind wings. Larvae green, with dark head and white hairs. **Biology** Adults perch on prominent twigs or leaves. Males make a clicking sound in flight when courting females. Larvae feed on Crossberry (*Grewia occidentalis*) and Natal Wild Pear (*Dombeya cymosa*). **Habitat** Bushveld and coastal forest. Adults occur year-round.

1 *Acleros mackenii* Macken's Dart

Small (wingspan 29–33mm). Upperside uniformly dark in male, with a few white spots in female; underside attractively mottled in cream and brown, becoming darker posteriorly. Tip of abdomen white. **Biology** Shade-loving. Flight skipping; settles frequently and often returns to the same perch. Males are territorial. **Habitat** Forest and dense bush.

2 *Leucochitonea levubu* White-cloaked Skipper

Small (wingspan 30–40mm), striking and unmistakable. Wings white, with black margins and veins, and chequered black blocks parallel to the outer margins. Conspicuous patches of yellow body hair under wings. **Biology** Males visit and perch conspicuously on the tops of koppies at midday, where they establish territories. Females often visit flowers. Larvae feed on Wild Raisin (*Grewia flava*). **Habitat** Savanna and rocky bushveld.

3 *Coeliades pisistratus* Two-pip Policeman

Large (wingspan 55–70mm). Wings greyish-brown above, underside of hind wings with diagnostic broad white band and 3 black spots. Orange fringe along back of hind wings. **Biology** Fast, powerful fliers, most active in the early morning and late afternoon. Larvae feed on a variety of host plants, including bush willow (*Combretum*) and Bug Tree (*Solanum mauritianum*). **Habitat** Edges of rainforest and coastal bush. **Related species** There are 4 other species of *Coeliades* in the region, 2 of which have a white band on the underside of the hind wings. The widespread Striped Policeman (*C. forestan*) has no black spots, while the One-pip Policeman (*C. anchises*), in the northeast of the region, has a single black spot on the white band.

Family Nymphalidae Brush-footed butterflies

A diverse family of butterflies that appear to have only 4 legs, because their fore legs are reduced and no longer used for walking. The component subfamilies were historically considered separate families and are treated sequentially below.

Subfamily Danainae Milkweed butterflies, monarchs

Large, conspicuously coloured butterflies. Black, with white markings, or orange and black, warning predators of their distasteful or poisonous properties. Flight slow and leisurely, with frequent gliding. Larvae have 2–4 pairs of fleshy outgrowths and feed by day, exclusively on milkweeds (Asclepiadaceae), Apocynaceae and Moraceae, the poisons of which accumulate in their bodies, rendering them distasteful to predators. These poisons persist into the adult stage. Pupae suspended by tail hooks from a silken pad attached to a support.

4 *Amauris niavius* Friar

Very large (wingspan 80–85mm), body black with white spots. Wings black around the edges, with large white patches and smaller white spots; veins black. **Biology** Slow fliers. Larvae feed on *Gymnema sylvestre* (Asclepiadaceae). **Habitat** Forested coastal areas of KwaZulu-Natal and the mountains of Mpumalanga. **Related species** There are 3 other species in the genus, all from the eastern parts of the region: the Novice (*A. ochlea*) is similar, but smaller, while both the Layman (*A. albimaculata*) (**4A**) and Chief (*A. echeria*) have yellow patches on the hind wings. *Papilio dardanus* females (p.446), both sexes of Variable Diadem (*Hypolimnas anthedon*) and several other butterflies and moths all mimic *A. niavius*.

5 *Danaus chrysippus* African Monarch

Large (wingspan 60–75mm) and 1 of the most common and familiar butterflies in the region. Sides of the thorax black, with conspicuous white spots. Wings orange-brown, with a black border. Apex of fore wings black, with a white bar and spots. Hind wings each with 4 (in male) or 3 (in female) black spots. Larva (**5A**) strikingly banded in yellow and black, with elongate fleshy filaments. Pupa (**5B**) green or yellow, short, fat and smooth, with gold spots. **Biology** Larvae feed mostly on *Asclepias* and other milkweeds. **Habitat** Common in almost all habitats. **Related species** No similar related species, but it is closely mimicked by several unrelated species, especially females of *Hypolimnas misippus*.

Subfamily Helioconiinae   Acraeas, leopards

Medium-sized butterflies with elongated fore wings. Wings conspicuously coloured, mostly in red or orange with black markings, sometimes with bare patches. Larvae covered with long, branched spines and feed mainly on granadillas (*Passiflora*), from which they extract toxins to deter predators. Pupae suspended head-down from a silken pad by tail hooks.

1 *Hyalites eponina*   Small Orange Acraea

Medium-sized (wingspan 35–40mm). In male (shown), wings are orange, with orange-dotted black margins and an orange blotch midway along the leading margin. In female, wings have a similar marginal pattern, but fore wings have black, white or transparent areas distally. **Biology** Weak fliers, often settling on low vegetation. Larvae feed on *Triumfetta*. **Habitat** Favours long grass at the edge of moist forest and other wooded areas. **Related species** There are several other similar species in this diverse genus.

2 *Hyalites esebria*   Dusky Acraea

Large (wingspan 50–60mm); highly variable in colour. Black, with a large yellow or sometimes orange patch on the hind wings, joining a similarly coloured patch on inner margin of the fore wings. Has yellow or white oblique bars below apex of fore wings. (Paler '*lycoides*' form shown here; wings can be much darker.) **Biology** Slow-flying, settling frequently. Larvae feed on *Laportea peduncularis*, Tree Nettle (*Obetia tenax*) and *Pouzalzia parasitica*. **Habitat** Woodland and forest fringes. **Related species** There are several similar species in this genus, including the similarly distributed White-barred Acraea (*H. encedon*) **(2A)** (wingspan 45mm), but that species has black spots on its wings.

3 *Acraea natalica*   Natal Acraea

Large (wingspan 55–65mm). Colour varies with sex and season. Males pinkish-red, females brownish-red. Wings with black spots and black bases. Fore wings with black apex and a group of contiguous black spots at end of the fore wing cell. Hind wings with broad black border on outer margin. **Biology** Zigzag flight, becoming rapid if disturbed. Larvae feed on various plants, including granadilla (*Passiflora*). **Habitat** Common in woodland and savanna. **Related species** This diverse genus contains several other species with similar colouring and distribution, including *A. oncaea*, *A. lygus* and the Clear-spotted Acrea (*A. aglaonice*) **(3A)** (wingspan 45–55mm), all of which have white or clear patches in the wings.

4 *Acraea horta*   Garden Acraea

Medium-sized (wingspan 45–53mm). Sexually dimorphic: male bright reddish, female paler reddish-orange **(4)**. Apical half of fore wings translucent, with prominent veins and a small dark bar at end of the fore wing cell. Hind wings with several black dots; outer margins chequered in black. **Biology** Larvae dark and covered with branched spines **(4A)**. They feed on Wild Peach (*Kiggelaria africana*) and some granadillas (*Passiflora*) and may be abundant in gardens. Pupate on or off the foodplant, often hanging on nearby walls and similar supports. Both larvae and pupae frequently parasitized by wasps such as *Charops* (p.462) and *Brachymeria kassalensis* (p.466). **Habitat** Woodland areas and gardens. **Related species** The genus includes several other species that are similar, but that can be distinguished by details in their wing pattern.

5 *Pardopsis punctatissima*   Polka Dot

Small (wingspan 33mm), easily recognized by pale orange colour, heavy black spotting, dark wing margins and rounded fore wings. Larvae blackish-brown or blue-black, with whitish protuberances along the back and 2 movable, antenna-like protuberances behind the head. **Biology** Flies weakly, close to the ground, often settling on low vegetation. Larvae feed on *Hybanthus capensis* (Violaceae). Mimicked by lycaenid butterflies of genus *Pentila* (p.452). **Habitat** Moist grassland.

1 *Phalanta phalantha* — Common Leopard

Medium-sized (wingspan 40–48mm), orange, with delicate black markings and 3 black lines along outer wing margin. The innermost black line is distinctly scalloped. Underside of wings yellowish-orange, with darker markings, a purplish tinge to the outer areas and a black spot near the inner angle of the fore wing. **Biology** Moves constantly, opening and closing the wings even when at rest. The larvae feed on various plants, including White Poplar (*Populus alba*), willow and Wild Mulberry (*Trimeria grandifolia*). **Habitat** Mostly wooded savanna. **Related species** The Forest Leopard (*P. eurytis*) (wingspan 45mm), which is very similar but paler, and the Eastern Blotched Leopard (*Lachnoptera ayresii*) (wingspan 50mm), which has squarer wings. Both are confined to the subtropical parts of the region.

Subfamily Satyrinae — Browns

Small to fairly large butterflies. Wings short and broad, tawny or various shades of brown or grey, often with several eyespots around the margins. Underside of wings often cryptically coloured. Identified by swollen main veins at base of each fore wing. Larvae green, yellow or brown, marked with longitudinal lines, densely covered in short hairs and with a forked rear segment. Foodplants mostly grasses and sedges. Pupae hang head-down from a foodplant or beneath stones or logs, or may lie free beneath debris on the ground.

2 *Aeropetes tulbaghia* — Table Mountain Beauty

Unmistakable and among the largest (wingspan 70–90mm) and most spectacular of satyrines. Dark brown, with 2 broken yellow bands parallel to outer margin of the fore wings and 1 on the hind wings. Row of vivid mauve eyespots on hind wing margin. Larvae brown, with prominent dark dorsal line. Pupae **(2A)** squat and white, with small black dots; hang head-down from a rock. **Biology** Adults strong fliers and frequently visit red flowers like red-hot pokers. The only pollinator of the Red Disa (*Disa unicornis*). Larvae feed on grasses. **Habitat** Rocky areas in mountainous and high-lying regions.

3 *Dira clytus* — Cape Autumn Widow

Medium-sized (wingspan 50mm), dark brown; each fore wing with a row of yellow patches on its outer margin and an eyespot near the wing tip. Hind wings with eyespots along outer margin. **Biology** Often very common, especially in autumn. Adults fly slowly just above grass, settling often on bare patches. Female scatters eggs in flight. The well-camouflaged, fat brown larvae feed on various grasses. **Habitat** Grassy areas on mountain slopes and on lower ground. Strays into urban areas. **Related species** There are several other, easily confused species in this genus and in the related genus *Dingana*.

4 *Bicyclus safitza* — Common Bush Brown

Medium-sized (wingspan 40–48mm), dark brown with 2 sometimes indistinct eyespots on the upperside of each fore wing. Hind wings 2-toned, with much more distinct eyespots of variable sizes along the paler outer margin. **Biology** Adults have slow, bouncy flight along the ground, often settling on low vegetation along a path or at the edge of a glade. **Habitat** Forests and well-wooded areas. Larvae feed on a variety of grasses. **Related species** There are 2 other similar-looking *Bicyclus* species, but they are less common; also, 1 *Henotesia* species.

5 *Cassionympha cassius* — Rainforest Brown

Medium-sized (wingspan 34–42mm), brown. Fore wings each with a large orange central patch and a yellow-ringed black eyespot containing 2 white dots. Hind wings each have 2 similar eyespots, but each contains just 1 white dot. Number of eyespots varies. A darker brown line runs parallel with outer margin of both fore and hind wings. **Biology** Hopping flight along edges of glades and roads. **Habitat** Coastal bush, wooded kloofs and forests. **Related species** There are many similar-looking species in several genera, especially *Pseudonympha*, which includes 13 difficult-to-distinguish but mostly range-restricted species, of which Cape coastal *P. magus* depicted **(5A)**.

6 *Melampias huebneri* — Boland Brown

Medium-sized (wingspan 36mm) with an elongated antennal club. Wings brown; fore wing with a large orange patch encircling a black eyespot that contains 2 small white spots. Hind wing has smaller orange patch, but no eyespots. **Biology** Locally common in low vegetation. Larvae feed on various grasses. **Habitat** In coastal fynbos or Nama karoo vegetation.

1 *Melanitis leda* — Common Evening Brown

Large (wingspan 65mm). Fore wings brown on upperside, with large orange eyespot near apex, enclosing a black comma-shaped patch marked with 2 white spots; outer margin near apex produced into a point; underside cryptically coloured. Hind wings with stout tail and up to 3 tiny, white-centred eyespots. **Biology** Remains hidden among dead leaves and debris below bushes and trees during the day. Reluctant to fly. If disturbed, flies a short distance before settling again. Larvae feed on *Cynodon* grass. **Habitat** Rainforest, coastal bush and bushveld. **Related species** The Yellow-banded Evening Brown (*Gnophodes betsimena*), which is similar in shape and habits, but lacks the eyespot on the upper wing.

Subfamily Limenitinae — Foresters, guineafowl, sailers, etc.

A small group of 14 regional species, the majority of which are rare and range-restricted. Their most constant characteristic is floating, sailing flight.

2 *Neptis saclava* — Spotted Sailer

Medium-sized (wingspan 40–50mm) with striking black wings marked with radiating white bars. Pale, mottled brown underside (other species of *Neptis* with black-and-white undersides). **Biology** Perches prominently and flies with slow, sailing flight, attracted to fermenting plant matter. **Habitat** Mostly coastal forests. **Related species** The are 5 other similar, but less widespread, regional *Neptis* species.

3 *Hamanumida daedalus* — Guineafowl

Large (wingspan 55–78mm), unmistakable, grey-brown, with numerous black-ringed white spots. Underside yellow-orange with white spots. **Biology** Flies low over rocky and sandy areas, often settling with wings open; always alert. Larvae feed on Velvet Bush Willow (*Combretum molle*) and Silver Cluster-leaf (*Terminalia sericea*). **Habitat** Common in thornveld and savanna.

4 *Euphaedra neophron* — Gold-banded Forester

Large (wingspan 55–78mm), unmistakable, upperside of wings metallic bluish, apical two-thirds of fore wings black with broad orange bar, extreme tip orange or yellow. Underside cryptically mottled in brownish-yellow. **Biology** Shade-loving. Glides low in glades or along pathways, often settling in patches of sun with wings open. Attracted to rotting fruit. Larvae feed on soapberry (*Deinbollia*). **Habitat** Coastal forest.

Subfamily Nymphalinae — Commodores, pansies, etc.

Small to large butterflies with coloration too diverse to allow for generalized description. As with other Nymphalidae, fore legs non-functional, formed into small brush-like structures. Powerful fliers, often attracted to fruit. Larvae diverse, usually cylindrical and covered with branched spines. Pupae angular or have tubercles and hang head-down, supported by their tail hooks.

5 *Hypolimnas misippus* — Common Diadem

Large (wingspan 60–80mm), strongly sexually dimorphic. Male **(5)** is larger and unmistakable: black, with a circular violet- or blue-ringed white patch on each wing, and a smaller white patch near fore wing tip. Female **(5A)** orange, with black borders, closely mimicking the African Monarch (*Danaus chrysippus*) (p.434), but distinguishable by hind wing cell that is open to the margin, and by single black spot (see arrow) on leading edge of hind wing. Larvae feed on the herbs Wild Foxglove (*Asystasia gagentica*), *Portulaca* and *Talinum*. **Biology** Strong fliers. Females often settle on foliage or on the ground. **Habitat** Open wooded country, parks and gardens.

6 *Precis octavia* — Gaudy Commodore

Medium-sized (wingspan 56mm), unmistakable but with remarkably different wet and dry season forms. Dry season form **(6)** bluish-purple, with black markings and a row of bright red spots. Wet season form brick red, with black patches, spots and wing margins **(6A)**. Several intermediate forms occur. **Biology** Dry season form may cluster in sheltered spots; wet season form is attracted to hilltops. Larvae feed on *Coleus* and *Plectranthus*. **Habitat** Bushveld, forest edges, grassland and gardens. Wet season form prefers grassy areas.

1 *Precis archesia*   Garden Inspector

Medium-sized to large (wingspan 45–60mm), ground colour brown. Dry season form has bright red wing bands parallel with outer wing margins. Red band forks at tip of fore wing, with red outer and white inner fork. Front margin of fore wings suffused with purple. Wet season form **(1A)** has yellow or orange wing bands. Intermediate forms occur. **Biology** Often settles on bare ground, rocks and paths. Larvae feed on the herbs *Coleus* and *Plectranthus*. **Habitat** Rocky or grassy slopes, forests and hilltops. Common in gardens.

2 *Junonia natalica*   Brown Pansy

Medium-sized (wingspan 45–55mm), brown and strikingly coloured, with orange to red bands and eyespots on wings; tips of fore wings dark, with 4 white spots. Larvae **(2A)** brown with a yellow head and rows of branched spines. One of many butterflies that feed on Wild Foxglove (*Asystasia gagentica*). **Biology** Often seen on the ground, slowly opening and closing the wings. Attracted to flowers. **Habitat** Forest edges, savanna, parks and gardens. **Related species** The Soldier Pansy (*J. terea elgiva*) **(2B)** (wingspan 50–60mm), which is similarly distributed and has an inward-curving yellow wing band on the fore wings and numerous orange ocelli on the hind wings.

3 *Junonia oenone*   Blue Pansy

Medium-sized (wingspan 40–52mm), ground colour black. Fore wings with broken white bar across apex and 3 white spots near wing tip. Hind wings each with a large blue spot and 3 broken white lines parallel to outer margins. Underside camouflaged in mottled browns. **Biology** Adults settle on the ground, slowly opening and closing their wings. Attracted to herbaceous borders and hilltops, where males establish territories. Larvae feed on Wild Foxglove (*Asystasia gangetica*), *Barleria* and *Phaulopsis imbricata*. **Habitat** Diverse. Prefers woodland and bushveld, but also frequents gardens and parks.

4 *Junonia hierta*   Yellow Pansy

Medium-sized (wingspan 40–50mm), black, with large yellow and orange patches on the wings; hind wings each with a violet-blue circular patch. Greyish-brown underside provides camouflage when wings are closed. **Biology** Often settles on the ground with wings open. Males establish and patrol territories. Attracted to rotten fruit and damp ground. Larvae feed on a variety of herbaceous plants, including Wild Foxglove (*Asystasia gangetica*), Zebra White (*Adhatoda densiflora*) and *Barleria*. **Habitat** Savanna, grassland, parks and gardens. Prefers open ground and grassy areas.

5 *Salamis parhassus*   Common Mother-of-Pearl

Large to very large (wingspan 65–90mm), greenish-white with mother-of-pearl lustre. Fore wings extended at tip into backward-curving point; hind wings with short tails and pointed rearmost angle. Colourful eyespot on each fore wing; 2 on each hind wing. **Biology** Strong fliers. Often settle on foliage, especially on the undersides of leaves in trees, sometimes gathering in swarms. Larval foodplants include Wild Foxglove (*Asystasia gangetica*) and Buckweed (*Isoglossa woodii*). **Habitat** Coastal and inland forest.

6 *Vanessa cardui*   Painted Lady

Medium-sized (wingspan 40–50mm); mostly orange with black markings. Tip of fore wings black, with several white markings and a broken black bar across orange area of wing. Hind wings with 3 rows of black spots parallel to outer margins. Underside of wings cryptically coloured, with a row of eyespots parallel to outer margins. Larvae black, with branched spines. **Biology** Rapid and powerful flier and among the most widely distributed of all butterflies. Adults often settle on the ground with wings open. Larvae feed on very wide range of plants, including *Arctotis*. **Habitat** Widespread in most habitats, especially in arid areas. Cosmopolitan.

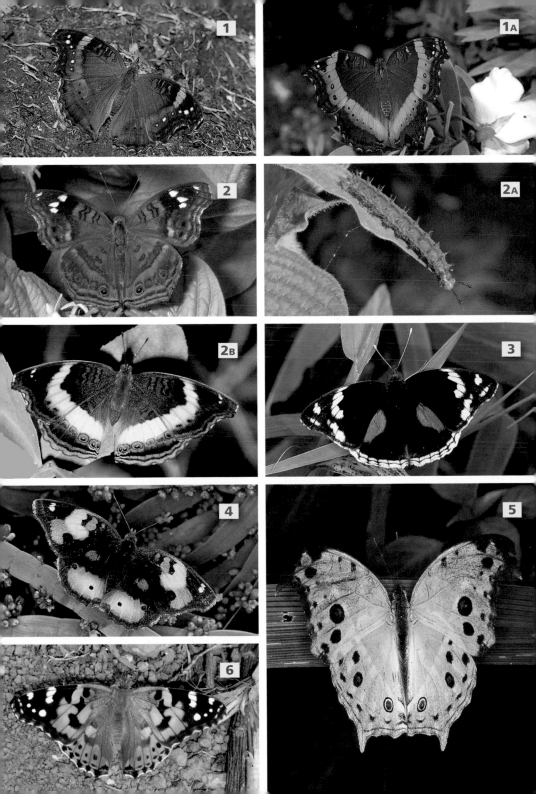

Subfamily Biblinae
Nymphs, jokers and pipers

A mostly tropical group with 3 genera in the northeastern parts of the region.

1 *Byblia ilythia* — Spotted Joker

Medium-sized (wingspan 40–45mm), rich orange, with bold black markings. A broad, black band along outer margin of hind wing encloses a row of orange spots. Orange area in mid hind wing has a row of small black dots. Underside paler orange, dramatically patterned with black bands and white spots. **Biology** Low, slow fliers, attracted to rotting fruit. **Habitat** Wide range of habitats. Favours grassland and open savanna. **Related species** The Common Joker (*B. anvatara*), which is very similar, but lacks black dots in the orange area of the mid hind wing.

2 *Eurytela dryope* — Golden Piper

Medium-sized (wingspan 40–55mm), with distinctive 2-toned colouring: dark brown anteriorly, with broad golden-orange posterior band. A dark scalloped line runs along wing margins. Underside mottled brown. **Biology** Male defends a territory from a leaf perch. Larvae feed on Castor Oil Plant (*Ricinus communis*) and Stinging Nettle Creeper (*Tragia durbanensis*). **Habitat** Forest edges and patches between forests.

Subfamily Charaxinae
Emperors

Medium-sized to large, robust butterflies that are fast and powerful fliers. Females larger than males. Coloration very diverse, with wings often exquisitely patterned. Ends of hind wings often with 1 or 2 mostly short tails. Adults attracted to rotting fruit. Larvae smooth and green, with slug-like bodies that taper posteriorly and end in 2 small projections. Head large, bearing 1 or 2 pairs of horns. Pupae angular and usually green, hanging head-down, supported by tail hooks.

3 *Charaxes brutus* — White-barred Emperor

Large to very large (wingspan 60–90mm), upperside unmistakable – bluish-black with a broad, white or yellowish diagonal band. Underside of wings brownish-red, with dark markings, a dark margin and a broad silver-white band. Hind wings with 2 longish, slender tails. **Biology** Adults aggressively defend territories; attracted to red flowers, sap flows and fermenting fruit. Green larvae **(3A)** have a large head and tail horns and feed on Forest Mahogany (*Trichilia dregeana*), among other plants. Pupae suspended from foodplant by tail hooks. **Habitat** Coastal and inland forest.

4 *Charaxes candiope* — Green-veined Emperor

Large to very large (wingspan 65–95mm). Fore and hind wings reddish-brown, with yellow at the base and a brownish-black band on outer margins enclosing a row of orange spots. Prominent green veins near front margins of fore wings. Underside brownish, with darker markings and yellow patches; green near main wing veins. Hind wings with 2 long tails in female; in male upper tail is shorter than lower tail. **Biology** Attracted to sap flows and fermenting fruit. Larvae feed on *Croton*. **Habitat** Common in woodland and savanna.

5 *Charaxes jasius* — Foxy Emperor

Large to very large (wingspan 65–90mm). Upperside with basal area of wings reddish-brown, followed by paler band and then a broad, brownish-black border, with a row of orange and blue dots on outer edge of the wings. Wings have long, slender tails, with upper tail shorter in male. Underside of wings **(5A)** with complex pattern of red, grey and orange, crossed by a broad white diagonal band. **Biology** Adults attracted to sap flows and fermenting fruit. Larvae feed on a large number of woody plants, including *Afzelia*, *Schotia*, *Xeroderris*, *Bauhinia* and Mopane (*Colophospermum mopane*). **Habitat** Thornveld, bushveld and savanna.

6 *Charaxes jahlusa* — Pearl-spotted Emperor

Medium-sized (wingspan 42–60mm), upperside orange, with a group of black spots on fore wings. Outer margin of hind wings narrowly bordered with orange, followed by a brownish-black border, with orange dots. Underside of fore wings light pink, with white markings and black dots; hind wings brown, with pearly spots and patches, and 2 short tails. **Biology** Fly around trees or congregate on hilltops. Larvae feed on Jacket Plum (*Pappea capensis*). **Habitat** Bushveld and karroid vegetation.

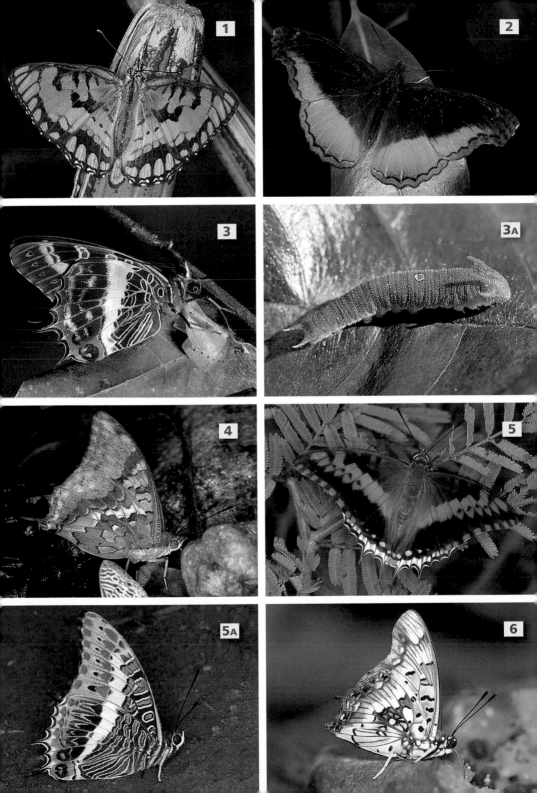

1 *Charaxes zoolina*
Club-tailed Emperor

Medium-sized (wingspan 40–58mm). Dry season form orange above, with dark orange wing borders, and a broken dark brown line parallel to the outer wing margin. Underside has white and light brown bands and spots. Wet season form greenish-white, with a broad black wing border; in female border is broken by greenish-white spots. Hind wings with 2 clubbed tails in female, 1 in the male. Larvae green, well camouflaged, and with long horns on the head. **Biology** Adults often fly around treetops. Larvae feed mainly on *Senegalia*. **Habitat** Rainforest and well-wooded savanna.

2 *Charaxes varanes*
Pearl Emperor

Large to very large (wingspan of male 68mm, female 78mm), easily recognized by pearl-white basal area of wings and orange-brown outer part. Fore wings with 2 rows of orange spots parallel to outer margin. Hind wings each have a row of dark brown spots and crescent-shaped brown marks along the outer margin and a slightly clubbed tail. Underside of wings orange-brown, resembling a leaf. **Biology** Adults attracted to sap flows and fermenting fruit. Larvae feed on *Allophytus, Cardiospermum* and *Rhus*. **Habitat** Thick bush and forest.

Subfamily Libytheinae
Snouts

A rare group of primitive butterflies characterized by their long labial palps. Only 1 genus worldwide, with a single species in South Africa.

3 *Libythea labdaca*
African Snout Butterfly

Medium-sized (wingspan 40–50mm). Easily recognized by the projecting snout (labial palps). Upperside dark brown, with paler patches and small white spots on the fore wing, which has a rectangular anterior projection. Underside mottled in whites and greys. **Biology** Mostly in the forest canopy, sometimes descending to lower branches or muddy ground; wary and difficult to approach. **Habitat** Coastal and Afromontane forest.

Family Papilionidae
Swallowtails

Large, spectacular butterflies, predominantly black, with spots, bands or patches of various other colours; hind wings usually end in a short or long tail. First-stage larvae are sometimes covered with small spines; later stages are smooth. A brightly coloured, forked scent gland (the osmeterium), concealed just behind the head of the larva, is a family characteristic and is everted, giving off a pungent smell, when the caterpillar is disturbed. Superbly camouflaged pupae are attached to the foodplant by tail hooks and held upright by a silk belt.

4 *Papilio demodocus*
Citrus Swallowtail

Very large (wingspan 100–130mm), black, with yellow bands, spots and speckling. Hind wings each with 2 black, blue and orange eyespots; no tails. Young larvae (**4A**) resemble bird droppings; later stages (**4B**) green, with brown and white markings and a red scent gland. Pupae (**4C**) resemble dead leaves. **Biology** Common throughout the region and 1 of the best-known local butterflies. Larvae are pests of citrus trees. **Habitat** Widespread in most habitats. **Related species** The Emperor Swallowtail (*P. ophidicephalus*) (**4D**), Constantine's Swallowtail (*P. constantinus*) and Forest Swallowtail (*P. euphranor*) are all superficially similar in colouring, but their hind wings have tails and all occur only in the east of the region.

5 *Papilio dardanus*
Mocker Swallowtail

Very large (wingspan 80–110mm). Male (shown) unmistakable, with black-bordered cream wings, and club-shaped hind wing tails. Female lacks tails and is extremely variable in appearance, mimicking distasteful Danainae (Nymphalidae), such as the African Monarch (*Danaus chrysippus*) (p.434), Chief (*Amauris echeria*) or Layman (*A. albimaculata*). **Biology** Males conspicuous in their leisurely flight through glades or along forest margins. Females remain in deep forest. Larvae feed on Rutaceae. **Habitat** Mountain and coastal forest, parks and gardens.

1 *Papilio nireus* Green-banded Swallowtail

Very large (wingspan 75–95mm) and unmistakable. Black, with an iridescent blue-green band down centre of each wing, blue-green apical spots on the fore wings, and a row of blue-green spots along outer margin of the hind wings. Hind wings lack tails. Larva resembles that of *P. demodocus* (p.446). **Biology** Fast fliers, often visiting flowers, muddy patches and, sometimes, fresh animal droppings. Larvae feed on citrus and other foodplants. **Habitat** Forest and bushveld, parks and gardens.

2 *Graphium colonna* Mamba, Black Swallowtail

Large (wingspan 55–65mm), black, with vivid turquoise lines and patches, mostly on fore wings, and straight lines across the fore wing cell. Hind wings with red spots on inner margin and long, straight, white-tipped tails. **Biology** Fly low through undergrowth and along forest edges. Often visit muddy places. Larval host plant is Red Hookberry (*Artabotrys monteiroae*). **Habitat** Coastal bush north of Richard's Bay. **Related species** *G. policenes* and *G. porthaon*, which have larger, more extensive turquoise markings; and the Large Striped Swallowtail (*G. antheus*) **(2A)**, which has wavy lines across the fore wing cell and may congregate in large numbers around wet mud.

Family Pieridae Whites, sulphurs, orange tips

A diverse family of medium-sized butterflies, usually white or yellow, with or without orange, red, gold, black or purple markings. The larvae are smooth, usually green or reddish-brown, sometimes with longitudinal stripes along their sides, and are often covered with short, fine hairs. Pupae are attached by tail hooks to various supports, away from the foodplant, and are held upright by a silk belt.

3 *Colotis auxo* Sulphur Orange Tip

Medium-sized (wingspan 35–40mm), variable. Male **(3)** yellow; apical half of fore wings orange, with a black border. Female with brown spots on the fore wings and outer margin of the hind wings; orange area sometimes reduced or absent. Underside yellowish with brown markings, resembling a dead leaf. **Biology** Fast fliers. Larvae feed on *Cadaba* (Capparaceae). **Habitat** Forest and drier savanna.

4 *Colotis danae* Scarlet Tip

Medium-sized (wingspan 35–55mm), very variable. Dry season form (shown here, male) white; apical half of fore wings red encircled with black band. Hind wings with black marginal spots. In wet season form, black areas are enlarged and more diffuse, and hind wing dots may become continuous. Female has more black markings and spots on red (or yellow) apical tip. Larval host plants are Bead-bean (*Maerua angolensis*) and Natal Worm-bush (*Cadaba natalensis*). **Biology** Low-flying, often settling on flowers or on the ground. **Habitat** Savanna and thornveld.

5 *Colotis euippe* Smoky Orange Tip

Medium-sized (wingspan 35–45mm), very variable. Male of dry season form **(5)** white, with large orange fore wing tip outlined in black, and black dots on the outer hind wing margin. Basal area of the wings is peppered with black scales, sometimes with additional, more extensive black areas **(5A)**. Larvae feed on Common Bush Cherry (*Maerua cafra*). **Biology** Flies low, often settling on flowers or on the ground. **Habitat** Widespread, except in grassland and the extreme southwestern Cape. **Related species** There are several similar orange-tipped species in this genus.

6 *Colotis eris* Banded Gold Tip

Medium-sized (wingspan 40–45mm), variable with sex and season. Male (shown here) distinctive, white, with gold-tipped fore wings edged with a broad black band along outer and hind margins; hind wings with smaller black markings. Female white or yellow, with blacker apex of fore wings and thinner black bar on inner fore wing margin. **Biology** Flight fast and erratic, close to the ground. Circles rapidly around foodplants, which include Shepherd's Tree (*Boscia oleoides*). **Habitat** Widespread in bushveld, preferring hot, dry areas.

1 *Eronia cleodora* — Vine-leaf Vagrant

Medium-sized to large (wingspan 45–60mm). Upperside white, with a broad black margin and 2 white spots at tip of the fore wing. Underside yellow, with brown patches and a wide brown hind margin; convincingly mimics a dead leaf. **Biology** A fast flier, stopping frequently to feed on flowers. Larvae feed on Zigzag Caper Bush (*Capparis fascicularis*). **Habitat** Wooded savanna and riverine forest.

2 *Pontia helice* — Meadow White

Medium-sized (wingspan 35–43mm). In male, upperside is white, with squarish brown or black markings at the tip of the fore wing, another at the end of the fore wing cell, and small dark markings on outer margin of the hind wing. Female has more extensive black markings and a broad dark outer margin to the hind wings, enclosing white spots. Undersides of hind wings have yellow and white markings and the veins are outlined in brown. **Biology** Slow-flying, settling on flowers. Larvae feed on sun flax (*Heliophila*). **Habitat** Common and widespread in open country and grassland.

3 *Belenois aurota* — Brown-veined White

Medium-sized (wingspan 40–50mm). Male upperside is white, with a black comma at end of fore wing cell, and white-spotted black outer margins to fore and hind wings. Underside with veins and black areas of fore wings outlined in brown, and with yellow basal patches. Female has more extensive black markings and no white on fore wing tip. **Biology** Flight rapid; often seen on flowers or at mud. Sometimes migrates in huge numbers. Breeds on, and may flock around, species of shepherd's tree (*Boscia*). **Habitat** Widespread; mostly in open country. **Related species** There are 5 similar species in the genus, including the African Common White (*B. creona*) (wingspan 43mm), which is found in the eastern parts of the region and has a yellow underside (**3A**) and a broad, yellow-spotted dark band along the wing margins.

4 *Catopsilia florella* — African Migrant

Large (wingspan 55–65mm). Upperside of male (**6**) white, with a faint greenish tinge, faint black front margin and single small black spot at the end of the fore wing cell; underside of wings yellow, speckled with brown, with 3 brown-ringed white spots at centre of hind wings. Female more yellow. **Biology** May migrate in huge numbers. Larvae feed on various species of *Senna* and on Sjambok Pod (*Cassia abbreviata*). **Habitat** Most habitats. Widespread.

5 *Pieris brassicae* — Cabbage White

Large (wingspan 55mm). Male (**5**) with white upperside, black tip to fore wing, and 1 black spot on underside. Female with 2 black spots on upperside of fore wings. Greyish-yellow speckles on underside in both sexes. **Biology** Black-and-yellow larvae (**5A**) feed mainly on crucifers, especially cabbages, and nasturtiums. A significant pest of vegetable gardens, but kept in check by the indigenous parasitic wasp *Pteromalus puparum* (p.468). Pupae (**5B**) often attach to walls. **Habitat** Gardens and smallholdings. Introduced accidentally into the Western Cape from Europe; first recorded in 1994.

6 *Mylothris agathina* — Common Dotted Border

Large (wingspan 50–65mm). Upperside of male creamish-white, of female pale yellow, becoming orange towards base. Undersides markedly orange at base of fore wing, fading to white or yellow, with black dots along the edge. **Biology** Slow-flying and readily attracted to flowers. Gregarious larvae drop off the foodplant if disturbed. Pupae resemble bird droppings, ornate, in brown and yellow, with projecting knobs. Larval foodplants include mistletoe species (*Tapinanthus*) and woody plants like Sour Plum (*Ximenia caffra*). **Habitat** Common and widespread in woodland, parks and gardens.

7 *Colias electo* — African Clouded Yellow, Lucerne Butterfly

Medium-sized (wingspan 35–40mm). Male orange, with broad brownish-black border to both wings and a black spot at the end of the fore wing cell. In female, orange areas paler, sometimes almost white, and suffused with brown on basal areas of wings; wing borders black, with a few yellow or white markings. Underside of wings green or greenish-yellow with a row of brown spots near the margins and a white central dot. Pupae pale green. **Biology** Found on a wide range of host plants, the larvae sometimes causing damage to lucerne crops. **Habitat** Most habitats. Widespread.

1 *Eurema brigitta* — Broad-bordered Grass Yellow

Medium-sized (wingspan 33mm). Upperside of wings bright yellow, with a broad black border along front and outer margins of fore wings (widest around tip) and along outer margin of hind wings. Marginal bands broader in wet season forms. Underside yellow, with a dusting of black dots and small black marginal spots. **Biology** Flies slowly, close to the ground, often settling on flowers. Larvae feed on St John's Wort (*Hypericum aethiopicum*) and English Tea Senna (*Chamaecrista mimosoides*). **Habitat** Widespread, except in the drier parts of the western Karoo and Namaqualand. **Related species** There are 2 other species in the genus, both with narrower black marginal bands on upperside of wing.

2 *Pinacopteryx eriphia* — Zebra White

Medium-sized (wingspan 40–55mm), unmistakable. Upperside black, with creamish-white bars and spots. On underside, black areas replaced with brown or pinkish mottling. **Biology** Low fliers, but capable of considerable speed. Larvae feed on Shepherd's Tree (*Boscia oleoides*) and Common Bush Cherry (*Maerua caffra*). **Habitat** Widespread in most habitats, preferring open country, especially bushveld, but also frequents parks and gardens.

Family Lycaenidae — Blues, coppers

A very diverse family, comprising about half of all regional butterfly species. Most are small, with the upperside often brilliantly coloured in iridescent blue, copper, purple or orange, and the underside usually duller, with spots or streaks. Hind wings often have narrow tails that, in combination with eyespots, create a false head, which distracts predators away from the real head. Larvae are flattened, often densely covered with short hairs (setae) of various types, and with diverse feeding habits. In most species larvae are herbivorous, but some prey on aphids, scale insects or ant larvae. In many species, feeding larvae are protected by ants, which are attracted to larval secretions; some larvae even live in ant nests, where they are fed by ants or prey on ant larvae.

3 *Ornipholidotos peucetia* — White Mimic

Medium-sized (wingspan 35–37mm), delicate. Wings thinly scaled, white with black margins, a black bar below apex of fore wings, and a few black spots on hind wings. Legs orange. **Biology** A shade-loving, weak-flying species that settles often on low plants. Larvae feed on blue-green algae growing on tree trunks. **Habitat** Thick bush and forest.

4 *Pentila tropicalis* — Spotted Buff

Small (wingspan 29–38mm). Wings orange-yellow, with a few black speckles on the upperside, especially along leading edge and tip of fore wing; underside more evenly and densely speckled. **Biology** Fly weakly through the forest understorey, settling often. Slug-shaped larvae feed on blue-green algae on tree bark. **Habitat** Warm coastal forest.

5 *Lachnocnema bibulus* — Common Woolly-legs

Small (wingspan 25mm) and easily recognized at rest by the short, very furry legs. Upperside of male dark brown; female paler, with diffuse white markings at centre of each wing. Underside of wings brownish-grey (whitish in parts in female), with various silver-dotted brown spots and bands, especially on hind wings. **Biology** Gregarious. Adults suck honeydew from various homopterans, including groundnut hoppers. Associated with ants. Larvae carnivorous, feeding on homopteran plant bugs, including coccids. **Habitat** Wooded and open savanna. **Related species** There are 3 other similar species in the region.

6 *Aloeides* — Coppers

A diverse genus of small butterflies (wingspan 22–36mm), with 49 easily confused regional species. Upperside brown to orange with dark borders; underside brown or orange, with white spots often circled in black. Hind wing colour usually cryptic. Larvae green and flattened, often with longitudinal stripes. *A. swanepoeli* (KwaZulu-Natal to Limpopo) depicted. **Biology** Adults defend territories on open ground. Larvae live associated with *Lepisiota* (p.482) or *Pheidole* ants (p.486), emerging to feed on plants at night; some species are fed by ants or prey on ant larvae. **Habitat** Members of the genus found across the region, but most species have a highly restricted range.

1 *Tylopaedia sardonyx* — King Copper

Medium-sized (wingspan 35–50mm), with sturdy body. Upperside of wings orange, with a broad black border and variable black bands and dots. Underside red-orange, fore wing lighter orange basally, with large black dots each with white centre. Grey area at base of hind wing, which is sometimes crossed with diagonal white lines. **Biology** Adults fast-flying around rocks or bushes. Larvae attended by *Crematogaster* ants and shelter under rocks by day, emerging at night to feed on *Aspalathus*, *Phylica* and *Euclea*. **Habitat** Rocky areas in the Karoo and Namaqualand.

2 *Phasis thero* — Silver Arrowhead

Medium-sized (wingspan 42mm). Upperside with dark brown margin and orange central area containing black dots, each with a white centre. Underside orange-brown, with silver-dotted brown spots and very broad greyish-brown band along outer and inner margins. Hind wings greyish-brown with characteristic hooked silver bar and 2 short tails. **Biology** Adults have a fast, undulating flight. Larvae occur inside the hollow stems of their foodplants, wild currant species (*Rhus*) and Honey Flower (*Melianthus major*), and are associated with ants. **Habitat** Sand dunes, coastal sandy wastes and hill slopes. **Related species** There are 3 other similar species in the genus.

3 *Chrysoritis* — Coppers and opals

Diverse genus of small (wingspans mostly 21–35mm) butterflies containing 42 regional species. Generally have orange upperside, with dark borders and black spots; base of wings often iridescent blue or silvery. Underside orange, with black and/or white markings, and with cryptic patterns on the hind wings. Common Opal (*C. thysbe*) from southwestern Cape depicted. **Biology** Fast-flying and territorial, often settling on flowers. Larvae nocturnal, sheltering under rocks or in ant nests by day. **Habitat** Genus occurs throughout the region, but most species have a very restricted range.

4 *Capys alphaeus* — Protea Scarlet

Medium-sized (wingspan 31–47mm) and spectacular. Upperside of both fore and hind wings dark brown, with a thin white margin and large, vivid red central patches; underside cryptic blue-grey, with a dark wavy band centrally and an orange flash on fore wing. **Biology** Fast-flying and territorial. Stout, dark larvae **(4A)** bore into large protea buds and consume the developing seeds. **Habitat** Mainly mountains; found wherever proteas occur.

5 *Cigarites* — Bars

Genus of relatively small (wingspan 25–34mm) but strikingly coloured butterflies, with 5 regional species; *C. natalensis* depicted. All have fore wings orange above, with dark bars; hind wings blue above, with 2 short tails. Underside with heavy red-orange diagonal bars across a paler ground. **Biology** Active group with rapid flight. *C. natalensis* larvae feed on Turkey Berry (*Canthium inerme*), Wild Plum (*Ximenia caffra*) and Smooth Tinderwood (*Clerodendron glabrum*). **Habitat** Grassland and savanna, especially in rocky and hilly areas. One rare species in northern Namaqualand, 4 others occur in warmer northeastern parts of the region.

6 *Iolaus* — Sapphires

Diverse genus of small (wingspan 25–40mm) but spectacular lycaenids. Upper wings usually bright blue with a black margin, often with red or black posterior spots. Underside usually white or grey, marked with red or yellow diagonal lines or bands. Posterior margin of wings with eyespots and long tails. One exceptional species (*I. pallene*) yellow on both upper and lower wing surface. Larvae flattened, green and cryptic when on foodplants. Pupae **(6A)** resemble plant galls. Genus includes 14 regional species; *I. silas* from Eastern Cape depicted. **Biology** Adults fast-flying and often chase each other. Larvae feed mostly on mistletoes. Not associated with ants. **Habitat** Some species rare and range-restricted; others are widespread in savanna and woodland.

1 *Myrina silenus* — Common Fig-tree Blue

Medium-sized (wingspan 27–41mm). Upperside mostly metallic blue, bordered by black and with a large brown patch on tip of fore wing; underside of wings rufous brown, resembling a dead leaf. Hind wings with long, twisted brown tails. Larvae **(1A)** bright green, velvety and slug-like. **Biology** Adults fly around very rapidly at midday. Larvae associated with various species of wild fig. Often attended by ants when boring into young figs. **Habitat** Adults found on and around koppies, larvae on the terminal growth of various trees and climbing figs.

2 *Tuxentius melaena* — Black Pie

Very small (wingspan 19–25mm), with black-and-white striped body. Upperside of wings black, with large white patches; underside white, with numerous large black spots. Larvae flattened, green and slug-like, with hairy margins. **Biology** Adults common around foodplants, feeding on their nectar and that of nearby flowers. Larvae feed on buffalo thorn species (*Ziziphus*). **Habitat** Savanna and forest, also gardens and parks. **Related species** The White Pie (*T. calice*) (tropical northeast of the region), which is very similar but distinguished by details of the dot pattern on its underside.

3 *Leptomyrina* — Black-eyes

Genus with 4 regional species of small to very small (wingspan 19–32mm) lycaenids characterized by a row of black eyespots or black dots encircled by a white ring on upper margin of fore wing. Upperside generally bronze-brown; underside silvery grey. Very short tails on hind wing. An exception is the Tailed Black-eye (*L. hirundo*) **(3A)**, which is silvery above, with only 1 eyespot, and has long white tails. **Biology** Fly slowly among low vegetation and rocks. Larvae feed on succulents, such as *Crassula* and pig's ears (*Cotyledon*). **Habitat** Widespread in savanna, grassland and mountains; also common in gardens.

4 *Tarucus thespis* — Fynbos Blue, Vivid Blue

Small (wingspan 20–27mm). Upperside of male sky blue. Upperside of female mostly black, with a blue sheen in basal areas of the wings and squarish white sub-apical dots on the fore wings. Both sexes have a black-and-white chequered fringe of hairs on outer wing margins. Underside of wings brownish-grey, with angular white spots. **Biology** Low- and slow-flying, settling frequently. Larvae feed on *Saxifraga* and *Phylica imberbis*. **Habitat** Bushes on hills or fynbos in valleys. **Related species** The Dotted Blue (*T. sybaris*), which has a non-overlapping range (occurs in the northeast) and a similar upperside; its underside is dramatically marked with large black dots, similar to those of the Black Pie (*Tuxentius melaena*) (above).

5 *Athene* — Hairtails

Diverse genus of small (wingspan 19–32mm) lycaenids. Upperside of wings blue or brown and underside variously banded in browns and white, often with 2 eyespots. Two pairs of thin, hairlike tails extend from hind wings and give the genus its common name. Larvae pale to dark green, with characteristic pale green tubercles along the back. Genus includes over 130 African species, 14 of which occur in South Africa. Common Hairtail (*A. definita*), Cape Town to northeastern parts, depicted. **Biology** Flies from ground level to treetops. Males often gregarious. Larvae feed on a wide variety of foodplants, including Sweet Thorn (*Vachellia karroo*), and are tended by ants. **Habitat** Widespread in bushveld.

6 *Virachola* — Playboys

Genus of small (wingspan 20–40mm) lycaenids, with 7 regional species. Upperside may be brown, red or blue; underside grey or brown, with several broken orange or brown crossbars. Eyespots and 1 pair of slender hind wing tails posteriorly. Lower eyespots project laterally when wings are closed. The widespread Brown Playboy (*V. diocles*) depicted. **Biology** Adults congregate around host plants, males defending territories. Larvae are seed-feeders, burrowing into seed pods. Host plants include invasive Balloon Vine (*Cariospermum grandiflorum*). **Habitat** Diverse, often forest edges.

1 *Cacyreus marshalli* — Geranium Bronze

Very small (wingspan 15–27mm). Upperside uniformly brown, with a brown-and-white chequered fringe along outer margin of wings. Hind wings with white-bordered black eyespot near base of single tail. Underside brown, with broken white banding. **Biology** Adults are weak fliers, settling frequently. May stay in an area for extended periods. No association with ants. Larvae feed on *Geranium* and *Pelargonium* species. **Habitat** Widespread, their distribution assisted by widespread use of host plants in gardens. Have been acidentally introduced to Spain and the UK. **Related species** The Water Bronze (*C. tespis*) **(1A)** (wingspan 20mm), which is very similar and occurs in the moister eastern coastal areas. Its larvae also feed on *Geranium* and *Pelargonium*.

2 *Azanus jesous* — Topaz-spotted Blue

Very small (wingspan 17–26mm). Male upperside lilac-blue tinged with pink; outer wing margins with a narrow brown border. Underside grey with complex pattern of black and white spots and bands. Female mostly brown. Central and inner areas of wing margins paler, with brown spot at end of fore wing cell and distinct black spot near rearmost angle of hind wings. **Biology** Common and widespread. Flies around flowering *Vachellia* trees and drinks at damp mud pools. Larvae feed on young *Vachellia* leaves. **Habitat** Most common in bushveld; also occurs in parks and gardens.

3 *Leptotes* — Common blues

Genus of small (wingspan 18–30mm) lycaenids, with 5 regional species that are almost impossible to distinguish in the field. Upperside violet-blue. Underside of wings with dark bands crossing a paler ground, a thin dark marginal line, 2 conspicuous eyespots on the hind wings, and a longish white-tipped tail. **Biology** Larvae feed on flowers, shoots and seeds of *Plumbago auriculata*. **Habitat** Widespread in most habitats, especially gardens, as *Plumbago* is a common garden plant.

4 *Lampides boeticus* — Pea Blue, Long-tailed Blue

Small (wingspan 24–34mm). Upperside of male violet-blue, of female paler blue with a brown border; both sexes with 2 black dots posteriorly and longish white-tipped tail. Underside with wavy white lines on a fawn background, 1 particularly conspicuous white band near the wing margin, and 2 eyespots circled in orange. **Biology** Fast, whirling flight, frequently returning to same perch. Males territorial. Larvae feed on a range of legumes, boring into pods and feeding on the seeds. **Habitat** Extremely widespread across Europe, Asia and Africa and occurs in all biomes in South Africa.

5 *Zizeeria knysna* — African Grass Blue

Very small (wingspan 18–26mm). Upperside of male violet, with a brown margin. Upperside of female brown, with some blue basally. Underside in both sexes grey, with a thin brown marginal line and scattered black dots. **Biology** Among the most common and widespread small butterflies in the region. Flight low and erratic, settling frequently on small flowers. Green larvae feed on Creeping Woodsorrel (*Oxalis corniculata*) and Devil's Thorn (*Tribulus terrestris*). **Habitat** Common across the region in all habitats, especially on suburban lawns.

6 *Oraidium barberae* — Dwarf Blue

Very small (wingspan 14mm). Upperside dark brown, with a margin of alternating white and brown hairs; underside brown, with darker brown spots bordered with white bars. End of hind wing with a row of white-centred brown spots. **Biology** Flies weakly, low above the ground. The smallest butterfly in South Africa, where it is very widespread. Foodplant Brakbos (*Exomis microphylla*). **Habitat** Common, especially in the fynbos and karoo regions.

7 *Orachrysops* — Blues

Endemic genus containing 11 species of small to medium-sized (wingspan 24–44mm) butterflies, with upperside brown to violet and underside grey, with a variety of pale bands and black spots. A pale band typically runs parallel to wing margins. Brenton Blue (*O. niobe*), only at Brenton-on-Sea, illustrated. **Biology** Low, erratic flight, close to the ground. Males patrol regular flight paths, searching for females. Larvae feed on various *Indigofera* species. **Habitat** Genus widespread within region, but each of the 11 species has a highly restricted range, and some, including the Brenton Blue, endangered.

ORDER HYMENOPTERA • SAWFLIES, WASPS, BEES & ANTS

Minute to large insects with 2 pairs of membranous wings (the front pair larger), with relatively simple venation. Wings are coupled in flight by a row of tiny hooks along the front edge of the hind wings.

Wasps, sawflies, bees and ants have biting mouthparts that may be modified for sucking and for lapping liquids. Mandibles are always present, and the egg-laying organ (ovipositor) is adapted for sawing, boring, piercing or stinging. All except sawflies have a waist of variable length between the thorax and abdomen. Bees are easily distinguished from wasps under magnification by the branched (not simple) hairs on their bodies. Most bees and many wasps lay their eggs in prepared nests. Bees, and a few wasps, feed their young on a mixture of pollen and nectar. All other wasps provide their young with paralysed insects or spiders for food, or parasitize other insects. Sawfly larvae feed on foliage. Some wasps and bees, and all ants, are social. Of about 198,000 species, over 6,000 are known from the region.

SUBORDER SYMPHYTA   Sawflies, woodwasps, horntails

This group contains the more primitive Hymenoptera, characterized by a lobe on each side of the last thoracic segment and the absence of a waist between the thorax and abdomen.

1 Family Tenthredinidae   Common sawflies

Usually smallish (body length 5–8mm), stout-bodied, and like other sawflies lack a 'wasp waist' between thorax and abdomen. Most are black, or black with yellow abdomen and legs. Antennae have 10–13 segments. *Anthelia* (1) are typical, while the slug-like larvae of the introduced Pear Slug (*Caliroa cerasi*) (1A) are perhaps most commonly seen. **Biology** Adults sluggish. Saw-like ovipositor used to insert eggs into leaves, stems or shoots. Larvae feed on foliage, *Anthelia* being pests of cabbages and turnips and *Caliroa* of plum and pear trees. **Habitat** A range of vegetation types (frequently grassland). Also in gardens and often on flowers.

2 Family Argidae   Argid sawflies

Smallish to medium-sized (body length 5–15mm), usually yellow or black and, like other sawflies, lack a 'wasp waist' between thorax and abdomen. Easily distinguished from Common sawflies (above) by having only 3 antennal segments, the apical segment being much elongated. Larvae (2A) resemble caterpillars, but often have more than 5 pairs of false legs on the abdomen. Only the diverse genus *Arge*, with approximately 45 species, found in the region. **Biology** Poorly known. *Arge taeniata* lays its eggs in slits cut into the underside of *Pelargonium* leaves, on which the larvae feed. **Habitat** Varied, including gardens.

3 Family Siricidae   Woodwasps, horntails

Characterized by a strong dorsal spike on the last abdominal segment. Females have a long, strong ovipositor projecting from end of abdomen. The only regional species, the introduced *Sirex noctilio*, is relatively large (20–30mm), with long antennae, orange legs and bluish-black body. **Biology** Drill into stressed pine trees to lay eggs, introducing a poisonous mucus and symbiotic fungus that kills the tree. Larvae feed on the fungus in the wood; adults feed on nectar. A parasitic wasp and a nematode worm have been introduced to curb the spread of this woodwasp. **Habitat** Became established in the Cape Peninsula in about 1994, spreading rapidly into the Eastern Cape and now KwaZulu-Natal, where it is a pest of pine plantations.

SUBORDER APOCRITA

Wasps, bees, ants

A diverse group of more advanced hymenopterans, distinguished from the Symphyta by the fusion of the first abdominal segment with the thorax. The first and sometimes second segment may be narrowed to form a long or short waist. Bees and a few wasps feed their young with a pollen-and-nectar mixture; others parasitize insects or provide their young with paralysed insects or spiders.

Family Ichneumonidae

Ichneumon wasps

Mostly slender-bodied wasps, with at least 16-segmented antennae (most wasps have 13 or fewer) that are more than half the length of the body. In females, the abdomen may end in a long ovipositor, used in some species to drill into plant stems to reach the host. Most species are black, yellow or brown, or patterns of these colours. Distinguished under magnification from the similar Braconidae by the absence of a square cell on the fore wing. The most diverse of all hymenopteran families, with over 500 regional species. All parasitize the larvae and pupae of Lepidoptera, Coleoptera (especially wood-borers), Diptera, Neuroptera or other Hymenoptera.

1 *Osprynchotus gueinzii*

Mud Dauber-mimicking Ichneumon

Medium-sized (body length of male 12–19mm, female 17–26mm, excluding ovipositor), entirely black, except for yellow bands on antennae and legs. Abdomen with long waist and, in female, a conspicuous ovipositor. Wings with metallic blue sheen. **Biology** Parasitizes nests of mud daubers and other wasps (e.g. *Vespa calida*) that seal their nests with mud. **Habitat** Associated with buildings and other man-made structures. **Related species** There are 5 other similar species in the region, including *O. violator*, which is very similar, but with a broader yellow band on the antennae, and the last 4 abdominal segments reddish-yellow.

2 *Enicospilus*

Medium-sized (body length 18mm), body and appendages reddish-brown except for the mostly creamish-white head. Waist narrow, abdomen expanding towards tip. Antennae as long as body. Wings clear. **Biology** Parasitizes the larvae of noctuid moths, including cutworms. **Habitat** Diverse, most often in buildings. Frequently attracted to lights.

3 *Gabunia ruficeps*

Medium-sized (body length of male 13–18mm; female 22mm, with ovipositor 31mm), black with bluish sheen, except for reddish-brown head, white bar midway along antennae and white second and third tarsal joints. Ovipositor shorter than abdomen. Wings dark brown, with transparent bar across fore wings near apex, and transparent tip to hind wings. **Biology** Parasitizes the larvae of long-horned beetles (Cerambycidae). **Habitat** Subtropical forest. **Related species** *Zonocryptus* species, which are similar, but lack white tarsal joints.

4 *Theronia lurida*

Medium-sized (body length 25mm), yellow, with 3 longitudinal black stripes on the pronotum, and black antennae. Ovipositor black, slightly shorter than abdomen. Legs yellow, mid and hind tarsi brown. Wings clear. **Biology** Larvae parasitize lepidopteran caterpillars. **Habitat** Diverse. Found in forest vegetation. **Related species** There are 2 other Afrotropical species in the genus. Also *Anomalon* wasps **(4A)** (body length 20mm), which are similar, but have a longer ovipositor and parasitize beetle or moth larvae.

5 *Charops*

Small to medium-sized (body length 10–14mm), black, with brownish abdomen that expands towards the tip. Ovipositor short. Wings clear, barely reaching halfway along abdomen. **Biology** Eggs are laid in later-stage caterpillars of various lepidopterans, including *Acraea* butterflies and the Codling Moth (*Cydia pomonella*) (p.372). Mature larvae leave the host and spin a cocoon **(5A)**, distinctively patterned with black patches and hanging on a silken thread. **Habitat** Forest, bushveld and gardens. **Related species** Numerous species in this and other genera, many of which are hard to identify, including *Odesia* species (body length 12mm) **(5B)**, which are similar, but have a very long, slender ovipositor and occur in grassland.

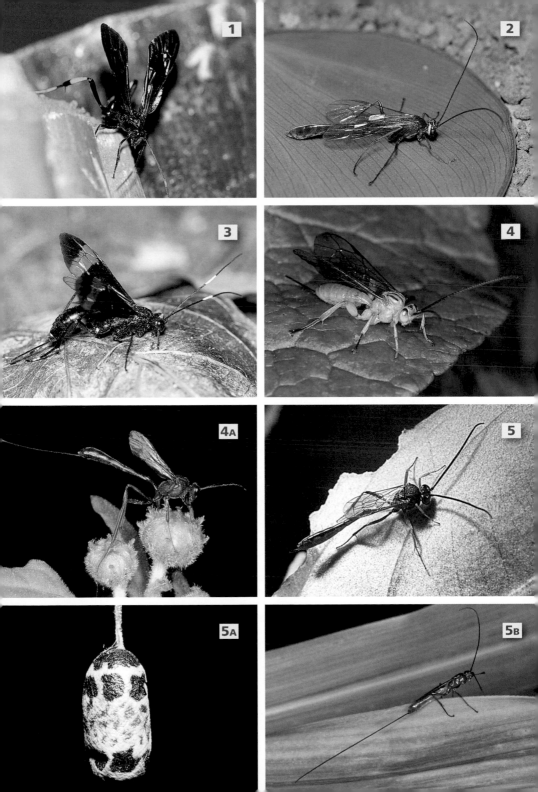

Family Braconidae
Braconid wasps

The most diverse hymenopteran family after Ichneumonidae, with several hundred known regional species and many others undescribed. Smaller and stouter than ichneumon wasps, with brown, black, orange or bright red coloration. Abdomen of female usually ends in a long ovipositor, used in some species to drill into plant stems to reach the larval host. Wing venation considerably reduced in some species, and there is a square cell in the fore wing. All parasitize the larvae and pupae of Lepidoptera, Diptera, Coleoptera or Hemiptera (especially aphids).

1 Archibracon servillei

Largish (body length 45mm), body bright orange, with last 5 abdominal segments black. Antennae black, almost as long as body. Legs black and orange. Ovipositor black, slightly longer than abdomen. Basal third of wings orange, remainder black, except for orange spot on pterostigma of fore wing. **Biology** Poorly known; some species in the genus parasitize cerambycid beetles and give off a strong ester-like odour when handled, probably to deter predators. **Habitat** Wooded savanna, subtropical forest and succulent karoo. **Related species** A number of similar species in the genera *Archibracon* and *Zombrus*.

2 Apanteles

Very small (body length 3mm), black, with brown-tipped antennae almost as long as the body. Legs brownish, with black on hind tibiae. Ovipositor about half the length of abdomen. Wings clear, nearly reaching tip of abdomen. **Biology** Females inject eggs into young *Acraea* caterpillars. Grubs feed internally and leave host's body when mature to spin white cocoons, covered with a net of golden silk, nearby. Host caterpillars of the moth subfamily Lymantriinae (p.418) are often seen covered with the silken cocoons of braconid wasps **(2A)**. Also, the related *Cotesia* species **(2B)**, which are very small wasps that have adapted to parasitizing the larvae of the introduced Cabbage White Butterfly (*Pieris brassicae*) (p.450) as a new host. **Habitat** Forest, bushveld and gardens. **Related species** Numerous other species in this and related genera, which are difficult to tell apart.

3 Bathyaulax

Medium-sized to large (body length 38mm). Head and thorax shiny black, abdomen red. Antennae black, longer than the body. Ovipositor black, equal in length to abdomen. Wings black, with a red spot near pterostigma of fore wing. A diverse genus with many relatively large and brightly coloured African species. **Biology** Parasitize the larvae of wood-boring longhorn beetles (family Cerambycidae). **Habitat** Wooded savanna.

4 Family Platygastridae
Scelionid wasps

Diverse, mostly minute to small (body length 0.5–10mm), often black wasps, with a well-sculptured surface and elbowed antennae that have 9–10 segmented flagellae (the second, 'post-elbow' part of the antenna). Some genera are wingless. Shown here is *Trissolcus*. **Biology** Parasitize the eggs of many insect orders and spiders. Some even attack the eggs of aquatic insects. Many are important biological control agents. *Trissolcus* parasitize the eggs of stink bugs, including those of the Southern Green Stink Bug (*Nezara viridula*) (p.160). **Habitat** Cosmopolitan, wherever hosts are found

5 Family Evaniidae
Ensign wasps

Small to medium-sized (body length 5–14mm), somewhat spider-like, usually entirely black, with relatively long legs and clear wings. Disproportionately small abdomen compressed sideways, triangular or semicircular in outline when viewed laterally; usually held upright. Abdomen arises from top of thorax via a long, narrow waist. *Evania* shown here. Wings folded in pleats when at rest. **Biology** Jerk the abdomen rapidly up and down as they move. Parasitize the eggs of cockroaches, including those of *Periplaneta americana* (p.54). Little is known about the species in the region. **Habitat** Widespread in wooded areas and gardens, usually on tree trunks and branches. Most often encountered on windows. *E. appendigaster* has spread globally from its putative Asian origin.

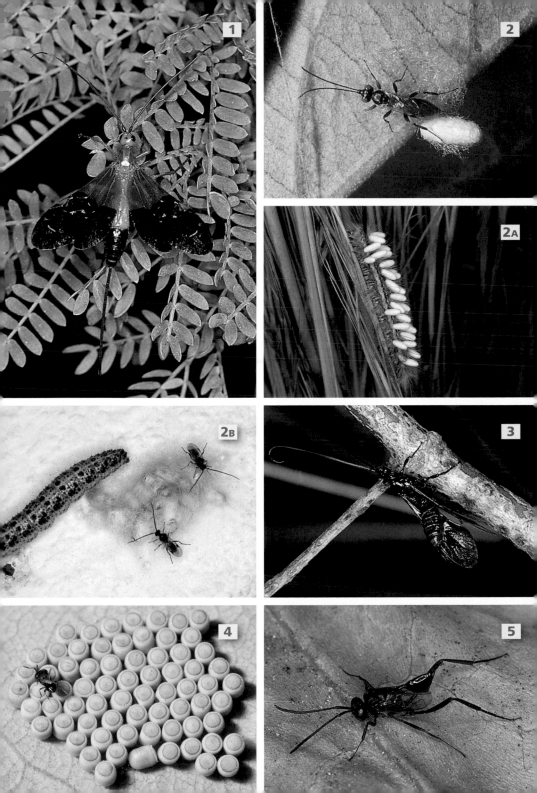

1 Family Gasteruptiidae
Carrot wasps

Small to medium-sized (body length 8–25mm, excluding ovipositor), usually black or reddish-brown, very slender and elongate, with abdomen arising from the top of the thorax and expanding gradually towards the tip. Ovipositor in females longer than the abdomen and sometimes tipped with white. Head directed forwards, with distinct neck and very large eyes. Hind tibiae characteristically narrow at base and swollen at the end. *Gasteruption* (shown here) are typical. About 40 similar species in the region. **Biology** Parasitize the nests of solitary wasps and bees, especially wood-nesting species such as carpenter and *Hylaeus* bees, feeding on their eggs, larvae and pollen stores. **Habitat** Widespread, usually seen flying in vegetation.

2 Family Agaonidae
Fig wasps

Very small (body length 1–4mm). In females **(2)** the head is flattened, forward-pointing, with a longitudinal groove on top and well-developed antennae; also have characteristic appendage on the mandibles to help them crawl into figs. Ovipositor is shorter than the body. Fore and hind legs more strongly developed than mid legs, and wings have almost no veins. Males **(2A)** have small antennae, usually no eyes or wings (sometimes reduced to mere filaments), and a long abdomen folded forwards beneath the thorax. **Biology** All development occurs within figs, which are pollinated by the females. The flightless male never leaves its fig. May be parasitized by *Apocrypta* wasps (p.468), which have much longer ovipositors, exceeding the body length. Commercially cultivated fresh fig varieties do not require pollination to produce fruits and thus do not contain fig wasps; but those varieties used to produce dry figs are normally wasp pollinated and may contain wasp remains. **Habitat** Wherever wild figs (*Ficus*) occur, each fig species having its own wasp pollinator species. **Related species** Some members of Family Pteromalidae (p.468) are also associated with figs and almost identical except for the absence of mandibular appendages in females.

3 Family Leucospidae
Leucospid wasps

Small to medium-sized (body length 3–16mm), easily recognized by fore wings that fold along a longitudinal pleat (as in vespid wasps). Waist relatively stout, about half the length of the rest of the abdomen. Ovipositor recurved over top of abdomen, sometimes reaching the thorax and often housed in a groove, a feature unique to this and closely related families. Hind femora swollen (as in chalcidoids) and toothed underneath. *Leucospis ornata* shown here. There are 8 *Leucospis* in the region. **Biology** Leucospidae parasitize solitary bees and wasps. Eggs are inserted through the wood or mud walls of the host's nest. Adults emerge by drilling holes through cell walls. **Habitat** Diverse.

4 Family Chalcididae
Chalcidid wasps

Small (body length up to 7mm), recognized by swollen hind femora combined with a robust, coarsely sculptured black body, sometimes with yellow or orange (never metallic) markings. Antennae 13-segmented in females, without differentiated club. Abdomen sometimes with waist. Ovipositor slightly protruding. Tarsi with 4–5 segments. Wings fully developed, clear or distinctly marked, with much reduced venation. **Biology** Parasitize larvae and pupae, mostly of Lepidoptera and Diptera. *Brachymeria kassalensis* **(4)** parasitizes the pupae of *Acraea* (p.436) and other butterflies. The larvae pupate within their host, the adults chewing their way out. *Hockeria* species **(4A)** are pupal parasitioids of moths. **Habitat** Forest, bushveld and gardens.

5 Family Eurytomidae
Seed chalcids

Very small (body length 2–4mm), narrow-waisted, usually black, sometimes yellowish or brown, never with metallic colours. Head and thorax coarsely punctate, abdomen shiny and smooth. Pronotum characteristically almost rectangular when viewed from above. Antennae with long, upright hairs in males. Wings fully developed. **Biology** Diverse. Includes parasites and hyperparasites of other insects, as well as species that feed on seeds or form galls on plants. *Eurytoma aloineae* **(5)** develops in the seeds of *Aloe ferox* and probably other aloes; the introduced *Quadrastichus erythrinae* **(5A)** forms the familiar galls on *Erythrina* leaves **(5B)**; *Sycophila* develop within wild figs; *Philolema* parasitize the egg sacs of Black Widow spiders. **Habitat** Diverse.

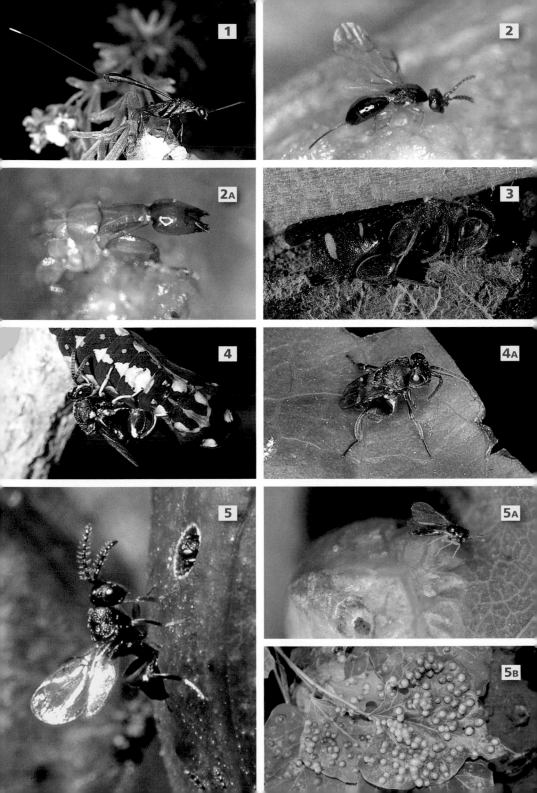

1 Family Eucharitidae
Eucharitid wasps

Very small (body length 3–4mm), often with bright metallic colours. Body smooth or coarsely punctate. Head small, with unusually large mandibles. Thorax bulbous, pronotum not visible from above, end of thorax sometimes with a projecting spine. Abdomen with narrow waist, sometimes joined to thorax by a long, thin stalk. Wings fully developed. **Biology** Among the few specialized parasites of ants. Eggs are laid in plant tissue. First-stage larvae attach themselves to passing ants and, if successfully transported to the ants' nests, bore into the ant larvae, completing their development in a pupal cocoon. **Habitat** A wide range of vegetation types.

Family Pteromalidae
Pteromalid wasps

A diverse and highly variable family, usually with well-armoured black or metallic green or blue bodies. Elongate and slender or relatively dumpy, with a large thorax. Winged and wingless species occur, the former with much-reduced wing venation. Fore and hind femora may be swollen. Parasitize a wide range of insects and arachnids, or form galls on various plants.

2 *Trichilogaster acaciaelongifoliae*
Acacia Gall Wasp

Very small (body length 3mm), shiny black, with yellow legs, long antennae and clear wings. **Biology** Introduced from Australia for biological control of the weedy Australian Long-leafed Wattle (*Acacia longifolia*), the seed production of which it can reduce by as much as 95 per cent. Eggs are laid in immature flower buds in mid-summer and hatch the following spring. Larval feeding causes distortion of buds and development of galls **(2A)**, which contain isolated chambers, each with 1 larva. Adults emerge in mid-summer. Larvae themselves often parasitized by other 'hyperparasitic' wasps, which then emerge from the same galls instead of *T. acaciaelongifoliae*. **Habitat** Diverse.

3 *Pteromalus puparum*

Very small (body length 2–3mm), metallic bluish-black, with short, elbowed black antennae. **Biology** Parasitizes various lepidopterans, including the introduced Cabbage White (*Pieris brassicae*) (p.450). Females lay eggs in final-instar larvae, and the grubs develop and pupate within the butterfly pupa, emerging through a small exit hole. **Habitat** Cosmopolitan; introduced to the USA as a biocontrol agent for Cabbage White butterflies.

4 *Apocrypta*
Fig wasp parasites

Female very small (body length up to 4mm), with an extremely elongated ovipositor, up to 4–5 times as long as the wasp's bronze-green body. **Biology** Found together with true fig wasps (p.466) with which they are often confused, but are distinguished by their much longer ovipositors. Females do not enter the fig via the ostiole, like true (pollinating) fig wasps do, but instead pierce the thick walls of the figs with their ovipositor, the tip of which has small teeth that may be hardened with a zinc metal 'drill bit'. The larvae develop as parasites inside the flower galls of figs, or may parasitize the larvae of the pollinating wasps. **Habitat** Wherever wild figs (*Ficus*) and their fig wasp hosts occur.

5 Family Chrysididae
Cuckoo wasps

A diverse family (32 genera and 428 species in sub-Saharan Africa). Small to medium-sized (body length 4–20mm), sometimes metallic blue or red, as in *Spintharina* **(5)**, but more usually bright metallic green, as in the cosmopolitan *Stilbum cyanurum* **(5A)** and many *Chrysis* species, or a combination of these colours, as in *C. concinna* **(5B)**. Body extremely hard and strongly sculptured to resist attack by host wasps and bees. **Biology** Most species lay eggs in the nests of solitary wasps and bees, their larvae feeding on the host's provisions or larvae. Roll up into a ball in self-defence. Adults visit flowers for nectar. **Habitat** Widespread in most terrestrial habitats. May be seen on mud walls or termitaria at midday, investigating nests of potential hosts.

H. Robertson

1 Family Scoliidae
Mammoth wasps

Medium-sized to large (body length 10–50mm), robust, bristly. Colour variable, some species black, sometimes with reddish hairs on the head and thorax as in *Campsomeriella caelebs* **(1)** or on the tip of the abdomen. *Scolia wahlbergii* **(1A)** has a bright yellow head, antennae and pronotum. *Aureimeris* **(1B)** (body length 46mm) has a golden thorax and dark abdomen. Other species are banded with yellow. Males are more slender, sometimes differently coloured from females, with 3 sharp spines projecting from the rear of the abdomen. Wings iridescent, dark brown, blue or clear, with or without brown tips, and finely corrugated longitudinally, especially towards the tip. **Biology** Parasitize beetle larvae, mostly of Family Scarabaeidae (p.238), which live in the soil or in rotting vegetable matter. Larvae feed externally and, when mature, pupate in a tough cocoon in soil. **Habitat** Fly around low-growing vegetation, manure and compost heaps.

2 Family Tiphiidae
Tiphiid wasps

Small to medium-sized (body length 4–30mm). Males elongate and slender, with very long antennae, colour variable, often wholly black or with a yellow-banded abdomen **(2)**; the abdomen may have a double constriction at the base and, in most species, has an upturned hook at the tip. Females stouter, with bristly legs; black, with or without a metallic blue sheen; sometimes with red head and legs as in *Mesa ruficeps* **(2A)** (body length 17mm). The red in some species extends onto the thorax, or the abdomen may be reddish. Wings clear, with or without brown tips, or dark brown with a metallic blue sheen. Females of some species are wingless, those of the subfamily Methochinae resembling ants. **Biology** Parasitic, mostly on subterranean beetle larvae (Scarabaeidae and Tenebrionidae). Some species parasitize tiger beetle larvae in their burrows. **Habitat** On the ground (wingless females) and on flowers (males), including those of shrubs and trees.

3 Family Mutillidae
Velvet ants

Small to medium-sized (body length 3–18mm), usually black, often with reddish-brown thorax and combination of red, yellow, white or silver spots or bands on the abdomen. Body extremely hard and coarsely punctured, covered with soft, velvety hairs. Females wingless, with box-like thorax, *Mutilla astarte* **(3)** being typical. *Ronisia* **(3A)** (body length 9mm) has a black hourglass pattern on a whitish abdomen. Males of most species, such as *Smicromyrme atropos* **(3B)**, have dark wings with a metallic sheen, and are often differently coloured from females; in some species males lack abdominal markings. *Dasylaboides philyra* male **(3C)** (body length 15mm) has a red thorax and black body. There are 30 genera and few hundred species in the region, which occur in diverse habitats. **Biology** Parasitize larvae and pupae of various solitary bees and wasps, pupae of certain flies, moths and beetles, and egg cases of cockroaches. *Dolichomutilla sycorax* **(3D)** mostly lays its eggs on and parasitizes larvae in the large, multi-celled mud nests of wasps such as *Sceliphron* (p.490) and *Synagris* (p.478), often found in buildings. It bites its way into a cell, lays an egg on the host's larva or pupa, and reseals the cell before leaving. Stings from female velvet ants are very painful. **Habitat** Females most often seen running about on the ground, males visiting flowers.

Family Pompilidae
Spider-hunting wasps

Active, small to very large, long-legged wasps, rear margin of pronotum reaching base of fore wings, antennae often curled. Most species have a glossy blue, black or brown body, sometimes with yellow or orange markings. Wings may be black with a blue sheen, orange to red, or transparent. A few species are wingless. Males are usually smaller, with longer antennae, and are sometimes differently coloured from females. Females run about on the ground, flicking and jerking their wings. They provision their nests with 1 paralysed spider per cell or, in some species, lay their eggs on prey caught by other pompilids. Frequently seen on flowers collecting nectar, and important pollinators of many milkweed (Apocynaceae) species. There are about 200 regional species.

1 Psammochares irpex

Females medium-sized (body length 10–24mm), short-waisted. Black, but covered extensively with grey pubescence, including the legs, but not the terminal segments of the abdomen. Antennae black. Terminal tarsal segments tend to be pinkish; fore tarsi equipped with comb-like sand rakes. Wings yellow, fore wings with dark tips. Face broad. Males smaller (body length 10–14mm), but with last abdominal segments also covered in fine grey hairs. **Biology** Females dig short single-celled burrows in sand and provision each of them with 1 large spider. **Habitat** Inhabits lake shores, sandy riverbeds and banks.

2 Tachypompilus ignitus

Large (body length 47mm), body black with reddish tip to abdomen. Face with red markings, antennae red. Legs red, femora mostly black. Wings red, with a black base and dark tip. **Biology** Most commonly provides young with rain spiders (*Palystes*). Usually seen dragging a paralysed spider along the ground. **Habitat** Ground-dwelling. Often enters buildings in search of prey. **Related species** *Hemipepsis* and *Cyphononyx* species (below), with which it is easily confused.

3 Cyphononyx optimus

Large (body length 50mm), wholly black, with black antennae and blue sheen to wings. Legs yellowish-red, black near base. **Biology** Females capture and paralyse large spiders, especially *Palystes*. **Habitat** Subtropical forest, bushveld and savanna. **Related species** *C. decipiens*, which has orange wings. Also, *C. atropos* and *C. flavicornis*, which are uniformly black, the latter with yellow antennae. May also be confused with like-coloured *Hemipepsis* species (below).

4 Hemipepsis tamisieri

Medium-sized (body length of males 19mm, females 23mm). Head and antennae orange. Thorax orange and black. Abdomen black, with a reddish tip. Legs orange. Wings metallic blue. **Biology** Preys mostly on baboon spiders (*Harpactira*) and rain spiders (*Palystes*). **Habitat** On the ground, usually searching among dead leaves or low vegetation. **Related species** There are several similar *Hemipepsis* species, including *H. vespertilio* and *H. dedjas*, which are very large, black with a violet sheen, and make loud rattling sounds with their wings when flying.

5 Batozonellus fuliginosus   Orb-web Spider Wasp

Large (body length 38–42mm). Females uniformly black, with a metallic blue sheen, especially on the wings; antennae yellow or black; first segment of abdomen clothed with recumbent black hairs. Males are smaller, banded brown and yellow, with brown-tipped yellow wings. **Biology** Females specialize in hunting orb-web spiders, which they capture in their webs **(5A)**. **Habitat** Diverse. Usually found digging nest burrows in sandy soil.

6 Episyron histrio

Smallish (body length of male 7–10mm, female 10–13mm), dark brown (male blackish), with 2 thin yellow bands across the abdomen and 1 on the front of the thorax. Wings dark brown. **Biology** Nests provisioned with orb-web spiders. **Habitat** Diverse. Usually in foliage. **Related species** There are 4 other similar regional species in this genus.

7 Dichragenia jacob

Medium-sized (body length 14–21mm). Head and front of thorax reddish, rest of body black. Legs and antennae red, without sand rakes (comb-like structures on the fore legs for digging). Wings dark brown, with a metallic blue sheen. **Biology** Characteristically (for the subfamily) removes legs of lycosid spider prey before transporting the body to the nest. **Habitat** Diverse. Most often found in buildings, trapped in windows. **Related species** There are 3 other similar species in *Dichragenia*. Some build retort-shaped mud turrets in the ground over the entrances to their burrows **(7A)**.

1 *Auplopus femoralis*

Female small to medium-sized (body length 13–16mm), male smaller (8–12mm). Female black, with brown on the face and front of thorax. Legs variable black to brown, but base of femur of fore legs usually brown. Antennae and wings dark brown. Male similar, but with paler brownish-red coloration. Wings lightly shaded to clear. **Biology** Females construct independent ovoid mud cells, arranged irregularly in cavities in wood or other substrates **(1A)**, and small enough to allow wasps to pass freely into the interior of the nesting cavity. Each cell is provisioned with 1 salticid spider. Nest cavities are occupied by several females and sometimes males as well, 1 of the females acting as a guard near or at the entrance to repel intruders, suggesting semi-social behaviour. **Habitat** Eastern coastal belt forest. **Related species** There are a large number of species in the region that are similar in appearance and biology.

Family Vespidae   Paper wasps, potter wasps, pollen wasps

A large family that includes most social wasps as well as solitary species. Nests are made of chewed plant fibres (paper wasps) or mud (potter wasps). In social species, colonies consist of a single queen and female workers. Important pollinators, and predators of other insects in agricultural settings. The four subfamilies occurring in the region are discussed below (Polistinae, Eumeninae, Masarinae, Vespinae).

2 Subfamily Polistinae   Paper wasps

Narrow-waisted (body length 8–37mm), brown, with black, yellow, grey or white markings on the body. Recognizable by fore wings that fold along a longitudinal pleat when at rest, eyes with strongly notched inner margins, and mid tibiae with 2 spurs and simple tarsal claws. Wings clear or dark brown. All species build papery multi-celled nests **(2)** of wood pulp and saliva, attached by a stalk to plants, rocks or buildings, generally in a rain shadow **(2A)**. All are social and defend their nests aggressively. The young are fed with chewed caterpillars. Nests are abandoned at the end of the season after males and reproductive females are produced. Fertilized females over-winter in sheltered places, founding new colonies in spring, their female offspring becoming workers.

3 *Polistes fastidiotus*   Umbrella Paper Wasp

Medium-sized to large (body length 18–26mm), distinctive and easily recognized, reddish-brown with 2 longitudinal yellow lines on last part of thorax. Abdomen with black base and 3 yellow bands. Wings brown, dark-tipped. **Biology** Predators of caterpillar pests, which they macerate prior to sucking out the body fluids. *Polistes* nest combs are constructed in the horizontal plane, with the openings of the cells facing downwards. **Habitat** Diverse. **Related species** There are several similar, but less conspicuously marked species that are widespread in the region, including *P. smithii* **(3A)** (body length 23mm), which has 2 thin yellow bands on the abdomen.

4 *Ropalidia*

Medium-sized (body length 8–17mm), usually reddish-brown or dark brown, some with white markings. Distinguishable from *Polistes* by swollen second segment of abdomen, into which remaining segments can be telescoped. **Biology** Nests are variable in shape, some very elongate. Comb constructed vertically, with cells opening horizontally (to the side) **(4A)**. **Habitat** Diverse, with large numbers of similar species in the genus.

1 *Belonogaster dubia*

Medium-sized to large (body length 19–37mm), brown, with a long, narrow waist similar to that of *Ammophila* (Sphecidae) species, with which it may be confused. Wings dark brown. **Biology** Over-wintering females sometimes congregate in large numbers in sheltered spots. Stings fairly readily. **Habitat** Most terrestrial habitats, and widespread in the rest of Africa. **Related species** *B. brachystoma* **(1A)** and *B. petiolata* **(1B)**, which have yellow markings on the abdomen.

2 *Polistes dominula*   European Paper Wasp

Medium-sized (body length 20–30mm), black, with yellow markings, reminiscent of the Yellowjacket (*Vespula germanica*) (p.480). Antennae orange. Wings pale brown. **Biology** Nest structure as for family, but nests significantly larger than those of indigenous species. Young fed with chewed-up caterpillars as well as a greater diversity of other insect prey compared with indigenous species. **Habitat** Exotic species introduced into South Africa in 2008. Inhabits forested areas, woodland, shrubland and grassland, also agricultural, suburban and urban areas where man-made structures provide suitable nesting sites. Currently confined to the southwestern Cape.

Subfamily Eumeninae   Potter wasps, mason wasps

Mostly slender, with a short or long waist. Highly variable in colour, but wings clear or brown. As with Polistinae, fore wings fold along a longitudinal pleat when at rest and inner margins of eyes are strongly notched. Distinguishable from Polistinae by long mandibles that cross each other when not in use, and by having only 1 spur on mid tibiae, and toothed or forked tarsal claws. All species solitary, with a range of nest types: from those made in existing cavities, to burrows in soil, to nests made with mud. Females suspend an egg from the roof of each nest cell before introducing paralysed caterpillars.

3 *Allepipona perspicax*

Smallish (body length 11mm), short-waisted. Head and thorax black, with yellow markings. Abdomen black, with yellow bands. Legs brown. Antennae brown, with dark tips. Wings clear, becoming brown at tip. **Biology** Females soften clay soil with water to construct burrows, using mud pellets to build a turret at the entrance and a surrounding wall. **Habitat** Paths and clearings in bush and forest. **Related species** There are a number of similarly marked regional species in the genus.

4 *Tricarinodynerus guerinii*

Medium-sized (body length of male 9–12mm, female 14–16mm), short-waisted. Head and thorax black, with brown markings. Abdomen usually red, sometimes reddish-brown with darker markings. Legs brown. Wings pale brown. **Biology** Females nest in existing cavities such as keyholes, open ends of piping and bamboo, holes in tree trunks, wood and walls, and empty cells in the disused mud nests of other wasps. Cells separated by mud partitions, entrances provided with characteristic curved mud turret. **Habitat** Diverse. Strongly associated with buildings, where nesting sites abound. **Related species** *Proepipona meadewaldoi*, which is smaller and has similar habits, but does not build a turret at the entrance to its nest. Also, *Odynerus* species **(4A)** (body length 22mm), which have a black-tipped abdomen and wings.

5 *Anterhynchium natalense*   Orange-tailed Potter Wasp

Medium-sized (body length 15mm), short-waisted, body uniformly black, with orange tip to abdomen. Legs black. Antennae red. Wings dark brown, with a metallic blue sheen. **Biology** Nests in existing cavities and the disused mud nests of other wasps, making cell partitions with mud. **Habitat** Diverse. **Related species** There are several similar species, including *A. fallax*, which is uniformly black, *A. grayi*, which is black, with a white band on the abdomen, and another *Anterhynchium* species, which has a broad yellow band that occupies over half the abdomen. Some *Synagris* species are also similar.

1 *Aethiopicodynerus capensis*

Medium-sized (body length 13–17mm), body black, with red markings on head and pronotum, narrow white bands on first and second abdominal segments, and a circular red patch on each side of second segment. Legs red. Wings brown. **Biology** Females construct burrows in clay soil with a funnel-shaped turret over the entrance, using water to soften the clay. Cells are stocked with paralysed bagworm caterpillars. **Habitat** Open areas of clay soil. May be found visiting pools of water. **Related species** The genus includes 5 other species in the region, probably with similar nesting habits.

2 *Afreumenes aethiopicus*

Medium-sized (body length 18–20mm), long-waisted. Dark brown, with paler markings mostly on thorax. Wings dark brown. **Biology** Females build 1-celled spherical mud pots resembling urns, with a lip around the entrance, attaching them to rock faces, buildings, tree trunks and roofing thatch in sheltered spots. After completion and sealing, no more mud is applied to the outside and the lip is left intact **(2A)**. **Habitat** Often associated with buildings. **Related species** Genus includes several other species with a similar appearance and habits.

3 *Delta lepeletierii*   Yellow-and-black Potter Wasp

Medium-sized (body length 17–25mm), long-waisted. Distinctively marked in black and yellow, with red markings on the waist. Antennae red, with dark tips. Legs yellow, with a red base and tips. **Biology** Females build multi-celled mud nests attached to plants. Water for making mud is collected from pools. **Habitat** Widespread in various vegetation types. **Related species** *D. caffrum*, which is very similar in appearance, except that its waist is black, with yellow markings.

4 *Delta hottentotum*

Medium-sized (body length 13–25mm), long-waisted, body mostly dark red, with black markings on thorax and abdomen. Antennae red, with dark tips. Legs red. Wings brown. **Biology** Females build 2-celled mud nests in crevices in tree trunks, rock faces and walls. The lip around the entrance to each cell is removed during sealing, and completed nests are daubed with more mud. **Habitat** Often associated with buildings. **Related species** Other *Delta* species, which include many similar-looking or very differently coloured wasps, with different nesting habits.

5 *Delta emarginatum*   Black Potter Wasp

Large (body length 24–35mm), long-waisted, dark brown to black, with dark red markings on head, thorax and waist. Antennae dark red, with dark tips. Legs black, with a reddish base. Wings dark brown. **Biology** Females build a series of contiguous 1-celled mud nests attached to plants, rock faces and buildings. Water for making mud is collected from pools. More mud may be daubed over the completed nest. **Habitat** Widespread in different vegetation types. Commonly seen visiting water or bare patches of ground, where it quarries soil.

6 *Synagris analis*

Medium-sized (body length 22–28mm), short-waisted, distinctive, black, with red-tipped abdomen. Wings black. **Biology** Females build massive, thick-walled mud nests attached to roof beams, walls and the branches of trees **(6A)**. Cells are provisioned progressively with caterpillars and guarded by the female. **Habitat** Diverse, most often found around buildings, collecting water or quarrying soil on bare patches of ground. **Related species** There are several similar-looking species, including *S. mirabilis*, which has a white tip to the abdomen. Also, some species of *Anterhynchium* (p.476), which, although smaller, have similar colour patterning.

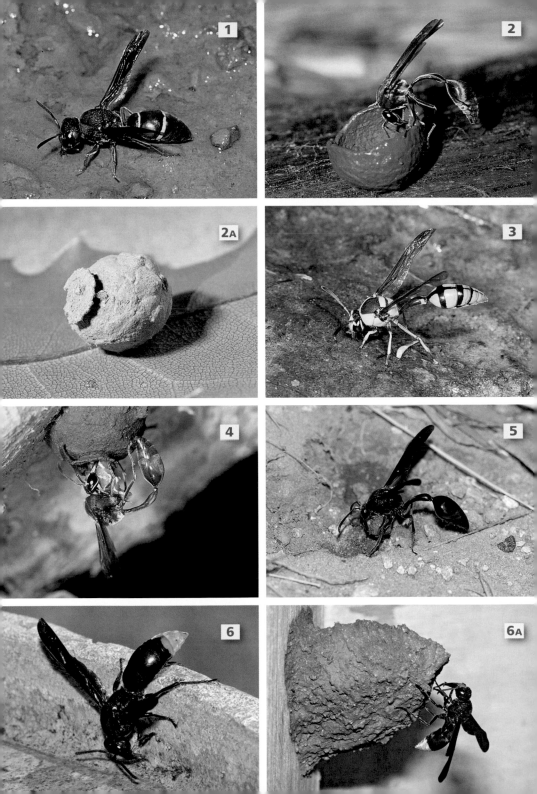

1 Subfamily Masarinae — Pollen wasps

Small to medium-sized (body length 3–19mm), mostly black (or brown) and yellow, distinguishable from other vespoid wasps by 2 (instead of 3) cells just below margin of fore wing and by absence of longitudinal fore wing pleat. An estimated 124 species occur in the region. **Biology** Some species nest in burrows in hard clay soil, never far from water, often in colonies. They collect water from pools for softening soil, often while floating on the surface. *Ceramius lichtensteinii* (shown here) (body length 15mm) uses excavated soil to form a straight or slightly curved vertical turret around the entrance to its burrow. Other species build mud cells on plants or stones. Uniquely (for wasps), cells are provisioned with pollen and nectar, hence the attraction to flowers and the common name. *C. lichtensteinii* larvae are parasitized by *Ceroctis* blister beetles (p.284). **Habitat** Dry fynbos and karroid vegetation.

Subfamily Vespinae — Hornets, yellowjackets

These are the largest members of the family. The group does not occur in sub-Saharan Africa, apart from the invasive German Yellowjacket.

2 *Vespula germanica* — German Yellowjacket

Medium-sized (body length 11–20mm). Head and thorax black, with yellow markings. Antennae black. Abdomen yellow, with black markings on each segment. Legs yellow. Wings clear. **Biology** Nests are made of several horizontal combs enclosed in a roughly spherical, greyish, papery envelope up to 300mm in diameter, and are built underground or in cavities in walls. Colonies may number several thousand. The young are fed with chewed insects, other fresh animal matter and sugary substances. Very aggressive, especially near the subterranean nest, where wasps can be seen flying in and out of the entrance hole. **Habitat** Exotic species, introduced from Europe in 1975. Impacts include damage to fruit and predation of local honey bees.

Family Formicidae — Ants

Ants have a conspicuous demarcation between head and thorax, a 1- or 2-segmented waist, and antennae elbowed at the first joint. All are social, with well-differentiated castes. Workers are wingless, usually sterile, females. Males and reproductive females generally have wings, which the females discard after mating and dispersing. The shape of the head and mandibles varies greatly with caste and even within the worker caste. Some ants have stings, sometimes modified for spraying defensive chemicals. Nests highly variable in structure and location. Some species are nomadic; others make small nests in hollow twigs, reeds or thorns, construct subterranean nests as deep as 6m or more, or nest under stones. Others build aerial nests in bushes and trees.

3 *Linepithema humile* — Argentine Ant

Small (body length 2–3mm), uniformly light brown, waist with 1 conical segment. **Biology** Runs in trails. Nomadic, occupying flimsy, unstable nest sites. Colonies huge and diffuse, with many queens, multiplying by budding. Nuptial flights rare. **Biology** Introduced from South America with horse fodder imported in 1870. An aggressive alien, eliminating most indigenous ant species. Strongly attracted to honeydew, protecting soft scales, aphids, etc. from parasites and predators. A pest in citrus orchards, vineyards and households. **Habitat** Prefers disturbed habitats but unable to survive in arid areas that do not have a constant water supply. May be displaced by the Small Black Sugar Ant (*Lepisiota capensis*), a common household pest in the southwestern Cape.

4 *Dorylus helvolus* — Red Driver Ant

Minute to very large (body length of workers 2–8mm, reproductives 40–50mm). Workers reddish-brown, without eyes, head squarish, sting non-functional. Queen similar, but with a distended abdomen. Males, sometimes referred to as 'sausage ants' **(4A)** – familiar, reddish-brown and wasp-like, with eyes, and furry head and thorax; abdomen and mandibles enlarged; wings clear. **Biology** Predatory and nomadic, living in very large colonies in temporary subterranean nests. Workers aggressive, biting vigorously. Males harmless, commonly attracted to lights. **Habitat** Diverse; common in well-watered lawns.

H. Robertson ▼

1 *Anoplolepis custodiens*   Pugnacious Ant
Small to medium-sized (body length of workers 2–10mm, queens 13mm), reddish-brown, head redder than rest of body, 1-segmented waist, abdomen appearing chequered because of 3 rows of refractive hairs. Workers with rounded abdomen. **Biology** Does not run in trails. Nests below ground, with no mound around the nest entrance. Colonies large, with several queens. Workers fast-running, very aggressive and predatory. Also tends aphids and coccids for their honeydew, often causing outbreaks of these pests. Seed disperser. **Habitat** Diverse. **Related species** *A. steingroeveri* **(1A)**, another ecologically important species, with similar behaviour and appearance, is darker and lacks the 3 rows of refractive hairs.

2 *Camponotus maculatus*   Spotted Sugar Ant
Large (body length of workers up to 20mm, queens 25mm). Workers yellow to reddish-brown, with slender head and thorax, 1-segmented waist, and somewhat bulbous abdomen. Queens **(2A)** dark brown, abdomen darker than enlarged thorax, sides of abdomen sometimes with creamish-yellow patches. **Biology** Does not run in trails. Nests in soil under stones. Colonies small, each with 1 queen. Workers do not sting, but spray formic acid at attackers by tucking the abdomen forwards under the thorax. Usually forages at night, sometimes entering houses. Tends aphids but is not a pest. **Habitat** Diverse. Common throughout Africa. **Related species** *C. importunus* **(2B)**, which is similar but less common, with a more limited distribution and which was once considered a subspecies of *C. maculatus*.

3 *Camponotus fulvopilosus*   Balbyter
Large (body length of workers 10–18mm, queens 20mm). Workers **(3)** black, with slender head and thorax, 1-segmented waist, and abdomen covered with reddish-fawn hairs. Major workers and soldiers have much enlarged heads. **Biology** Does not run in trails. Nests in soil below large rocks, fallen trees or at the base of a small bush. Colonies small, with 1 queen in each nest. Workers do not sting, but spray formic acid at attackers by tucking abdomen forwards under the thorax. Workers visit mealy bugs on bushes for honeydew, prey on termites and other insects, and collect bird droppings. **Habitat** Arid savanna and woodland with sandy soil. Often on melkbos species (*Euphorbia*). **Related species** The Desert Balbyter (*C. detritus*) **(3A)**, which is very similar in appearance and biology, but is confined to the Namib/Naukluft region of Namibia.

4 *Camponotus cinctellus*   Shiny Sugar Ant
Medium-sized (body length of major workers 9mm, minors 5.5–6mm). Body black, legs dark reddish-brown, except for black femora. Sparse creamy pubescence on head and thorax. Apices of abdominal segments yellowish. Winged reproductive shown. Abdomen covered with shimmering golden pubescence, sometimes absent from first segment. **Biology** Usually seen running along the ground or among low foliage. **Habitat** Eastern Afrotropics down to the Wild Coast, in forest, moist savanna, gardens, urban parks and dwellings.

5 *Camponotus vestitus*   Shimmering Sugar Ant
Medium sized (body length of major workers 10mm, minors 6mm). Head, antennae, mandibles, legs and thorax pale to dark brick red. Abdomen dark brown, covered with greyish-gold pubescence arranged in longitudinal bands, creating a distinctive chequered appearance as a result of the alternating lay of the 'pile'. Resembles Pugnacious Ant (above) but less common and not aggressive. **Biology** Diurnal, often seen running rapidly across the ground. **Habitat** Widespread throughout sub-Saharan Africa, inhabiting hot, sandy areas.

6 *Lepisiota capensis*   Small Black Sugar Ant
Small (body length 2.5–3.5mm). Shiny black. Mandibles, antennae and legs reddish-brown, with darker brown femora. Sparse pilosity mostly on abdomen. **Biology** Runs in trails, nesting under stones, in walls and under floors, and forms large colonies. Appears capable of displacing the invasive Argentine Ant (p.480). **Habitat** Widespread in diverse habitats, including homes in the southwestern Cape, where it is a pest. Also occurs across Africa and as far afield as India.

1 *Oecophylla longinoda* — Weaver Ant, Tailor Ant

Medium-sized (body length of workers 7–11mm). Workers reddish-brown, eyes large, thorax slender, abdomen reflexible over thorax. **Biology** Binds leaves with silk spun by larvae to make nests, which may enclose colonies of coccids. Workers very aggressive, biting intruders and injecting formic acid into bites. **Habitat** Coastal bush and forest.

2 *Polyrhachis gagates* — Shiny Spiny Sugar Ant

Medium-sized (body length of workers 12mm). Workers black, covered with short silvery hairs, with 1-segmented waist. Front and last part of thorax of workers armed with sharp spines. Abdomen globular and down-turned as in *Mymicaria*. **Biology** Tends aphids, coccids and membracids for honeydew. Often found on shoots and stems or in foliage. **Habitat** Savanna. **Related species** About 13 species. *P. schistacea* is distinguishable by hairs all over top of thorax (not only on front) and is more common; sometimes a pest in citrus orchards.

3 *Crematogaster peringueyi* — Black Cocktail Ant

Small (body length of workers 3–6mm, queens 10mm). In workers, head, thorax, 2-segmented waist and legs reddish-brown. Workers reflex the black, heart-shaped, pointed abdomen over thorax when alarmed. Sting replaced by a gland that secretes irritant defensive fluid. **Biology** Runs in trails. Builds carton nests **(3A)** in bushes and trees, from vegetable fibres glued with salivary secretions. Workers tend aphids and coccids for honeydew. Larvae of various *Chrysoritis* butterflies (p.454) develop in the nests, where they feed on ant larvae, and supply honeydew to the ants via a gland. **Habitat** Diverse. **Related species** There are many similar species in the genus, including some ground-dwellers that nest under stones and logs. Also, *C. acacia*, the Thorn Cocktail Ant **(3B)**, which is slightly smaller, reddish-brown in colour, and nests in the hollow thorns of *Vachellia* species.

4 *Carebara vidua* — African Thief Ant

Minute to large (body length of workers 2mm, queens 25mm, males 18mm). Workers yellow, queens dark brown or black with bulbous thorax and huge, inflated abdomen. Males resemble queens, but with yellow abdomen, legs and antennae, body covered with pale hairs, and smaller thorax and abdomen. Reproductives with pale brown wings. **Biology** Nests in walls of termite mounds. Colonies of tiny workers release huge winged females (shown here) and males after the first heavy summer rains, a few workers carried with the queen. The subterranean and seldom-seen workers raid termite colonies to prey on the eggs and brood. **Habitat** Savanna.

5 *Messor capensis* — Common Harvester Ant

Medium-sized (body length of workers 5–12mm, males 8mm, queens 14mm). Thorax narrow, widest at front. Waist 2-segmented, attached to last part of the thorax by a narrow stalk. Abdomen large and somewhat globular, shiny black. Workers with dark reddish-brown head and thorax, darker abdomen, and very large head, especially in major workers. **Biology** Walks slowly in trails, and makes large nests underground. Feeds mostly on seeds, leaving husks lying in heaps around the nest entrance. Assists in seed dispersal, since some seeds may be dropped or left uneaten. Often forages at night. **Habitat** Dry areas, where its nests of mounded white seed husks, surrounded by a bare apron, stand out in the Karoo landscape.

6 *Myrmicaria natalensis* — Drop-tail Ant

Small (body length of workers 7mm). Workers (shown here) with reddish-brown head and thorax, and 2-segmented waist. Abdomen black, characteristically globular and down-curved. Last part of thorax armed with 2 backward-pointing spines. **Biology** Slow-moving, but aggressive. Like others in the genus, builds nest mounds of soil and organic material thatched with leaves, stems and other materials, often at the base of bushes. Workers tend aphids and coccids. Sometimes damages the leaves of young citrus trees. **Habitat** Forest and woodland; also gardens.

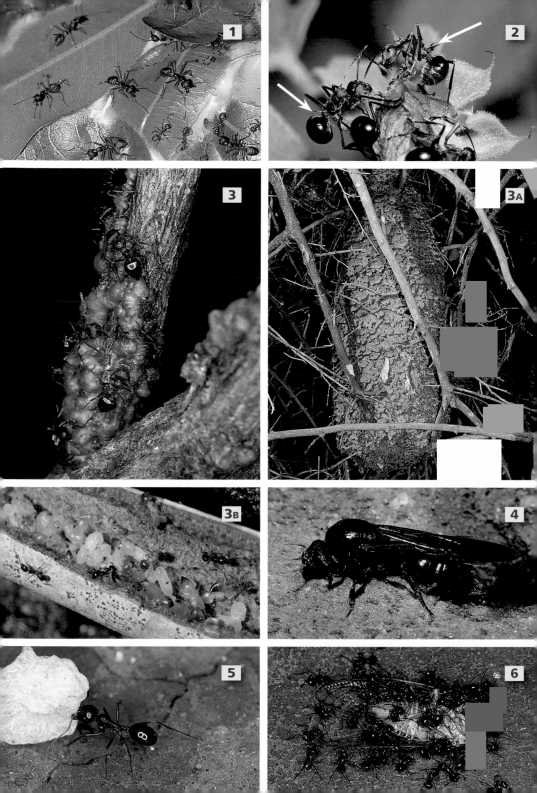

1 *Pheidole* — House ants, Big-headed ants

Small to medium-sized (body length of major workers 3–6mm, minors 1.8–2.4mm depending on species). Easily identified by the disproportionately large head and huge, heavy mandibles of the major workers. Antennae 12-segmented, with a club of 3 segments. Body colour brown or black, depending on species. The Big-headed House Ant (*P. megacephala*) is invasive, now cosmopolitan, probably originating in the warmer parts of Africa. **Biology** Mostly omnivorous; major workers use their large mandibles mainly to cut up prey. Nests in the soil under various objects such as logs, stones, paving, etc. **Habitat** Widespread in diverse habitats, especially where the soil has been disturbed.

2 *Solenopsis punctaticeps* — Fire Ant

Small (body length of workers 2–3mm, queens 6mm), with a 2-segmented, bead-like waist. Recognizable by 2-segmented club at the end of the antennae and 4 protruding spines (the outer 2 larger than the inner 2) on front margin of upper lip. Workers yellowish, with bulbous abdomen and relatively large head. Queens with bulbous thorax and inflated abdomen. Clusters of small yellow ants of different sizes under rocks are likely to be of this species. **Biology** Sometimes builds mounds. Preys on termites and invades colonies of other ants, preying on their broods. Painful, burning sting may cause allergic response. **Habitat** Subterranean in diverse habitats.

3 *Tetramorium capense* — Cape Fierce Ant

Small (body length of workers 2–3mm). Workers reddish-brown, with dark brown abdomen, 2-segmented waist, last part of thorax with 2 backward-pointing spines. Queens similar but slightly larger. **Biology** Does not run in trails. Nests in a rotten tree stump or below ground, with the entrance surrounded by a crater of soil. Predatory, mainly on termites. **Habitat** Diverse, in both moist and dry areas.

4 *Megaponera analis* — Matabele Ant

Medium-sized to large (body length of major workers 18mm, minors 9–10mm), shiny black ants. Queens permanently wingless. **Biology** Run in large swarms or columns, attacking termites at their feeding sites. Relatively large colonies (500–2,000 workers) in subterranean nests up to 0.7m below surface, often near trees, rocks or in deserted termite mounds, and sometimes in rotting logs. When disturbed, they release an unpleasant smell and produce loud squeaking sounds. **Habitat** Widespread in sub-Saharan Africa, wherever termites occur.

5 *Paltothyreus tarsatus* — African Ant, Foul Stink Ant

Large (body length of workers and males 17mm, queens 22mm), black, with a shiny abdomen and fine grooves on the head and thorax. Waist has 2 segments, the anterior segment being knobbed and narrow transversely, with a flattened, grooved top. The posterior segment has 2 small knobs on the front. Tibiae with golden hairs. **Biology** Does not run in trails. A scavenger and predator, mostly of termites. Lives in relatively small colonies, and gives off a strong, fetid smell from the mandibular glands when alarmed. **Habitat** Sandy savanna. **Related species** *Plectroctena mandibularis* **(5A)**, which is large (body length of workers 19mm) and shiny black, with a 2-segmented waist, the anterior segment being knobbed and rounded. It does not run in trails. Preys on termites, millipedes and beetles, which it grasps in its massive mandibles **(5B)**, and nests in deep tunnels with a few chambers; colonies relatively small. Widespread, but absent from the Western and Northern Cape.

6 *Streblognathus peetersi* — Smooth Ringbum Ant

Very large (body length of workers 17–20mm), the largest worker ant in the region. Head and thorax matt black due to fine sculpturing, abdomen shiny black, mandibles reddish-brown. Waist 1-segmented, the sides of the anterior segment converging dorsally into a pointed, keel-like crest. Last part of thorax armed with a pair of short spines. **Biology** Does not run in trails. Predatory. Colonies small, often only a few dozen. Nest entrance with mud shelf. Stridulate when disturbed. There is no queen, and egg-laying is undertaken by a single mated worker. **Habitat** Areas of hard clay soil.

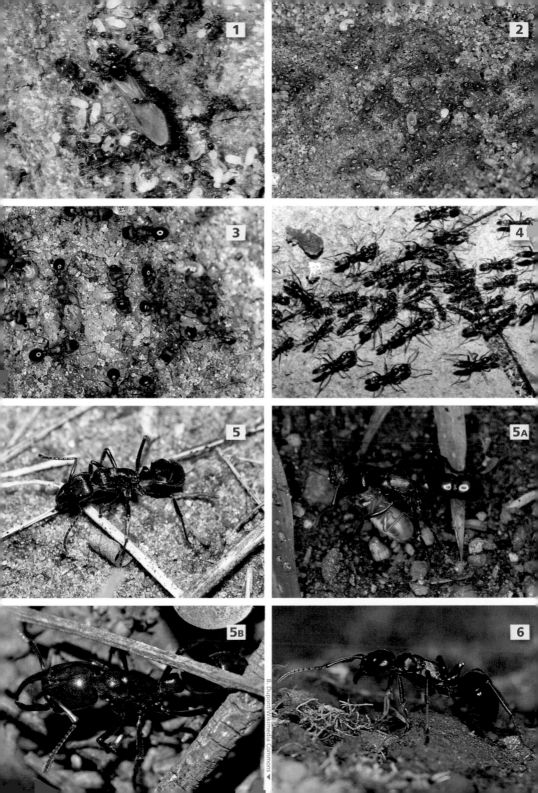

1 *Tetraponera* — Slender ants

Medium-sized (body length of workers 4–10mm). Workers long, slender, short-legged, reddish-brown, with relatively large eyes, and last part of thorax unarmed. Males and queens also long and thin. **Biology** Workers have a large, strongly developed, painful sting used to capture prey and defend the colony. Nest in hollow twigs, reeds or thorns. Small colonies, defended aggressively. Selectively prune neighbouring plants to reduce invasion of host plant by other ants. **Habitat** Diverse, but absent from grassland. *T. ambigua* and *T. natalensis* nest in *Senegalia* trees, *T. bifoveolata* on trunks of Marula (*Sclerocarya caffra*).

2 Family Ampulicidae — Cockroach wasps

Small to large (body length 3–33mm), elongate, usually metallic green, blue or purple, sometimes black or with red head, thorax and tip of abdomen. Pronotum relatively long, narrower than and distinct from rest of thorax, the last part of the thorax usually with longitudinal ridges converging posteriorly. Legs long and slender, with drumstick-shaped femora. Antennae long. *Ampulex* shown here (body length 15mm) is typical. *Dolichurus* species are dark with a black-and-brown abdomen and are similar in biology to *Ampulex*. These 2 regional genera contain about 22 species. **Biology** Nests provisioned exclusively with cockroaches, including domestic species. Once the victim is lightly paralysed with a sting, it is drawn by its antennae **(2A)** to a suitable cavity, an egg is laid on it, and the entrance to the cavity blocked up with plant fragments and other debris. **Habitat** Subtropical forest and woodland. Usually on tree trunks.

Family Sphecidae — Mud dauber, digger and sand wasps

A diverse group of solitary wasps, varying greatly in size (body length 1–40mm), with relatively broad head and a pronotum that forms a more-or-less raised collar, its rear margin not reaching the base of the fore wings. The base of the abdomen may form a long, thin waist. Females dig using a comb-like structure (sand rake) on the tarsi of the fore legs. Nests consist of burrows excavated in the ground, a plant stem, an existing cavity or the disused burrow of another insect species, or else are mud structures attached to plants, rocks or buildings. They are provisioned with paralysed insects or spiders. Some species lay their eggs in nests provisioned by other wasps or are parasitic as larvae. Adults visit flowers to feed on nectar. Females can sting but are seldom aggressive. Some species have economic potential as pest-control agents, depending on their prey preferences.

3 *Chlorion maxillosum* — Cricket Hunter

Medium-sized to large (body length 24–40mm), relatively short-waisted, body and wings black with a metallic-blue sheen. Head and legs dark reddish-brown or black. Mandibles exceptionally large and adapted for grasping crickets. Well-developed comb-like sand rakes on fore tarsi. **Biology** Does not construct nests. Digs for large crickets, especially *Brachytrupes*, which are stung into paralysis. An egg is then attached to the victim, which later recovers and resumes normal activities until killed by the wasp larva. **Habitat** Open, sandy areas in subtropical forest and savanna.

4 *Chalybion spinolae* — False Mud Dauber

Medium-sized (body length 15–30mm), body black, with a metallic blue sheen (distinguishing it from *Sceliphron spirifex* (p.490), which is matt black) and a long yellow waist. Hind legs black, with yellow bands, fore and mid legs reddish. Antennae reddish, with black tips. Wings clear. **Biology** Females excavate short, single-celled burrows in vertical banks, using water collected from pools to soften the soil and to make mud to seal the nest. Cells are mass provisioned with button spiders (*Latrodectus*) and false button spiders (*Steatoda*). **Habitat** A wide variety of vegetation types. **Related species** *C. laevigatum* and the similar *C. tibiale*, which are metallic blue and nest in existing cavities in wood and other substrates.

Colin Ralston ▼

1 *Sceliphron spirifex*　　　　　　　　　　　　　　　　Mud Dauber

Medium-sized (body length 17–27mm), body dull black, with long yellow waist. Legs black, with yellow bands. Antennae black, distinguishing it from the False Mud Dauber (*Chalybion spinolae*) (p.488). Wings clear. **Biology** Females build large, multi-celled mud nests attached to cliffs, rocks, tree trunks, bridges and buildings. Cells are mass provisioned with several spiders (mostly orb-web spiders) and sealed with mud. **Habitat** Diverse. Strongly associated with buildings and other man-made structures. **Related species** There are 3 other similar-looking species, including *S. fossuliferum*, which makes nest cells from dung, attaching them end-to-end to grass stems.

2 *Isodontia pelopoeiformis*　　　　　　　　　　　Charcoal Digger Wasp

Medium-sized (body length 18–30mm), dark brown, with brown wings. Petiole about as long as the rest of the abdomen, which is particularly bulbous. **Biology** Nests in existing cavities in the branches of trees, sometimes in man-made cavities such as pipes. Cells are separated by partitions of plant material and are mass provisioned with katydid nymphs, crickets and, rarely, cockroaches. The cavity is plugged with stones, soil clods, seeds and/or pieces of charcoal. **Habitat** Widespread in wooded areas. **Related species** The grass-carrying wasps *I. stanleyi* and *I. simoni*, which both have red on the basal half of the abdomen and clear or slightly cloudy wings. *I. stanleyi* nests in wood crevices and seals the nest with dry grass inflorescences, while *I. simoni* nests in cavities in the ground.

3 *Sphex tomentosus*　　　　　　　　　　　　　Yellow-back Digger Wasp

Medium-sized (body length 23–35mm), body black, with conspicuous yellow 'fur' on the face and rear of thorax. Abdomen with short, straight waist. Hind femora and tibiae red, legs otherwise black, with conspicuous comb-like sand rakes on fore tarsi. Wings clear, with pale brown tips. **Biology** Females dig extensive multi-celled nest burrows up to 1.3m deep, each cell provisioned with several, usually adult, katydids. Excavated soil forms a conspicuous mound at the entrance to the nest. **Habitat** Areas with deep friable sand in subtropical forest and savanna. **Related species** There are 10 other regional species in this genus that provision their nests with grasshoppers or katydids.

4 *Podalonia canescens*

Medium-sized (body length 14–24mm), narrow-waisted with waist about as long as the rest of the abdomen. Black, with red basal half of abdomen. Legs black, with well-developed comb-like sand rakes on fore tarsi. Wings clear. **Biology** Females hunt cutworms and other caterpillars that live in soil or among leaf litter. These are cached while a single-celled nest is dug in the soil. Each nest is provisioned with a caterpillar. **Habitat** Diverse, occurring across Africa. **Related species** *Ammophila* species (p.492), which are similar, but have a longer petiole; also *Prionyx* species (below), which have fairly short, rounded abdomens.

5 *Prionyx kirbyi*　　　　　　　　　　　　　　　　Kirby's Hunting Wasp

Medium-sized (body length 12–20mm), slender-bodied with waist about as long as the rest of the rounded abdomen. Black, with red basal half of abdomen, each segment of abdomen with a fine white bar. Legs black. Wings clear. **Biology** Females hunt acridid grasshoppers, which are cached while a single-celled nest is dug in soil. Each nest is provisioned with 1 grasshopper. **Habitat** Open ground in areas with friable soil in a wide variety of vegetation types. **Related species** *P. subfuscatus* (black with yellow wings), which occurs in Zimbabwe and preys on locust hoppers, and *P. funebris*, which is a very large (body length 37mm) black species that occurs in arid areas.

1 *Ammophila rubripes*   Thread-waisted Wasp

Medium-sized (body length 19–28mm), slender, with very long waist at least twice as long as the rest of the abdomen. Comb-like sand rakes on fore tarsi. Mostly black, with some red markings on thorax and more-or-less red basal half of abdomen. Wings clear. **Biology** Females hunt caterpillars among foliage. These are carried, usually in flight, to pre-prepared, single-celled nests dug in the soil. Several prey are provided per nest in a series of provisioning events. **Habitat** Widespread in areas with bare, sandy soil. **Related species** There are 17 other similar regional species in this genus.

2 *Ammophila beniniensis*   Benin Thread-waisted Wasp

Medium-sized to large (female body length 23–35mm, male smaller 18–23mm), slender, both sexes with a very long waist, almost 1.5 times the length of the rest of the abdomen. Comb-like sand rakes on fore tarsi. Entirely dark brown, but sometimes reddish-brown on the head and junction of the waist and on rear section of the abdomen. Wings fuscous. **Biology** Females hunt caterpillars among foliage. These are carried to prepared, single-celled nests dug in the soil. One prey item is provided per nest. **Habitat** Widespread in areas with friable, bare or grass-covered soil below tree canopies or surrounded by rank grass or herbaceous vegetation.

3 *Ammophila insignis*   Conspicuous Thread-waisted Wasp

Females medium-sized to large (body length 26–35mm), males smaller (20–28mm), both slender with a very long waist, almost 1.4 times the length of the rest of the abdomen. Comb-like sand rakes on fore tarsi. Black, except for basal half of the abdominal bulb, which is red. Legs red, except for the black femora. Wings pale brown. **Biology** Females dig single-celled nests in the face of a vertical bank, the floor of a cave or a disused animal burrow. Several caterpillars are provided per nest, the entrance being temporarily sealed with a clod of earth or a stone in between provisioning events. Prey is carried in flight unless unusually heavy. **Habitat** Widespread in areas with bare, sandy soil and where vertical banks or otherwise sheltered sites are available for nesting.

Family Crabronidae   Bee wolves, sand wasps

This very large family includes a diverse assemblage of wasps, many of which are probably not closely related to one another. However, they all sting and paralyse some kind of insect or spider on which their young develop. All solitary, but some nest in aggregations in sandy cliffs and banks. There are at least 150 regional species

4 *Tachytes observabilis*   Golden Cricket Wasp

Medium-sized (body length 16–22mm), short-waisted. Head and thorax black, with fine white hairs. Abdomen covered with flat-lying golden hairs arranged in a chequered pattern. Legs black. Wings clear, with brown outer margins. **Biology** Females dig multi-celled nests in sandy soil and provision them with acridid grasshoppers. **Habitat** Subtropical forest and savanna. **Related species** There are many similar species, some with a striped abdomen and others, like *T. nigropilosellus*, totally black.

5 *Tachysphex aethiopicus*

Small (body length of females 8–12mm, males 6–8mm), body black, except for the first 3 segments of the abdomen, which are red. Legs red, antennae black. Compound eyes greenish. Comb-like sand rakes on fore tarsi. Wings pale fuscous. **Biology** Females dig short, sloping entrance tunnels with up to 3 lateral shafts, each ending in a single cell. One or several grasshoppers are carried to the nest in flight or on foot, depending on their size. **Habitat** Areas with horizontal to sloping, bare, friable soil.

1 Cerceris iniqua

Medium-sized (body length 14–17mm), body black with a faint blue sheen, except for the abdomen, which is orange, except for the first segment. Legs and antennae black. Fore legs with comb-like sand rakes. Wings dark brown. **Biology** Females dig more or less vertical shafts branching into a number of secondary shafts, each terminating in a single cell. Excavated soil is deposited around the entrance, forming a characteristic tumulus of 'sand sausages'. Each cell is provided with several weevils (Curculionidae) carried in flight to the nest. **Habitat** Areas with bare, more or less friable soil such as roadsides, dry watercourses, floodplains and stable, vegetated, coastal dunes.

2 Trypoxylon   Keyhole wasps

Smallish (body length 14mm), body black to dark brown. Abdomen longer than the rest of the body, narrow at the base, expanding gradually to the rear. Inner margins of compound eyes with a distinct notch. Wings clear, slightly brown towards tip. At least 12 species in the region. **Biology** Females nest in disused bee burrows and almost any other existing cavity, including old mud nests and man-made holes. Mud is collected and used to partition cavities into cells, each provisioned with several spiders. **Habitat** Diverse. Often found collecting mud at muddy pools.

3 Dasyproctus bipunctatus   Watsonia Wasp

Small (body length 12mm), long-waisted, body black, usually with yellow bars on the pronotum and abdomen, much reduced in some varieties. Head black, large and squarish, basal segment of antennae yellowish. Legs red, with some black at base. Wings pale brown. **Biology** Females excavate nests in live flower stems of monocotyledons, mostly bulbs in the families Amaryllidaceae, Iridaceae and Liliaceae. Nests are entered at the side, their numerous cells separated by chewed pith and mass provisioned with flies, mostly Muscidae. **Habitat** Diverse, common in gardens. **Related species** There are about 12 other similar species in the region.

4 Stizus fuscipennis

Medium-sized (body length 14–24mm), short-waisted, body dark brown, covered with fine whitish hairs, especially on last part and underside of 'thorax'. Abdomen has 2 yellow lateral patches on the second and third segments. Wings clear. **Biology** Females dig burrows with 1 or more cells in sandy soil. Cells are provisioned with several mantid nymphs. **Habitat** Bare, sandy areas in coastal bush. **Related species** There are at least 20 other species in the region with yellow markings, some, like *S. imperialis*, almost entirely yellow. Many provision their young with grasshoppers.

5 Stizus imperialis

Small to medium-sized (body length 13–21mm). Head and thorax black, with extensive yellow markings. Pronotum with 2 longitudinal, teardrop-shaped yellow markings. Abdomen with bold yellow bars and a narrow black central line. Antennae rufous, yellowing at the tip. Legs yellow. Wings clear pale fuscous, with a small dark brown patch on the leading edge towards the apex. **Biology** Females burrow into the vertical face of a sandbank or into the roof of a similarly situated, abandoned animal burrow. The initial burrow branches into 2 or more shafts, each ending in a single cell. Cells are provisioned with a variety of grasshoppers, up to 12 in a cell. **Habitat** Arid areas with low vegetation.

6 Gorytes natalensis   Flowerpot Wasp

Medium-sized (body length 20mm), short-waisted, body black with brown markings on the thorax. Second abdominal segment bright yellow, apex brown. Legs brown. Wings clear, yellowish, with a brown patch near the tip. **Biology** Females dig nests in soil, commonly in flowerpots, provisioning them with cercopid nymphs such as those of *Ptyelus grossus* (p.184). **Habitat** Diverse, including gardens. **Related species** There are 2 other *Gorytes* species in the region.

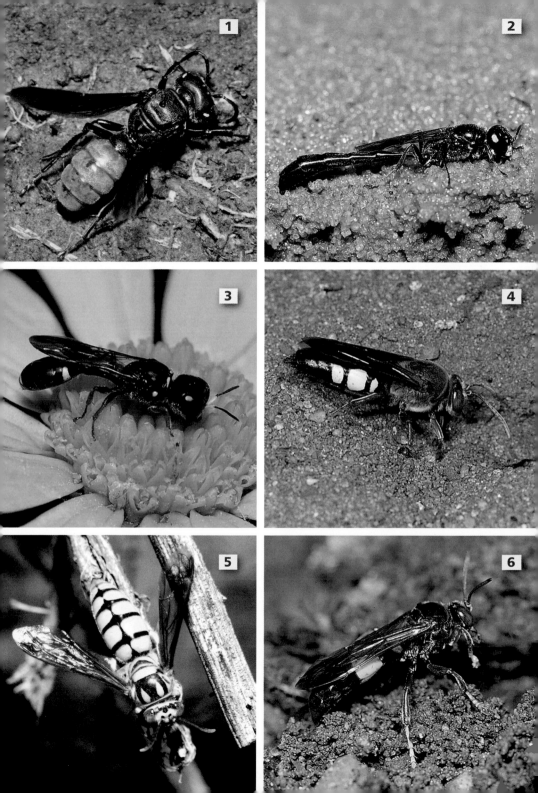

1 *Bembecinus laterimacula*

Medium-sized (body length 16mm), short-waisted. Thorax dark brown with paler markings. Abdomen dark brown with yellow patches on each side, 1 on each segment. Legs and wings brown. **Biology** Females hunt for cercopid nymphs, or sometimes tettigid grasshoppers, which they supply progressively to their larvae. Nests are dug in sandy soil, often in colonies. **Habitat** Subtropical forest and savanna. **Related species** There are 39 other regional species of *Bembecinus*. May be confused with some species of *Stizus* (p.494).

2 *Bembix* — Sand wasps

Medium-sized (body length 13–27mm), short-waisted. Head and thorax usually black, labrum conspicuously long. Abdomen barred in black and yellow. Legs yellow, with black markings, females with large comb-like sand rakes on the fore tarsi. Wings clear. At least 43 species in the region. **Biology** Females dig multi-celled nests in sand, provisioned progressively with flies from several families and sealed only when larvae are almost mature. **Habitat** Diverse. Nest in colonies on sandy beaches and banks and in other open, sandy areas.

3 *Philanthus triangulum* — Bee Wolf

Small to medium-sized (body length 10–17mm), short-waisted, black, with white markings on face and thorax. Head broad, with inner eye margins sharply notched. Abdomen and legs bright yellow, with comb-like sand rakes on fore tarsi. Wings clear. **Biology** Females dig multi-celled nests in bare, sandy soil. Cells are mass provisioned with bees, commonly honey bees ambushed at flowers. **Habitat** Diverse. **Related species** There are 13 other species in the region, which are similar, with a black-and-yellow, black-and-red or red abdomen. Also, *Cerceris* species (p.494), which are similar, but lack eye notches and have ridges along the rear margin of the first 4 abdominal segments. They provision their nests with beetles, especially weevils. Another similar species is the Bee Pirate (*Palaris latifrons*), which has yellow legs, a yellow abdomen banded in black, and hunts bees at their hives, causing the collapse of bee colonies in the sandy areas that it favours.

4 Family Colletidae — Membrane bees

Small to medium-sized (3–13mm), recognizable by the unique forked tongue. Mostly black or dark-coloured, with a hairy or smooth body and wasp-like appearance (species of *Colletes* resemble honey bees). Hairs on the abdomen often arranged in conspicuous bands. Some species (e.g. *Colletes*) use a brush of hairs on the hind tibiae to carry pollen; others use their crop. The Brown Membrane Bee (*Hylaeus heraldicus*) **(4)** is typical. Some 70 species are found in the region. **Biology** Non-social. Nests are made in existing cavities or by burrowing into pithy plant stems, rotting wood or soil. Uniquely, nests are lined with a transparent, cellophane-like material secreted by the female from an abdominal gland (hence the family's common name), and the cells are provisioned with regurgitated pollen and nectar. **Habitat** Diverse.

5 Family Halictidae — Sweat bees, flower bees, alkali bees

Sweat bees, so-called because they are commonly attracted to sweat, are small to medium-sized (body length 2–10 mm) and short-tongued. Most are brown or black, but some are metallic blue or metallic green. Basal vein of the fore wing is strongly curved in several genera. Subfamily Halictinae is the largest and most diverse, and a typical example, *Seladonia* sp., is shown here **(5)**. They range from solitary, through several grades of semi-sociality, to true sociality. Several species, mostly *Sphecodes*, parasitize bees, including other halictids. Multi-celled nests are made in burrows in the ground, sometimes in weakened wood or existing cavities. Nest architecture very diverse. Cells are lined with a water-repellent material and provisioned with pollen and nectar gathered from a wide variety of flowers.

6 *Nomia amabilis* — Blue Bee

Smallish (body length 13mm), body black, with white hairs on face, parts of the thorax and fore and mid tibiae. Abdomen with white or blue lateral bars. Concave, W-shaped plate on top of last section of thorax. Wings dark brown. **Biology** Aggregates to dig nests in clay soils in shady areas, including old termite mounds. In some areas pollinates cucurbit crops. **Habitat** Diverse.

1 *Nomioides*

Small (body length 5mm), black, with a metallic sheen, yellow at front and back of thorax. Broad yellow bands across abdomen. Legs yellow, with black basal half of femora. Body covered with long, relatively sparse white hairs. **Biology** Collects pollen from a few plant species, including Devil's thorn (*Tribulus*). Nests dug in mostly sandy or stony soils. **Habitat** Arid areas, rarely in moist, forested habitats.

2 **Family Melittidae**   Melittid bees

Small to medium-szed (body length 3–22mm), short-tongued, distinguishable from halictids by basal vein of fore wings, which is never strongly curved. Mostly hairy, black, reddish-brown or greyish, often with banded abdomen. Brush of hairs on hind tibiae used for carrying pollen. May superficially resemble honey bees. Some, such as the large *Meganomia binghami*, have yellow markings and abdominal bands (due not to hairs but to a coloured integument). *Melitta* species shown **(2)** is typical. **Biology** Solitary. Construct burrows in soil, almost exclusively in open ground. *Rediviva* collect oils (to mix with pollen) from certain flowers, assisted by unusually long fore legs that are specially adapted for reaching and adsorbing oils. **Habitat** Diverse. Especially numerous in the Western Cape.

Family Megachilidae   Leafcutter bees, mason bees

Long-tongued bees. Species with non-parasitic females have a brush of pollen-holding hairs on the underside of the abdomen. All are solitary, and some (cuckoo bees) feed on pollen collected by other megachilids. Nest structures are highly diverse: some construct cells in existing cavities in wood or in the ground, using mud, plant fibres, chewed leaf material or round pieces cut from living leaves; others build nests in the open from combed plant material (resembling cotton wool), resin, mud and small stones. Like all other bees, they provision their nests with a mixture of pollen and nectar.

3 *Fidelia major*   Fideliine Bee

Medium-sized (body length 10–23mm), pale, with a dense covering of greyish or yellowish hair. Upper lip often yellow, and abdomen often banded. Hind tibiae and basal tarsal segments particularly hairy in female, and used to shift excavated soil. **Biology** Nests in loose aggregations in flat areas. Nests consist of oblique burrows with several branches, each ending in an unlined cell, some reaching 2m in depth. Pollen collected from a variety of flowers, especially Asteraceae and Mesembryanthemaceae. **Habitat** Arid areas. **Related species** There are 10 other regional species in the genus, including *F. friesei*, a common visitor to *Sesamum* flowers.

4 *Anthidium cordiforme*   Carder Bee

Small (body length 6–8mm), robust, with wide, flat head. Head and thorax black, with yellow markings, top of head and pronotum covered with short golden hairs. Compound eyes white in female. Abdomen black, with yellow bands interrupted in the middle (only on first 2 segments in female). Legs black and yellow, with white hairs. Wings brown. **Biology** Nests **(4A)** made of cotton-like plant fibres. Species in the genus generally dig cavities in loose soil or build multi-celled nests of plant fibres in existing cavities in soil, wood or a stem. **Habitat** Savanna and bushveld. **Related species** There are several other related genera that make similar nests, including *Serapista*, which build large, white, fluffy, multi-celled nests using plant fibres and soft animal hair and attach them to the stems of plants.

5 *Megachile combusta*   Red-tailed Leafcutter Bee

Medium-sized (body length 24mm), head and thorax black, with short black hairs. Base of abdomen black, the rest covered with red hairs. Legs black. Wings dark brown. **Biology** Nests by constructing a series of mud cells in existing cavities, hollow canes or disused insect burrows. **Habitat** Diverse. **Related species** *M. bombifrons* **(5A)** (body length 34mm), which is very similar, and *M. felina*, which has greyish hairs, apical bands of whitish hairs on the abdominal segments, and builds resinous cells.

1 *Megachile chrysorrhoea* — Lesser Red-tailed Leafcutter Bee

Medium-sized (body length 17mm), black, with white hairs on sides and rear of thorax. Base of abdomen white, the last 2 or 3 segments covered with red hairs. Wings dark brown. **Biology** Nests in existing cavities in wood, hollow canes (shown here) or disused insect burrows; cell walls and partitions constructed from very tough resinous material. Cells provisioned with a mixture of pollen and nectar. **Habitat** Diverse. **Related species** *M. maxillosa* **(1A)**, which is slightly larger (body length up to 25mm), uniformly black, with a band of dense white hairs on the rear half of the thorax and base of abdomen, and with only the apex of the fore wing fuscous. It builds multi-cellular aerial mud nests attached to stones, buildings or plant stems.

2 *Coelioxys* — Cuckoo bees

Small to medium-sized (body length 11–14mm), black, with a punctured cuticle. White hairs cover the face, eyes and parts of the thorax. Often have 2 sharp, backward-pointing projections on either side of the thorax. Abdomen tapers to a sharp point and is banded with scale-like white hairs. Legs partially covered with white hairs. Wings clear, with pale brown tips. **Biology** Parasitize the nests of *Megachile* (above) and *Anthophora* (p.502), using the narrow abdomen to penetrate the nest cells of the host bee. **Habitat** Diverse.

Family Apidae — Honey bees, carpenter bees, digger bees, stingless bees

Members of this family, along with Megachilidae, are the only long-tongued bees in the region. Non-parasitic species have a fringe of hairs around the smooth, concave pollen baskets on the hind tibiae. Nests are built in the ground (carpenter bees, Xylocopinae), in wood (digger bees, Anthophorini), or in existing cavities (honey bees, Apinae and stingless bees, Meliponini).

Subfamily Xylocopinae — Carpenter bees

Small to large, ranging from 5–35mm. Nest in holes bored into dead wood or pithy stems. There are solitary, semi-social and social species, including some social parasites.

3 *Xylocopa caffra* — Banded Carpenter Bee

Large (body length 20–24mm), stout. Female (shown here) black and hairy, with 2 bands of yellow hairs on the last part of the thorax and first segment of the abdomen. Male **(3A)** entirely covered with yellow hair. **Biology** Excavates nest burrows (often H-shaped) in dry, pithy *Aloe* and *Agave* inflorescences, tree branches or structural timber. Burrows are subdivided into linear cells, separated by partitions made of wood scrapings and saliva **(3B)**. Males patrol fixed areas around flowers, twigs or leaves of particular bushes and trees. As in all carpenter bees, some females house special mites in a chamber on the abdomen. **Habitat** Fynbos, succulent karoo, dry savanna woodland and coastal bush. **Related species** There are many other *Xylocopa* species, including the Giant Carpenter Bee (*X. flavorufa*) (body length 52mm) **(3C)**, which is black with a rufous thorax, guards brood cells and is common in the southern and Eastern Cape and KwaZulu-Natal.

4 *Xylocopa nigrita* — Black-and-white Carpenter Bee

Large (body length 28–33mm). Female black, with white pilosity on face, and white tufts on the sides of the thorax and abdominal segments. Fore legs with white, black, or black-and-white pilosity on tibiae and tarsi. Mid and hind legs black. Antennae black. Wings dark brown. Male entirely yellow, pale on head and legs, pale or brownish-yellow on thorax and abdomen. **Biology** Females drill burrows into dead branches, untreated roof timbers and woody or pithy plant stems, in which a series of cells are arranged linearly. Each cell is provisioned with a food mass of pollen and nectar. Pollinates a range of plants, including cucurbits. **Habitat** Widespread and extends up to Central Africa.

1 *Ceratina nasalis*   Small Blue Carpenter Bee

Small (body length 7mm), metallic blue or metallic green, with heavily punctured cuticle. Covering of white hairs especially towards rear of abdomen. Tibiae with white lines. Wings pale brown. **Biology** Shows many features of social and subsocial behaviour, nesting in hollow stems, twigs and dead, pithy aloe inflorescences, the nest partitioned into individual cells using wood shavings. The oldest larvae are at the bottom of the nest, the youngest at the top. Collects pollen from a wide range of plants. **Habitat** Various vegetation types, agricultural fields and gardens. **Related species** There are 39 other species of *Ceratina* in southern Africa.

2 *Allodapula variegata*   Small Carpenter Bee

Small (body length 7mm), black, with yellow markings and white hairs around last part of the thorax. Abdomen reddish-brown, last 3 segments with flattened whitish hairs. Legs black and reddish-brown, with distinct brushes of long hair on hind legs. **Biology** Subsocial. Excavates cavities in dry, pithy and soft woody stems and herbaceous plants **(2A)**. As there are no cells in the nests, several larvae are reared together and fed regurgitated pollen and nectar. Larvae feed on a central food mass, and larvae of some *Allodapula* species form a ring around the food mass. The adult bee uses the end of its abdomen to block the nest entrance against intruders. **Habitat** Vegetation suitable for nesting. **Related species** Other *Allodapula* species, some of which parasitize the larvae of other species in the genus.

Subfamily Apinae   Honey bees, stingless bees and other bees

Small to medium-sized bees (4–20mm). A diverse group of social species, cleptoparasites, oil collectors, pollen and nectar collectors (including honey bees) and robbers. There are no simple characteristic features defining this subfamily of bees.

3 *Amegilla atrocincta*   Black-ringed Bee

Medium-sized (body length 14–19mm), with orange hairs on top of the head and thorax and white hairs on the face and on sides and underside of thorax. Last part of thorax and base of abdomen black, with broad orange bands on rest of abdomen. Fore legs white, mid and hind legs black. Wings dark brown. **Biology** Females construct burrows in clay, which they soften with water, the entrance surmounted by a turret up to 70mm long, made of pellets of excavated clay. Sleeps in groups (comprised mainly of males) with jaws embedded in grass stems. **Habitat** Nests among low vegetation in bare, flat, clayey soils, generally close to a waterbody. Widespread in sub-Saharan Africa.

4 *Amegilla caelestina*   Green-banded Bee

Medium-sized (body length 14mm), head, thorax and, especially, last part of thorax covered with long white and dark brown hairs. Abdomen with metallic bluish-green hairs. Legs with white hairs, a brush of white hairs on the outside of the tibiae, and first tarsal segment reddish on the inside. **Biology** Collects nectar from sage plants (*Plectranthus*). **Habitat** Subtropical woodland and savanna. **Related species** *A. cymatilis*, which is similar but smaller.

5 *Anthophora braunsiana*

Small to medium-sized (body length 12–14mm). Entirely black, head and thorax densely covered with grey hairs, sometimes brownish on thorax. Abdomen with bands of grey hairs. Legs fringed with grey hairs. Antennae black. Wings clear. **Biology** Nests in aggregations in soil, termite mounds, the surfaces of dirt roads **(5A)** and similar sites. Nest entrances are surrounded by a ring of excavated soil pellets. Cells clustered in linear series at the ends of burrows, each lined with a coating of white wax. Cells provisioned with a moist paste of pollen and nectar. The larvae are parasitized by *Thyreus* bees, mutillid wasps and meloid beetles. **Habitat** Diverse karroid vegetation. **Related species** *Anthophora vestita* **(5B)** (body length 22mm), which is a yellowish, furry bee. As is typical for *Anthophora* species, the female excavates her own nest in the soil, often doing so gregariously. An important pollinator of crops such as melons.

1 *Thyreus delumbatus* White-barred Cuckoo Bee

Medium-sized (body length 13mm), black, with conspicuous white hairy markings on face, thorax and legs, and white lateral bars on each segment of abdomen. W-shaped plate on top of last section of thorax. Wings pale brown. **Biology** Parasitizes *Anthophora* and *Amegilla* bees (p.502), entering nests and depositing eggs in cells. Their larvae kill host larvae and feed on nest provisions. **Habitat** Diverse. **Related species** There are 38 other regional species of *Thyreus*, including several that look similar to *T. delumbatus*, but which may have blue, not white, markings **(1A)**.

2 *Meliponula* Stingless bees

Very small (body length 5–6mm), black, covered with white hairs, segments of abdomen with reddish-brown basal bands. Wings clear. **Biology** Nest in existing cavities lined with a mixture of wax, mud, resin and plant material. Adults feed on plant sap, fruit juice, nectar, honeydew from coccids and membracids, and pollen. One of the most important pollinators in tropical parts. Truly social, with queen and worker castes. **Habitat** Subtropical bushveld, forested areas and savanna. **Related species** Other stingless bees, which occur in 4 other genera in the region. Many nest in hollow trunks and branches or in the ground.

3 *Apis mellifera* Honey Bee

Medium-sized (body length of worker 13mm). Workers black, with pale hairs on head, thorax and base of abdomen. Basal segments of abdomen more or less banded in reddish-brown. Legs black. Wings clear. Queens **(3)** (see marked individual) resemble workers but are larger. Males (drones) are also larger than workers, but darker, the abdomen lacking banding and squared off, rather than pointed, at the tip. **Biology** Nests in existing cavities and builds vertical parallel combs of wax **(3A)** with 2 layers of hexagonal cells opening at opposite ends. Larvae are fed progressively by workers. Workers of the Cape Honey Bee (subspecies *Apis mellifera capensis*) invade and eventually overwhelm colonies of the African Honey Bee (subspecies *Apis mellifera scutellata*). Its workers lay unfertilized eggs that develop into females, thus reproducing at a faster rate than the African Honey Bee. Although very important as crop pollinators, honey bees may deprive more specialized and efficient indigenous bees of pollen and nectar, effectively reducing pollination of wild flowers. **Habitat** Occurs in all habitats, but needs hollow trees or rock crevices and overhangs in which to nest.

4 *Hypotrigona* Sweat bees

Very small (body length 2–4mm), black, with pollen-collecting baskets (corbicula) on the hind legs. There are 3 regional species in the genus. **Biology** Nest entrance is usually surmounted by waxen tubes guarded by the workers. They are well known for trying to collect moisture from the eyes and mouths of humans, and are incapable of stinging. In some parts of Africa humans collect the pollen, honey stores and wax from their nests, which are located underground, in dead wood crevices or even in window frames. They are important pollinators of a range of crops including coffee. **Habitat** Open woodland and savanna.

GLOSSARY

Abdomen: Hindmost of the three major body divisions, after the head and thorax.
Antennae: Paired 'feelers' on the head.
Apical: At the tip or end of a structure.
Apterous: Lacking wings.
Binomial: Literally two names: the system of double names (genus and species name) given to animals and plants.
Caste: Form or type of adult individual found in social insects, such as worker or soldier.
Cerci: Paired, usually segmented, appendages at the tip of the abdomen.
Cleptoparasite: Parasite that feeds by 'stealing' food from the host.
Cocoon: Covering or case, usually of silk, spun by larva as protection for the pupa.
Compound eyes: Large lateral eyes of most insects, formed of numerous facets or lenses, each representing a separate ommatidium.
Coxa: First segment of the leg, where it joins the body.
Cuticle: Hard outer covering of an insect, made of chitin.
Dimorphic: With two different forms within the same species (usually male and female forms).
Diurnal: Active during the day.
Ectoparasite: Parasite living on the external surface of the host.
Elytra: Hardened fore wings of beetles, not used in flight.
Endemic: Occurring naturally only in a particular region.
Endoparasite: Parasite living within the body of its host.
Femur: Third and usually largest segment of an insect leg ('thigh' in humans).
Filiform: Thread-like.
Frass: Solid faecal pellets produced by insects.
Halteres: Small club-shaped balancing organs derived from and replacing the hind wings in flies (Diptera).
Hemimetabolous: Undergoing incomplete metamorphosis, in that immatures resemble and gradually develop into the adult body form. Wing pads present in older nymphs.
Holometabolous: Undergoing complete metamorphosis, with clearly differentiated larval, pupal and adult stages.
Imago: Adult life stage in insect metamorphosis.
Instars: Larval or nymphal stages.
Labium: Mouthpart behind the mouth opening, comparable to a lower lip.
Labrum: Mouthpart in front of mouth opening, comparable to an upper lip.
Larva: Immature stage of a holometabolous insect.
Mandibles: Paired 'jaws', or biting and chewing mouthparts (sometimes highly modified).
Maxillae: Paired mouthparts situated behind the mandibles, assisting them to hold and chew food.
Mesothorax: Middle thoracic segment.
Metathorax: Third and last thoracic segment.
Nocturnal: Active at night.
Nymph: Young stage of an insect that undergoes incomplete metamorphosis.
Ocelli: Simple eyes, comprised of a single ommatidium.
Ommatidia: Individual units that together make up the compound eyes of insects.
Ovipositor: Egg-laying apparatus at posterior end of abdomen of female insects.
Palp: Segmented sensory appendage of maxilla or labium.
Parasitoid: Parasitic insect that feeds and lives (usually) internally on another insect, eventually killing it.

Parthenogenetic: Females producing offspring without being fertilized by a male.

Petiole: Thin stalk connecting the thorax and abdomen of some insects, mainly wasps.

Pheromone: Chemical released from the body of one insect to influence the behaviour or physiology ofanother.

Proboscis: Mouthparts modified into an elongate tube or trunk-like structure.

Proleg: 'False' leg found in addition to the three pairs of true legs in insect larvae, especially caterpillars.

Pronotum: Dorsal part of prothorax.

Prothorax: First thoracic segment.

Pterostigma: Dark spot on the front or leading edge of the wing.

Pupa: Inactive non-feeding stage between larval and adult phases of life cycle of holometabolous insects.

Raptorial: Legs adapted to seize and hold prey.

Rostrum: Beak-like piercing and sucking mouthparts, typically in hemipterans (bugs).

Scutellum: Triangular extension of thoracic segment (seen from above), prominent in metathorax of many bugs.

Setae: Bristles or hair-like structures.

Stridulation: Sounds produced by rubbing together parts of the body.

Tarsus: Fifth and last jointed section or 'foot' of an insect leg, bearing claws.

Tegmina: Hardened fore wings of grasshoppers and their relatives.

Thorax: Second of the three main body divisions, between the head and abdomen, bearing the wings and legs.

Tibia: Fourth segment of an insect leg, between the femur and tarsus.

Trochanter: Small second segment of an insect leg.

Venation: Pattern of strengthening veins on an insect wing, important in identification.

FURTHER READING

Evans, AV & Bellamy, CL. 2000. *An inordinate fondness for beetles*. University of California Press, Berkeley. 209pp.

Gess, SK & Gess, FW. 2014. *Wasps and bees in southern Africa*. SANBI Biodiversity Series 24. South African National Biodiversity Institute, Pretoria. 327pp.

Griffiths C, Day, J & Picker, M. 2015. *Freshwater life – A field guide to the plants and animals of southern Africa*. Struik Nature, Cape Town. 368pp.

Henning, GA, Pringle, ELL & Ball, JB. 1994. *Pennington's butterflies of southern Africa* (2nd ed). Struik-Winchester, Cape Town. 800pp.

Henning, GA, Henning, SF, Joannou, J & Woodhall, SE. 1997. Living butterflies of South Africa: biology, ecology and conservation, Vol 1, *Hesperiidae, Papilionidae and Pieridae*. Umdaus Press, Hatfield. 397pp.

Holm, E. 2017. *Insectopedia. The secret world of southern African insects*. Struik Nature, Cape Town. 208pp.

Holm, E & Marais, E. 1992. *Fruit chafers of southern Africa*. Ekogilde, Hartebeespoort. 328pp.

Marshall, SA. 2012. *Flies: A natural history of Diptera*. Firefly Books. Richmond Hill, Canada. 616pp.

Martins, DJ. 2014. *Pocket guide: Insects of East Africa*. Struik Nature, Cape Town. 152pp.

McGavin, GC. 1993. *Bugs of the world*. Blanford, London. 192pp.

Picker, M & Griffiths, C. 2015. *Pocket guide: Insects of South Africa*. Struik Nature, Cape Town. 152pp.

Pinhey, ECG. 1975. *Moths of southern Africa*. Tafelberg Publishers, Cape Town. 347pp.

Prinsloo, GL & Uys, VM (eds). 2015. *Insects of cultivated plants and natural pastures in southern Africa*. Entomological Society of Southern Africa, Hatfield. 786pp.

Samways, MJ. 2008. *Dragonflies and damselflies of South Africa*. Pensoft Publishers, Sophia, Bulgaria. 297pp.

Schoeman, AS. 2002. *A guide to garden pests and diseases in South Africa*. Struik Nature, Cape Town. 160pp.

Scholtz, CH & Holm, E. 1985. *Insects of southern Africa*. Butterworths, Durban. 502pp.

Skaife, SH. 1979. *African insect life* (2nd ed). C. Struik, Cape Town. 279pp

Slingsby, P. 2017. *Ants of southern Africa*. Slingsby Maps, Cape Town. 256pp

Tarboton, W & Tarboton, M. 2015. *A guide to the dragonflies & damselflies of South Africa*. Struik Nature, Cape Town. 224pp.

Uys, VM & Urban, RP. 1996. *How to collect and preserve insects and arachnids*. ARC – Plant Protection Research Institute, Pretoria. 73pp.

Weaving, A. 2000. *Southern African insects and their world*. Struik Nature, Cape Town. 88pp.

Woodhall, S. 2005. *Field guide to butterflies of South Africa*. Struik Nature, Cape Town. 440pp.

Woodhall, S. 2013. *Pocket guide: Butterflies of South Africa*. Struik Nature, Cape Town. 152pp.

Woodhall S & Gray, L. 2015. *Gardening for butterflies*. Struik Nature, Cape Town. 188pp.

Yeo, PF & Corbet, SA. 1983. *Solitary wasps, Naturalist's handbook No. 3*. Cambridge University Press, Cambridge. 70pp.

WEB-BASED RESOURCES

The following sites provide checklists, key taxonomic references for South African insects and notes on the biology of selected species:

Dragonflies & Damselflies of Southern Africa
www.dragonflies.co.za

WaspWeb: Wasps, Bees and Ants of Africa and Madagascar
www.waspweb.org

Animal Demography Unit Virtual Museums, featuring identified images of regional dung beetles (Coleoptera), Neuroptera, Lepidoptera and Odonata
vmus.adu.org.za

A Catalogue of the Insects of Southern Africa
www.ru.ac.za/media/rhodesuniversity/resources/martin/main.html

African Moths
www.africanmoths.com

Caterpillar Rearing Group, Facebook
www.facebook.com/groups/caterpillarrg

iNaturalist
www.inaturalist.org/lists/702812-Insects---South-Africa

iSpot Share Nature
www.ispotnature.org/communities/southern-africa/view/observation/544122/insects

ACKNOWLEDGEMENTS

A project of this size would not have been possible without the generous help and support of our families, friends, entomological colleagues and students.

For assistance with identifications we are indebted to the following (several of whom also aided us by providing published or unpublished biological notes, and by checking the relevant sections of the text for accuracy):

Blattodea – Philippe Grandcolas (Paris Museum of Natural History); Riaan Stals (ARC – Plant Protection Research Institute); Vivienne Uys (ARC — Plant Protection Research Institute)

Coleoptera – Erik Arndt (Fachhochschule Anhalt Fachbereich Loel, Germany); Chuck Bellamy (Natural History Museum of Los Angeles County); Vasily Grebennikov; James Harrison (Transvaal Museum); Chi-Feng Lee (Ohio State University); Vin Whitehead

Diptera – Dave Barraclough; Maureen Coetzee (South African Institute for Medical Research); Peter Jupp; Jason Londt & Ray Miller (Natal Museum, Pietermaritzburg); Brian Stuckenberg; Diedrich Visser (Plant Protection Research Institute)

Embioptera – Edward Ross

Ephemeroptera – Helen James (Albany Museum, Grahamstown)

Hemiptera – Ian Miller (ARC – Plant Protection Research Institute); Patrick Reavell (University of Zululand); Michael Stiller (ARC – Plant Protection Research Institute); Martin Villett (Rhodes University)

Hymenoptera –Sarah and Fred Gess (Albany Museum, Grahamstown); Hamish Robertson; Simon Van Noort & Vin Whitehead (Iziko Museums)

Lepidoptera – Henk Geertsema (University of Stellenbosch); Martin Kruger (Transvaal Museum)

Neuroptera – Mervyn Mansell (ARC – Plant Protection Research Institute) Odonata – Michael Samways (University of Natal, Pietermaritzburg)

Orthoptera – Peter Johns (University of Canterbury); Dan Otte (The Academy of Natural Sciences, Philadelphia); Rob Toms (Transvaal Museum); Moira van Staaden (Bowling Green State University)

Phasmatodea – Paul Brock

Psocoptera – Courtney Smithers (Australian Museum)

Siphonaptera – Joyce Segerman (South African Institute of Medical Research)

Zygentoma, Archaeognatha – John Irish (National Museum, Bloemfontein)

Trichoptera – Ferdie de Moor (Albany Museum, Grahamstown).

Staff of the Entomology Department of the Iziko Museums kindly allowed us access to their entomology collection.

A special vote of thanks is due to our families. The Picker family provided us with accommodation, transport and logistical support during several collecting expeditions out from Johannesburg, while the Griffiths family accompanied us on numerous photographic excursions and endured having their kitchen regularly turned into a photographic studio and bug-culture area.

Finally, we are grateful to our editor Emily Donaldson and our designer Neil Bester for their careful and patient editing and formatting of this guide.

INDEX TO SCIENTIFIC NAMES

CAPITAL LETTERS DENOTE SUBCLASSES, ORDERS, INFRAORDERS AND SUBORDERS.

INDEX TO COMMON NAMES

DEMIBOLD CAPITAL LETTERS DENOTE ORDERS AND SUBORDERS. BOLD DENOTES FAMILIES.

519

Springtails
COLLEMBOLA p.24

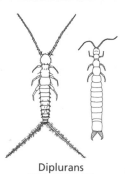

Diplurans
DIPLURA p.26

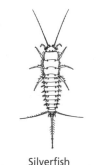

Silverfish
ZYGENTOMA p.28

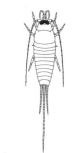

Bristletails
ARCHAEOGNATHA p.30

Mayflies
EPHEMEROPTERA p.32

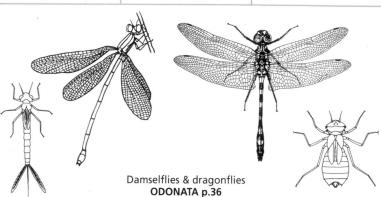

Damselflies & dragonflies
ODONATA p.36

Cockroaches & termites
BLATTODEA p.52

Mantids
MANTODEA p.70

Earwigs
DERMAPTERA p.80

Web-spinners
**EMBIOPTERA
(EMBIIDINA) p.82**

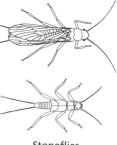

Stoneflies
PLECOPTERA p.84

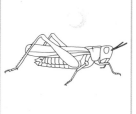

Crickets, katydids,
grasshoppers & locusts
ORTHOPTERA p.88

Stick insects
PHASMIDA p.128